MASTERING MATHEMATICS

FOR OCR GCSE

FOUNDATION 1

Assessment Consultant:
Keith Pledger

FOUNDATION 1

MASTERING MATHEMATICS FOR OCR GCSE

Series Editor: Roger Porkess

Gareth Cole, Heather Davis, Sophie Goldie,
Linda Liggett, Robin Liggett, Andrew Manning,
Richard Perring, Rob Summerson

HODDER
EDUCATION
AN HACHETTE UK COMPANY

The publisher would like to thank the following for permission to reproduce copyright material:

Photo credits:

p. 5 © Gina Sanders – Fotolia.com; **p. 10** © Brebca – Fotolia.com ; **p. 17** © Kadmy – Fotolia.com; **p. 26** © annacurnow – Fotolia.com; **p. 27** © Caitlin Seymour; **p. 28** © Ulrich Müller – Fotolia.com; **p. 30** © Kate Crossland-Page (t); © Sue Hough (b); **p. 35** © HugoCCampos – Fotolia.com; **p. 42** © Scott Griessel – Fotolia.com; **p. 46** © Alexandr Mitiuc – Fotolia.com; **p. 51** © Food photo – Fotolia.com; **p. 51** © Coprid – Fotolia.com; **p. 54** © Kate Crossland-Page; **p. 55** © Sandy Officer; **p. 59** © Rich Lindie – Fotolia.com; **p. 65** © Yali Shi – Fotolia.com (t); © Kate Crossland-Page (b); **p. 72** © Brian Jackson – Fotolia.com; **p. 80** © Imagestate Media (John Foxx) / Vol 18 Golddisc I (t); © Sue Hough (b); **p. 88** © yanlev – Fotolia.com; **p. 94** © Pavel Losevsky – Fotolia.com; **p. 100** © cherries – Fotolia.com; **p. 107** © Florin Capilnean – Fotolia.com; **p. 108** © Kate Crossland-Page; **p. 112** © Dave Gale; **p. 123** © olgataranik – Fotolia.com; **p. 130** © Warren Goldswain – Fotolia.com; **p. 131** © Sue Hough; **p. 136** © klikk – Fotolia.com; **p. 153** © GaryBartlett – Thinkstock; **p. 161** © Photodisc/Getty Images/Retail, Shopping and Small Business 21; **p. 171** © Ingram Publishing Company / Ultimate Business 06; **p. 179** © Kautz15 – Fotolia.com; **p. 189** © GoodMood Photo – Fotolia.com; **p. 201** © Oleg Prikhodko / iStockphoto; **p. 205** © Kate Crossland-Page; **p. 208** © absolut – Fotolia.com (t); © GoldPix – Fotolia.com (b); **p. 215** © Photodisc/Getty Images/ Business & Industry 1; **p. 224** © Marc Tielemans / Alamy; **p. 225** © Sandy Officer (l); © Kate Crossland-Page (r); **p. 226** © Kate Crossland-Page; **p. 230** © Eric Farrelly / Alamy; **p. 238** © Stockbyte/ Photolibrary Group Ltd/ Big Business SD101; **p. 248** © Heather Davis ; **p. 254** © Heather Davis; **p. 265** © George.M – Fotolia.com; **p. 277** © Anne Wanjie; **p. 288** © Heather Davis; **p. 298** © Doug Houghton / Alamy; **p. 306** © Elaine Lambert; **p. 315**, **p. 321** © Pumba – Fotolia.com; **p. 325** © moodboard – Thinkstock.com; **p. 335** © Andres Rodriguez – Fotolia.com; **p. 347** © sytilin – Fotolia.com; **p. 356** © auryndrikson – Fotolia.com; **p. 366** © Sean Gladwell – Fotolia.com; **p. 373** © Dawn Hudson – Fotolia.com.

Orders: please contact Bookpoint Ltd, 130 Milton Park, Abingdon, Oxon OX14 4SB. Telephone: (44) 01235 827720. Fax: (44) 01235 400454. Lines are open 9.00–17.00, Monday to Saturday, with a 24-hour message answering service. Visit our website at www.hoddereducation.co.uk

First published in 2015 by

Hodder Education
An Hachette UK Company
Carmelite House
50 Victoria Embankment
London EC4Y 0DZ

Impression number	5	4	3
Year	2022	2021	2020

Cover photo © science photo – Fotolia

Typeset in ITC Avant Garde Gothic Std Book 10/12 by Integra Software Services Pvt. Ltd., Pondicherry, India

Printed in Dubai

A catalogue record for this title is available from the British Library

ISBN 978 1471 840012

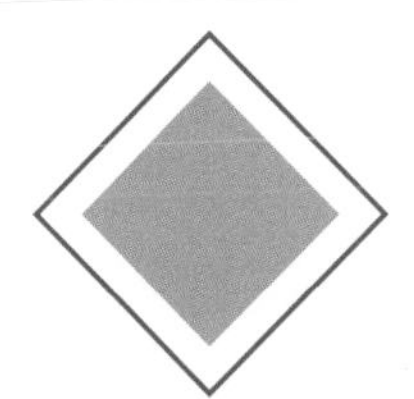

Contents

ALGEBRA

GEOMETRY AND MEASURES

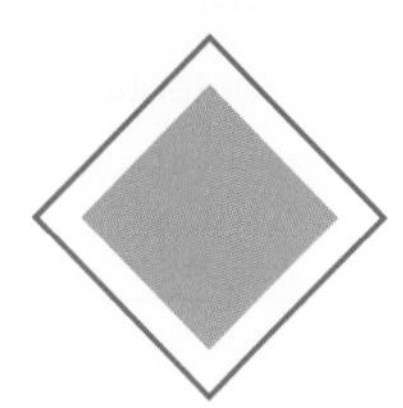

How to get the most from this book

This book covers the initial content for the OCR Foundation Tier GCSE.

Content common to both Foundation and Higher can be found in the Foundation 2/Higher 1 book.

The material is split into 20 **'strands of learning'**:

Number strands	Algebra strands	Geometry and Measures strands	Statistics and Probability strands
Calculating	Starting algebra	Units and scales	Statistical measures
Using our number system	Sequences	Properties of shapes	Statistical diagrams
Accuracy	Functions and graphs	Measuring shapes	Probability
Fractions	Algebraic methods	Constructions	
Percentages		Transformations	
Ratio and proportion		Three-dimensional shapes	
Number properties			

Each strand is presented as a series of units that get more difficult as you progress (from Band b to Band k). This book mainly deals with units in Bands e to h. In total there are 46 units in this book.

Getting started

At the beginning of each strand, you will find a '**Progression strand flowchart**'. It shows what skills you will develop in each unit in the strand. You can see:

- what you need to know before starting each unit
- what you will need to learn next to progress.

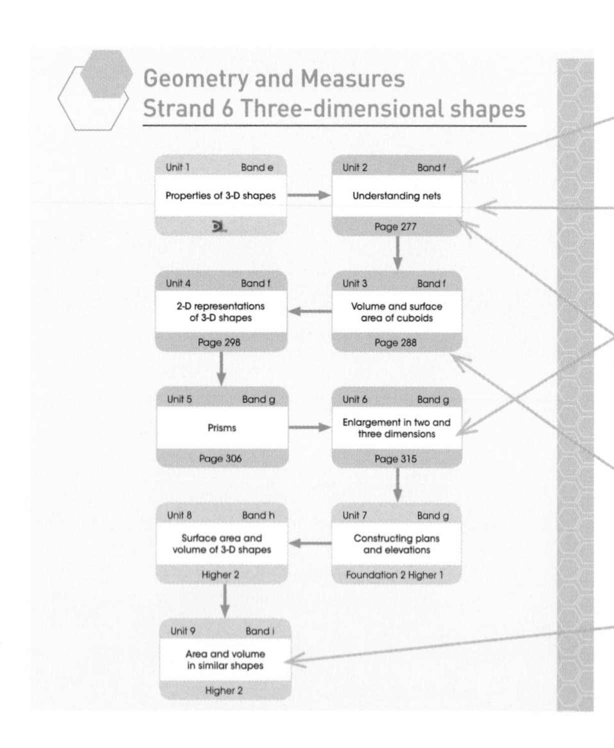

The '**Band**' tells you how difficult the unit is. Bands go from Band b to Band k.

The flowchart tells you the order you need to study the units in.

Assumed knowledge is reviewed and extended in the '**Moving on**' section at the beginning of the strand. Students not yet confident in the knowledge assumed in this book will find support online.

The page number tells you where to find the start of that unit in the book.

Units that build on to the work in this book are shown so it is easy to identify next steps.

When you start to use this book, you will need to identify where to join each strand. Then you will not spend time revisiting skills you have already mastered.

If you can answer all the questions in the **'Reviewing skills'** section of a unit then you will not have to study that unit.

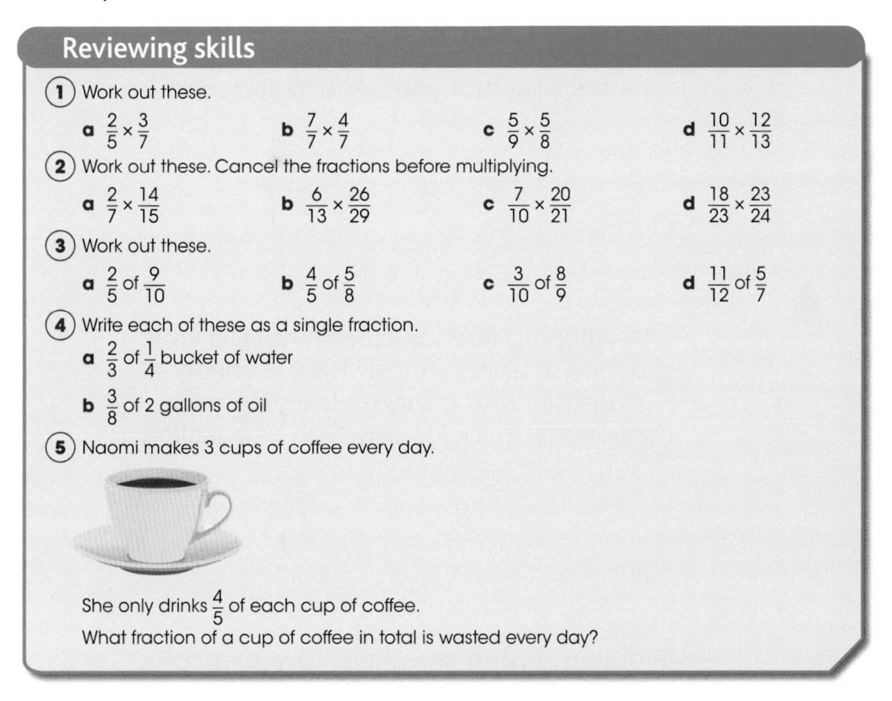

Reviewing skills

1. Work out these.
   a $\frac{2}{5} \times \frac{3}{7}$ b $\frac{7}{7} \times \frac{4}{7}$ c $\frac{5}{9} \times \frac{5}{8}$ d $\frac{10}{11} \times \frac{12}{13}$
2. Work out these. Cancel the fractions before multiplying.
   a $\frac{2}{7} \times \frac{14}{15}$ b $\frac{6}{13} \times \frac{26}{29}$ c $\frac{7}{10} \times \frac{20}{21}$ d $\frac{18}{23} \times \frac{23}{24}$
3. Work out these.
   a $\frac{2}{5}$ of $\frac{9}{10}$ b $\frac{4}{5}$ of $\frac{5}{8}$ c $\frac{3}{10}$ of $\frac{8}{9}$ d $\frac{11}{12}$ of $\frac{5}{7}$
4. Write each of these as a single fraction.
   a $\frac{2}{3}$ of $\frac{1}{4}$ bucket of water
   b $\frac{3}{8}$ of 2 gallons of oil
5. Naomi makes 3 cups of coffee every day.
   She only drinks $\frac{4}{5}$ of each cup of coffee.
   What fraction of a cup of coffee in total is wasted every day?

When you know which unit to start with in each strand you will be ready to start work on your first unit.

Starting a unit

Every unit begins with some information:

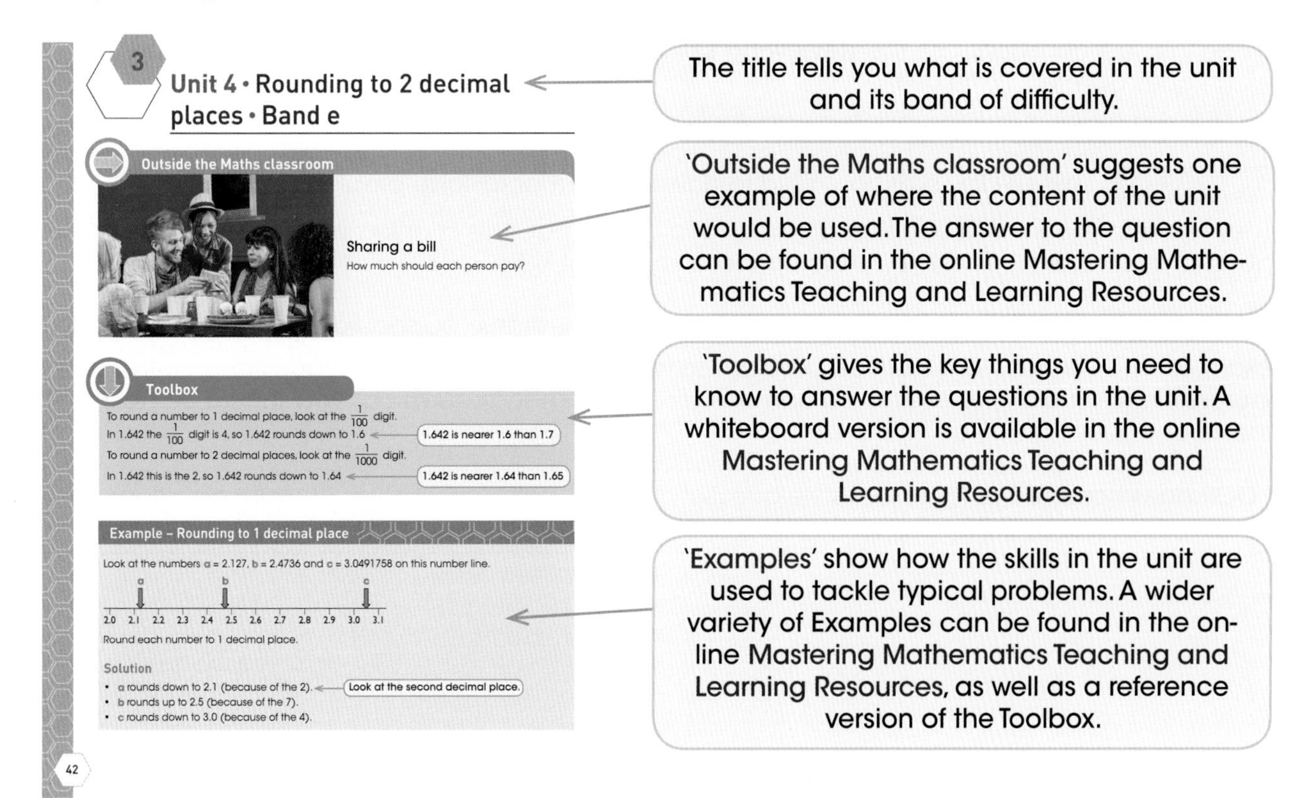

3

Unit 4 • Rounding to 2 decimal places • Band e

Outside the Maths classroom

Sharing a bill

How much should each person pay?

Toolbox

To round a number to 1 decimal place, look at the $\frac{1}{100}$ digit.

In 1.642 the $\frac{1}{100}$ digit is 4, so 1.642 rounds down to 1.6 ← 1.642 is nearer 1.6 than 1.7

To round a number to 2 decimal places, look at the $\frac{1}{1000}$ digit.

In 1.642 this is the 2, so 1.642 rounds down to 1.64 ← 1.642 is nearer 1.64 than 1.65

Example – Rounding to 1 decimal place

Look at the numbers a = 2.127, b = 2.4736 and c = 3.0491758 on this number line.

Round each number to 1 decimal place.

Solution

- a rounds down to 2.1 (because of the 2). ← Look at the second decimal place.
- b rounds up to 2.5 (because of the 7).
- c rounds down to 3.0 (because of the 4).

42

Now you have all the information you need, you can use the questions to develop your understanding.

Practising skills

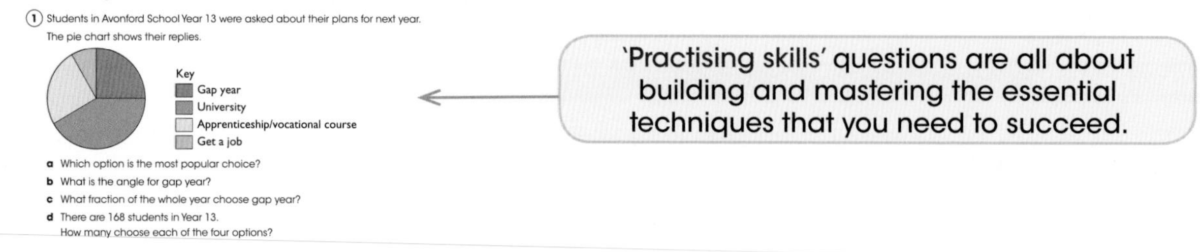

1 Students in Avonford School Year 13 were asked about their plans for next year.
The pie chart shows their replies.

a Which option is the most popular choice?
b What is the angle for gap year?
c What fraction of the whole year choose gap year?
d There are 168 students in Year 13.
How many choose each of the four options?

'Practising skills' questions are all about building and mastering the essential techniques that you need to succeed.

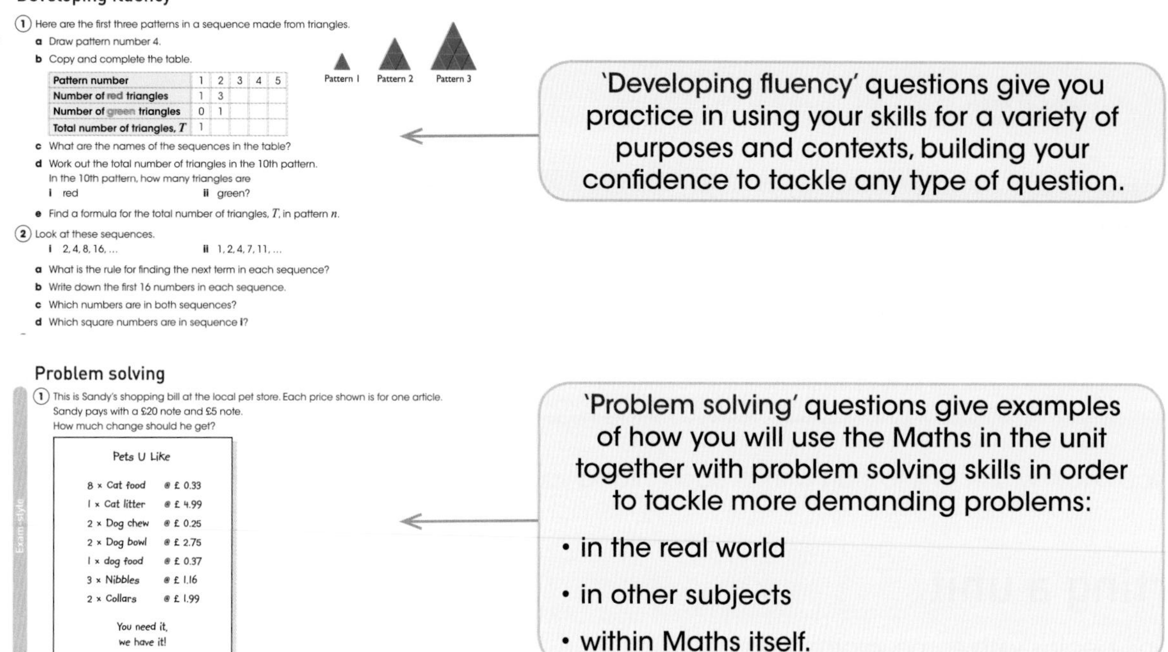

Developing fluency

1 Here are the first three patterns in a sequence made from triangles.
a Draw pattern number 4.
b Copy and complete the table.

Pattern number	1	2	3	4	5
Number of red triangles	1	3			
Number of green triangles	0	1			
Total number of triangles, T	1				

c What are the names of the sequences in the table?
d Work out the total number of triangles in the 10th pattern.
In the 10th pattern, how many triangles are
i red ii green?
e Find a formula for the total number of triangles, T, in pattern n.

2 Look at these sequences.
i 2, 4, 8, 16, … ii 1, 2, 4, 7, 11, …
a What is the rule for finding the next term in each sequence?
b Write down the first 16 numbers in each sequence.
c Which numbers are in both sequences?
d Which square numbers are in sequence i?

'Developing fluency' questions give you practice in using your skills for a variety of purposes and contexts, building your confidence to tackle any type of question.

Problem solving

1 This is Sandy's shopping bill at the local pet store. Each price shown is for one article.
Sandy pays with a £20 note and £5 note.
How much change should he get?

'Problem solving' questions give examples of how you will use the Maths in the unit together with problem solving skills in order to tackle more demanding problems:

- in the real world
- in other subjects
- within Maths itself.

When you feel confident, use the '**Reviewing skills**' section to check that you have mastered the techniques covered in the unit.

You will see many questions labelled with Reasoning or Exam-style

Reasoning skills are key skills you need to develop in order to solve problems.

They will help you think through problems and to apply your skills in unfamiliar situations. Use these questions to make sure that you develop these important skills.

Exam-style questions are examples of the types of questions you should be prepared for in the exam. Exam-style questions that test problem solving will often be unfamiliar questions, so they are impossible to predict. However, if you can answer the exam-style questions you should be able to tackle any question you get in the exam.

OCR does not endorse the exam-style questions in this book as examples of exam questions that may be set by OCR.

About 'Bands'

Every unit has been allocated to a Band. These bands can be used as a general indication of the level of difficulty of the Maths that you are working on, but are not to be related directly to the GCSE grades.

Each Band contains Maths that is of about the same level of difficulty.

This provides a way of checking your progress and assessing your weaker areas, where you need to practise more.

Moving on to another unit

Once you have completed a unit, you should move on to the next unit in one of the strands. You can choose which strand to work on next but try to complete all the units in a particular Band before moving on to the next Band.

A note for teachers

Lower Bands have been assigned to units roughly in line with the previous National Curriculum levels. Here they are, just to help in giving you a reference point. Bands I, j and k have been assigned to the units that address the very hardest skills in the GCSE specification (the bold content).

Band	Approximate equivalence in terms of old National Curriculum levels
b	Level 2
c	Level 3
d	Level 4
e	Level 5
f	Level 6
g	Level 7
h	Level 8
i, j, k	No equivalent levels

Answers and Write-on sheets

Write-on sheets to aid completion of answers are denoted by . These and answers to all the questions in this book are available via **Mastering Mathematics 11-16 Teaching and Learning Resources for OCR GCSE** or by visiting www.hoddereducation.co.uk/MasteringmathsforOCRGCSE

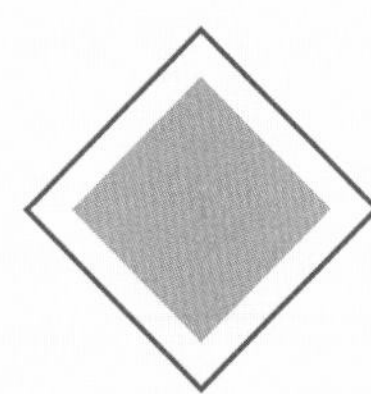

Number Strand 1 Calculating

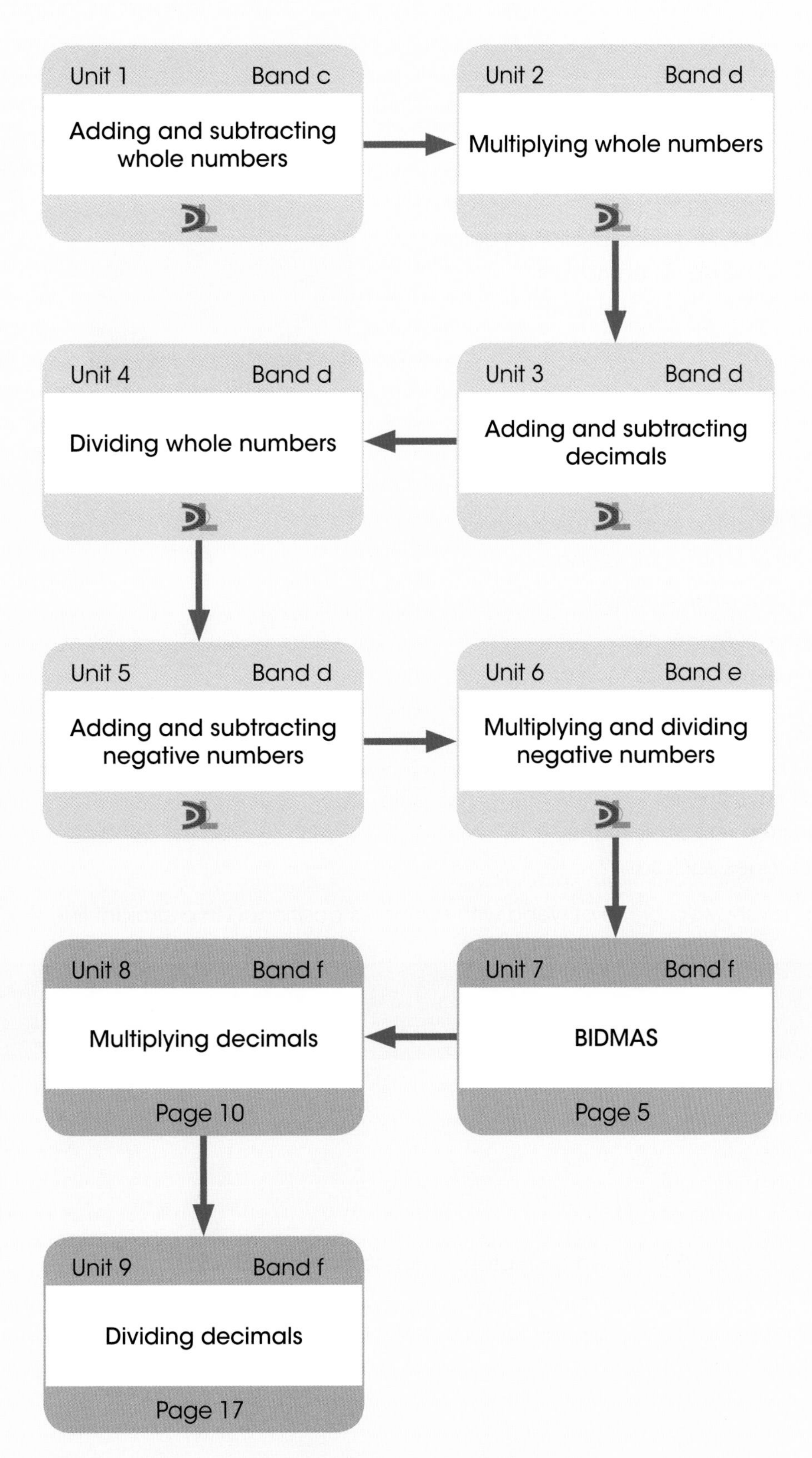

Units 1–6 are assumed knowledge for this book. They are reviewed and extended in the Moving on section on page 2. Knowledge and skills from these units are used throughout the book.

Units 1–6 • Moving on

The questions in this section should be answered *without* the use of a calculator.

Exam-style

1 A supermarket has a special offer on cereal.
Lucy buys two boxes of cornflakes.
How much money does she save?

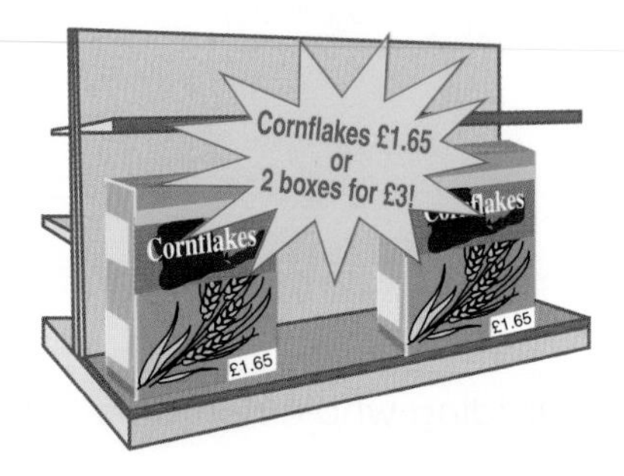

Exam-style

2 Jack has £50 at the start of the weekend.
Here is a list of what Jack spends:

£7.50 for a cinema ticket
£4.25 for burger & fries
£1.50 for a drink
£3.40 for bus fares
£2.50 for a magazine
£14.99 for a t-shirt

Jack saves the rest of his money.
How much does Jack save?

Exam-style

3 The diagram shows a piece of wood with five holes positioned in a straight line, ABCDE.

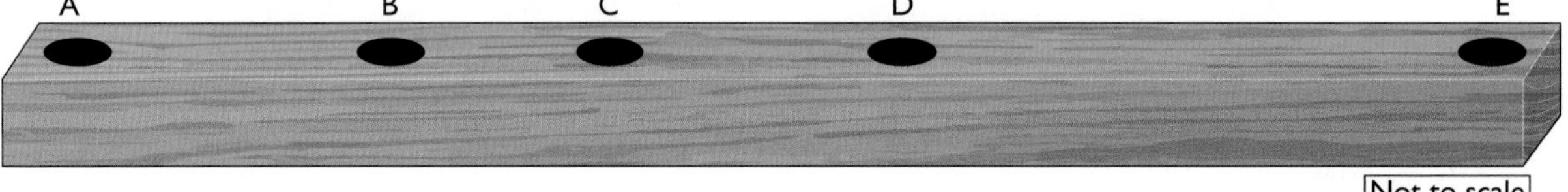

$AD = 9.5$ cm
$DE = 6.7$ cm
$CE = 10.9$ cm
Work out the length of CD.

Exam-style

4 Mrs Jones has £8000 to spend on laptops. Each laptop cost £325.
There are 25 students in Mrs Jones' class.
Will the students have to share the laptops or will there be enough for one each?

Exam-style

5 Ged buys and sells second hand sports equipment.
The table shows some information about some of Ged's sales at a local auction.

Item	Cost price in £	Selling price in £	Profit or loss £
Football boots	35.75	18.50	
Snooker cue	23.50	42.00	
Golf clubs and bag	112.00	108.44	
Set of weights	98.00	72.50	
Cricket bat	15.10	24.30	
Snooker table	58.00	93.60	

How much profit or loss did Ged make overall?

Exam-style

6 A theatre has seating for 500 people.
Last night the total amount of money taken in ticket sales was £3081.
£345 was taken in the sale of children's tickets.
Work out the number of seats that were vacant.

Ticket prices	
Adults	£8
Children	£5

Exam-style

7 The table shows a list of Joe's fixed monthly expenditure.

Rent	£525
Food	£230
Utilities	£186
Travel	£292

Joe earns £24 816 each year.
Each month £517 is deducted from his earnings for taxes.
Is Joe able to save more or less than £3000 each year?

Exam-style

8 Waqas is going to tile the walls in his kitchen.
The tiles are sold in boxes.
Each box contains 24 tiles and costs £15.
Waqas has £275 to spend on tiles.
He estimates that he will need 18 boxes of tiles.

a Does Waqas have enough money to buy 18 boxes of tiles?

Waqas will use in total 425 tiles to tile the walls in his kitchen.

b Work out the number of tiles that Waqas will have left over.

Exam-style

9 The diagram shows a rectangular field.

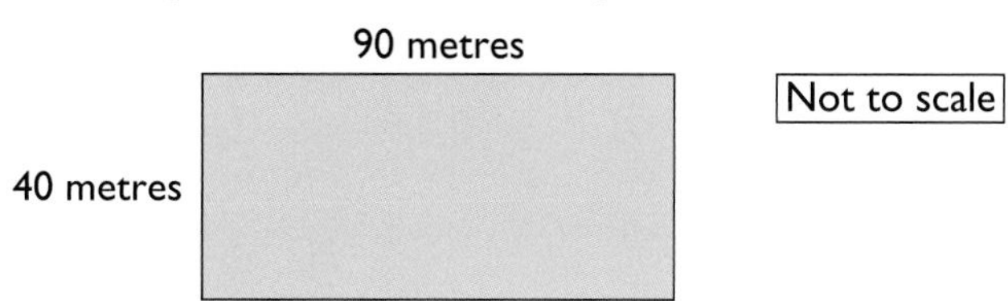

Bill has 28 goats.
He wants the goats to be able to graze in this field.
Each goat needs 125 square metres of field in which to graze.
Show that the field is large enough for Bill's 28 goats.
Would there be enough space for another goat?

Exam-style

10 In an aeroplane there are 31 rows of Economy class seats.
There are 6 seats in each row.
In addition there are 36 seats in Business class.
On one flight to Montego Bay, Jamaica, all the seats are taken.
Work out the total cost of all the tickets sold.

Exam-style

11 Mrs Green has £150 in her bank account. She receives a cheque for £50 from an insurance company. She has a phone bill of £95 to pay.
Mrs Green records her finances on this account sheet.

Credits (+)	Debits (-)	Balance (£)
		150
50		200
	95	?

a Copy the account sheet and fill in the missing balance.
Extend your copy of the account sheet. Write these items in the correct column.

Payment for dress £85
Payment to hairdresser £40
Competition win £100
Payment for car repairs £150
Payment for food £25
Sale of old car £400

b What is the final balance?

Exam-style

12 In a hotel there are 15 floors, the ground floor is floor 0.
There are 9 floors above ground level, floors 1 to 9.
There are 5 floors below ground level, floors −1 to −5.
In the hotel there are two lifts, lift A and lift B, both positioned at floor 0.
The table shows a sequence of movements for each lift.

Lift A	Up 5 floors	Down 2 floors	Up 4 floors	Down 7 floors	Up 8 floors
Lift B	Down 3 floors	Up 10 floors	Down 6 floors	Up 1 floor	Down 6 floors

a On which floors are the two lifts now?
b How many floors are there between the two lifts now?
Jenny is waiting for a lift on floor 2.
c Which lift is Jenny likely to get?

Unit 7 • BIDMAS • Band f

Outside the Maths classroom

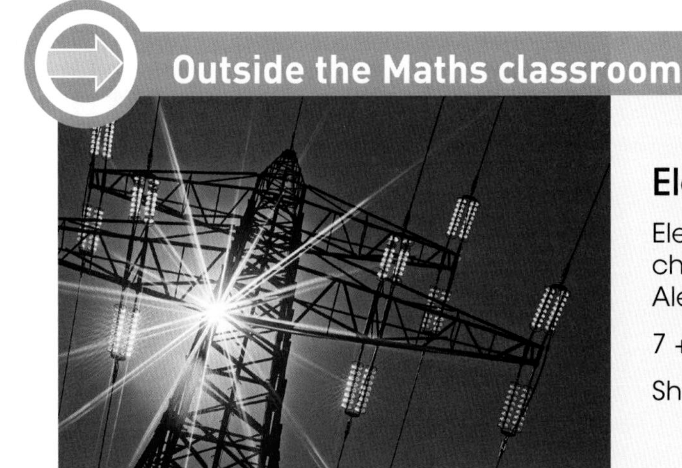

Electricity bills

Electricity bills come in two parts – a standing charge and a price per unit of electricity used. Alex uses this calculation to work out her bill.

$7 + 0.1 \times 50$

Should her bill be £12 or £355? Why?

Toolbox

The order that operations should be carried out in is

- **B**rackets
- **I**ndices (powers, such as cubes or square roots)
- **D**ivision
- **M**ultiplication
- **A**ddition
- **S**ubtraction.

You can use the word **BIDMAS** to help you remember the right order.

Example – Using brackets

Work out these.

a $(8 + 5) \times 4$ **b** $50 - 2 \times (9 + 6)$

Solution

Remember BIDMAS.

a $(8 + 5) \times 4$ ← Brackets

$= 13 \times 4$

$= 52$ ← Multiplication

b $50 - 2 \times (9 + 6)$

$= 50 - 2 \times 15$ ← Brackets

$= 50 - 30$ ← Multiplication

$= 20$ ← Subtraction

Example – Using powers

a Work out $(20 + 7) \div 3^2 + 4$ **b** $(10 - 3) \times 2^3 - 4$

Solution

a $(20 + 7) \div 3^2 + 4$ — Brackets

$= 27 \div 3^2 + 4$ — Indices

$= 27 \div 9 + 4$ — Division

$= 3 + 4$

$= 7$ — Addition

b $(10 - 3) \times 2^3 - 4$

$= 7 \times 2^3 - 4$ — Brackets

$= 7 \times 8 - 4$ — Indices

$= 56 - 4$ — Multiplication

$= 52$ — Subtraction

Example – Using fractions (division)

Work out $\dfrac{6 \times 8}{14 - 2}$

Solution

The fraction bar counts as a bracket.

So this is the same as $(6 \times 8) \div (14 - 2)$

$= 48 \div 12$ — Brackets

$= 4$ — Division

The questions in this unit should be answered *without* the use of a calculator, unless otherwise stated. (However, you may wish to use a calculator to check some of your answers.)

Practising skills

Work these out.

1. **a** $(5 + 2) \times 3$ **b** $5 + 2 \times 3$ **c** $5 + (2 \times 3)$
2. **a** $7 \times (10 - 4)$ **b** $7 \times 10 - 4$ **c** $(7 \times 10) - 4$

3 a $(20 + 15) \div 5$ b $20 + 15 \div 5$ c $15 + 20 \div 5$

4 a $20 - 12 \div 2$ b $(20 - 12) \div 2$ c $20 - (12 \div 2)$

5 a $(21 + 12) \times 3$ b $21 + 12 \times 3$ c $21 - 12 \div 3$

6 a $(13 + 5) \times (7 - 3)$ b $13 + 5 \times 7 - 3$ c $(13 + 5) \times 7 - 3$

7 a $(8 + 12) \div (4 + 1)$ b $8 + 12 \div 4 + 1$ c $(8 + 12) \div 4 + 1$

8 a $13 + 5 - 7 - 3$ b $(13 + 5) - (7 - 3)$ c $13 + (5 - 7) - 3$

9 a 4×3^2 b $4 + 3^2$ c $(4 + 3)^2$

10 a $8^2 - 5^2 + 1^2$ b $(8 - 5)^2 + 1^2$ c $(8 - 5 + 1)^2$

11 a $2^3 + 2^2 + 2 + 1$ b $2^3 - 2^2 - 2 - 1$ c $2^3 \times 2^2 \times 2 \times 1$

12 a $5^2 - 4^2 - 3^2$ b $5^2 - (4 - 3)^2$ c $(5 - 4 - 3)^2$

13 a $5 + 2 \times 2 - 2$ b $(5 + 2) \times (2 - 2)$ c $5 + 2^2 - 2$

14 a $4 - 6 - 8 \div 2$ b $(4 - 6 - 8) \div 2$ c $4 - (6 - 8) \div 2$

15 a $7 \times 7 - 7$ b $7^2 - 7$ c $7 \times (7 - 7)$

Developing fluency

1 Find the value of $(a + b) \times (c - d)$ when

a $a = 3, b = 2, c = 4, d = 1$ b $a = 3, b = 7, c = 1, d = 0$ c $a = 1, b = -1, c = 4, d = 1$

Exam-style

2 Find the value of $p^2 - (q^2 + r^2)$ when

a $p = 3, q = 2, r = 1$ b $p = 10, r = 8, q = 6$ c $p = 6, q = 5, r = 4$

3 Find the value of $\sqrt{x \times (y - z)}$ when

a $x = 8, y = 12, z = 4$ b $x = 100, y = 4, z = 3$ c $x = 10, y = 5, z = 5$

4 Find the value of $r^3 + r^2 + r + 1$ and of $\frac{r^4 - 1}{r - 1}$ when

a $r = 2$ b $r = 5$ c $r = 9$

Exam-style

5 Insert brackets in each calculation to make these correct.

a $6 + 3 \times 2 = 18$ b $5 + 9 - 2 \times 2 = 19$ c $8 + 3 \times 2 + 1 = 23$

d $4 + 3 \times 3 + 2 = 35$ e $6 + 8 - 2 + 1 = 11$ f $13 - 5 + 4 - 2 = 6$

Exam-style

6 Jim and Ted are contestants in a TV quiz show. They each have to make a target number.
They can use the operations +, −, × and ÷ as many times as they wish.
Jim's target number is 240. The numbers he has are 100, 75, 5, 3 and 1.
Ted's target number is 151. The numbers he has are 200, 8, 5, 4 and 1.

a Show how Jim can make his target number using all 5 numbers.

b Show how Ted can make his target number using all 5 numbers.

In questions 7 to 9, work out these.

7 **a** $(5+1) \div (2+1)$ **b** $\frac{(5+1)}{(2+1)}$ **c** $\frac{5}{2}+\frac{1}{1}$

d $\frac{5+1}{2+1}$ **e** $5+1 \div 2+1$ **f** $(1+5) \div (1+2)$

8 **a** $3^2+4^2+12^2$ **b** $\sqrt{(3^2+4^2+12^2)}$ **c** $\sqrt{3^2}+\sqrt{4^2}+\sqrt{12^2}$

d $\sqrt{3^2+4^2}+\sqrt{12^2}$ **e** $(3+4+12)^2$ **f** $\sqrt{(3+4+12)^2}$

9 **a** $\frac{3^2+4^2}{36+64}$ **b** $\frac{(3+4)\times 2}{\sqrt{36}+\sqrt{64}}$ **c** $\frac{3^2+4^2}{\sqrt{36+64}}$

d $\frac{\sqrt{3^2}}{\sqrt{36}}+\frac{\sqrt{4^2}}{\sqrt{64}}$ **e** $\frac{\sqrt{3^2+4^2}}{\sqrt{36+64}}$ **f** $\sqrt{\frac{3^2+4^2}{36+64}}$

Reasoning

10 **a** Fill in the answers to questions **i** to **ix** in a copy of this square. What do you notice about the completed square?

i	**ii**	**iii**
iv	**v**	**vi**
vii	**viii**	**ix**

i $4 \times (4-4)+4$ **ii** $4+4 \div 4+4$ **iii** $\frac{4 \times 4}{4+4}$

iv $(4 \times 4-4) \div 4$ **v** $\frac{4 \times 4+4}{4}$ **vi** $44 \div 4-4$

vii $\sqrt{4}+\sqrt{4}+\sqrt{4}+\sqrt{4}$ **viii** $4^4 \div 4^4$ **ix** $(4+4) \div 4+4$

b Now find your own ways of using four 4s to get the same answers for **i** to **ix**.

Problem solving

Exam-style

1 Mo bought a new hammer for £10 and 5 bags of nails at £2 per bag.
He paid £30 and continued with his work.
Later that week, his partner Joe, was checking the accounts,
'Why did you pay £30 for this, Mo? You have been overcharged!'
Mo said 'No, that's right.' He got out his calculator. 'Look…'
Sure enough – the total was £30.

a What buttons did Mo press on the calculator?

b Who is right and what is the correct bill?

c What will Mo have to be careful to do in the future so this does not happen again?

d Will Mo get the items cheaper if he buys the hammer last? Is that a rule he can work with in future?

Exam-style Extension

2 Put brackets where appropriate to make the following statements true.

a $10-1 \div 3+6=9$

b $3+7 \times 2-5=-30$

c $2 \times 1+3^2=32$

Exam-style

3) Michelle worked out the value of $6x + 3x^2$ when $x = 2$.
Hannah worked out the value of $4y^2 - 52$ when $y = 5$.
Finlay worked out the value of $4z^2 \div 3$ when $z = 3$.
They all got an answer of 48.
Whose answer of 48 is correct?
Use brackets to show how the other two got their answers.

Extension

Exam-style

4) Here are four different numbers.
2 3 4 11

a Put one number in each box to make a correct statement.
You may only use each number once.

Answer box

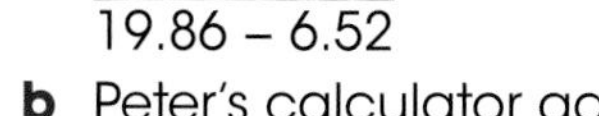

$\square + \square \times \square = \square$

b Using just three of these numbers, what is the greatest answer that could appear in the answer box?

Exam-style

5) Here are four different symbols.

(+ ×)

a Use each symbol once only to make this statement correct.

6 1 7 = 48

Here are four more symbols.

(÷ −)

b Use each symbol once only to make this statement correct.

9 7 4 = 3

Exam-style

6) **a** Use your calculator to work out

$$\frac{5.36 + 62.87}{19.86 - 6.52}$$

Extension

b Peter's calculator gave an answer of 2.005659617
Write, in order, the buttons Peter may have pressed on his calculator to get this answer.

Reviewing skills

Work these out.

1) **a** $(7 - 2) \times 3$ **b** $7 - 2 \times 3$ **c** $7 - (2 \times 3)$

2) **a** $100 \div (10 - 6)$ **b** $100 \div 10 - 6$ **c** $\sqrt{100} \div 10 - 6$

3) **a** $5^4 \div (5^3 - 5^2)$ **b** $5^4 \div 5^3 - 5^2$ **c** $(5^4 \div 5^3 - 5)^2$

4) **a** $\dfrac{8 + 4}{2 + 1}$ **b** $(8 + 4) \div (2 + 1)$ **c** $\dfrac{8}{2} + \dfrac{4}{1}$

5) Find the value of $(a - b) \div c$ when

a $a = 10, b = 5, c = 2$ **b** $a = 5, b = 5, c = 2$ **c** $a = 5, b = 10, c = 2$

6) Insert brackets in these calculations to make them correct.

a $12 + 4 \div 5 - 1 = 4$ **b** $7 - 3 \times 4 - 2 = 1$

Unit 8 • Multiplying decimals • Band f

Outside the Maths classroom

Laying a lawn

Why does a gardener need to use decimals when buying turf?

Toolbox

When you multiply decimals you have two things to think about:

- getting the correct digits
- getting the decimal point in the right place.

Here are three methods you can use.

- rewriting the decimals so you are working with whole numbers.
  For example
  $0.03 = 3 \div 100$
  $8 \times 0.03 = 8 \times 3 \div 100 = 24 \div 100 = 0.24$
- the grid method
- long multiplication.

Example – Rewriting the decimals

Work out these calculations.

a 6×0.3

b 12×0.005

Solution

a $6 \times 0.3 = 6 \times 3 \div 10$ (0.3 is rewritten as $3 \div 10$)

$= 18 \div 10$

$= 1.8$

b $12 \times 0.005 = 12 \times 5 \div 1000$ (0.005 is rewritten as $5 \div 1000$)

$= 60 \div 1000$

$= 6 \div 100$

$= 0.06$

Example – The grid method

Calculate 4.36 × 23

Solution

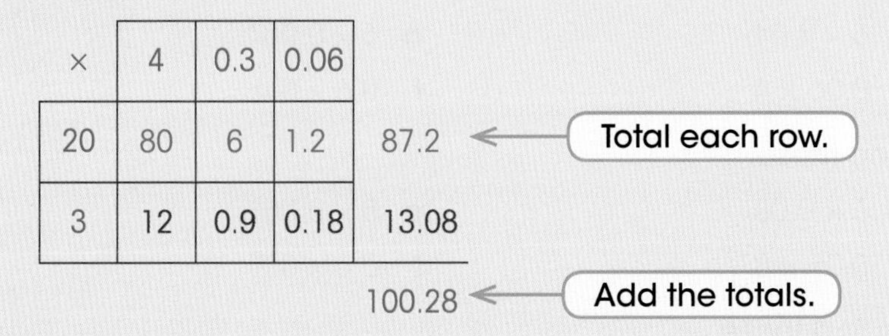

×	4	0.3	0.06	
20	80	6	1.2	87.2
3	12	0.9	0.18	13.08
				100.28

Example – Using long multiplication

Work out 12.36 × 1.5

Solution

We can treat 12.36 × 1.5 as 1236 × 15 as long as you adjust the number of decimal places at the end.

There are 2 + 1 decimal places here… → …so put 3 decimal places in your answer.

Work out 1236 × 15

	1	2	3	6
	×		1	5
1	2	3	6	0
	6	1	8	0
1	8	5	4	0

1 × 1236 → 12360

This zero is a placeholder.

Now locate the decimal point.

12.36 × 1.5 is between 12 × 1 = 12 and 12 × 2 = 24.

So the answer should have two digits before the decimal point.

It is 18.540 or 18.54

Another way to locate the decimal point is to count the number of digits to the right of the decimal point. This should be the same in the question and the answer.

12.36 × 1.5 has 3 digits to the right, so the answer is 18.540
(12 3 321)

The questions in this unit should be answered *without* the use of a calculator.

Practising skills

1. You know that $6 \times 4 = 24$. Use it to find these.
   - **a** 6×0.4
   - **b** 6×0.04
   - **c** 6×0.004
   - **d** 0.6×4
   - **e** 0.06×4
   - **f** 0.6×0.4
2. You know that $5 \times 8 = 40$. Use it to find these.
   - **a** 5×0.8
   - **b** 5×0.08
   - **c** 5×0.008
   - **d** 0.5×8
   - **e** 0.05×8
   - **f** 0.5×0.8
3. You know that $15 \times 8 = 120$. Use it to find these.
   - **a** 15×0.8
   - **b** 15×0.08
   - **c** 15×0.008
   - **d** 15×0.0008
   - **e** 1.5×0.8
   - **f** 0.15×0.8
4. Find these.
   - **a** 25×4
   - **b** 25×0.4
   - **c** 25×0.04
   - **d** 25×0.004
   - **e** 25×0.000004
   - **f** 0.25×0.04
5. You are given that $19 \times 53 = 1007$. Find these.
   - **a** 1.9×53
   - **b** 1.9×5.3
   - **c** 190×5.3
   - **d** 0.19×0.53
   - **e** 0.19×5300
   - **f** 0.0019×0.00053
6. Find these.
   - **a** 0.2×0.3
   - **b** 0.4×0.2
   - **c** 0.6×0.7
   - **d** 0.9×0.8
   - **e** 0.5×0.4
   - **f** 0.9×0.9
7. Find these.
   - **a** 25×0.064
   - **b** 41×0.022
   - **c** 99×0.00011
   - **d** 90×0.8
   - **e** 500×0.000004
   - **f** 125×0.008
8. Find these.
   - **a** $12 \times 0.5 \times 0.5$
   - **b** $600 \times 0.5 \times 0.4$
   - **c** $0.1 \times 0.2 \times 0.3$
   - **d** $0.2 \times 0.04 \times 0.008$
   - **e** $0.02 \times 50 \times 0.71$
   - **f** $0.12 \times 2.4 \times 0.5$
9. Find these.
   - **a** 3^2
   - **b** 0.3^2
   - **c** 0.03^2
   - **d** 0.3^3
   - **e** 0.03^3
   - **f** 0.3^4
10. Work out the cost of these.
    - **a** 8 apples at £0.46 each
    - **b** 5 books at £2.58 each
    - **c** 12 pens at £1.09 each
    - **d** 18 drinks at £0.79 each

Developing fluency

1. Find these.

   a $3 \times (-0.4)$ b $0.3 \times (-0.4)$ c $(-0.3) \times 0.4$

2. Find these.

   a $9 \times (-1.2)$ b $(-9) \times (-1.2)$ c $(-900) \times (-0.0012)$

3. Petrol costs £1.42 per litre. Calculate the cost of 55 litres.

Exam-style

4. Turkey costs £8.10 per kilogram and chicken costs £6.45 per kilogram.
   Alana buys 0.6 kg of turkey and 1.2 kg of chicken.
   What is the total cost?

5. An electricity bill shows the following information.

Electricity bill	
Present reading	12768 units
Previous reading	11988 units
Cost per unit	£0.152
Standing charge	£19.50

   Work out the total charge.

Exam-style

6. The table shows the calories used up when doing some types of exercise.

Exercise	Calories per minute
Aerobics	10.5
Running	12.4
Swimming	6.8

   a Jana has a 40 minute workout. She spends 15 minutes doing aerobics and the rest swimming. How many calories does she burn?

   b Jack goes running for 45 minutes. How many more calories does he burn than Jana?

7. Zack's pay is £9.28 per hour, Monday to Friday. On Saturdays he is paid time and a half.
   Last week he worked 6 hours on Monday, 7.2 hours on Tuesday and 8.5 hours on Saturday.
   Calculate his pay for the week.

Exam-style

8. The diagram shows a water tank.

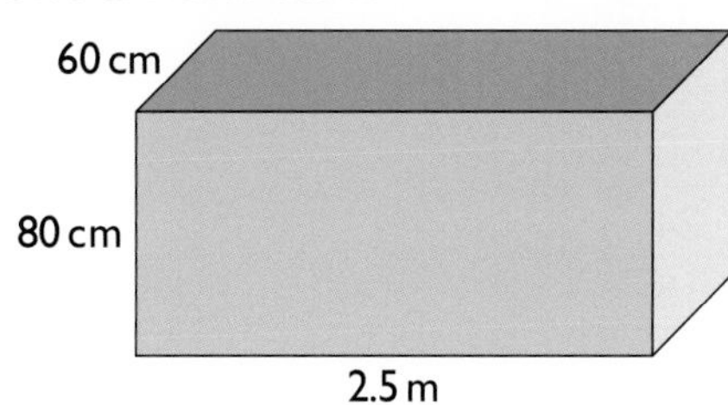

   Find its capacity in cubic metres.

Problem solving

1 This is Sandy's shopping bill at the local pet store. Each price shown is for one article.
Sandy pays with a £20 note and £5 note.
How much change should he get?

Pets U Like

8 × Cat food @ £ 0.33
1 × Cat litter @ £ 4.99
2 × Dog chew @ £ 0.25
2 × Dog bowl @ £ 2.75
1 × dog food @ £ 0.37
3 × Nibbles @ £ 1.16
2 × Collars @ £ 1.99

You need it,
we have it!

2 Five of the rectangular tiles pictured are used to make the pattern below.

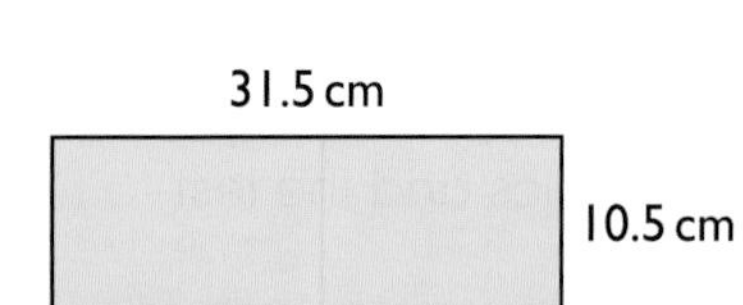

What is the area, in cm^2, of this pattern?

3 The table below shows the monthly payments per person for an insurance policy.
The payments depend upon the age of the person.

Age (in years)	Monthly payment per person	
	Standard rate	Premium rate
0–16	£8.15	£9.58
17–20	£11.45	£13.80
21–30	£15.30	£17.24
31–40	£20.42	£23.15
41–50	£35.46	£41.98
Over 50	£62.85	£82.50

Tanvir is aged 17 and his father is 49.
Tanvir pays for his insurance at the premium rate.
His father pays for his insurance at the standard rate.
Work out how much more money Tanvir's father pays each year than Tanvir.

Exam-style

4 David and Elaine both work at the sports centre.
David works from 6 p.m. to 9.30 p.m. from Monday to Friday each week.
He earns £7.80 per hour.
Elaine works on Saturdays and Sundays.
She works from 8.30 a.m. to 12.30 p.m. and from 1.30 p.m. to 4.30 p.m.
She earns £9.50 per hour.
Who earns more money each week?
You must show clearly how you got your answer.

Exam-style

5 It costs £12.50 for a rail ticket lasting one day from Ashton to Manchester.
It costs £2700 for an annual rail ticket lasting one year from Ashton to Manchester.
Becky travels from Ashton to Manchester on 220 days in every year.
Is it cheaper for Becky to buy a £12.50 ticket each day or to buy a £2700 ticket for the year?

Exam-style

6 Caroline works in a factory.
The table shows how long she works each day one week.

Monday	Tuesday	Wednesday	Thursday	Friday
8.5 hours	8.4 hours	8.5 hours	10.2 hours	7 hours

Caroline is paid for the total time she works in a week.
Her pay is £9 per hour for the first 40 hours she works in the week.
She is paid £12 per hour for any extra hours she works in the week.
Work out Caroline's pay for the week.

Exam-style

7 A double sheet of a newspaper measures 72.5 cm by 60 cm.
The newspaper contains six double sheets.
Work out the area of paper in one copy of the newspaper.
Give your answer in square metres.

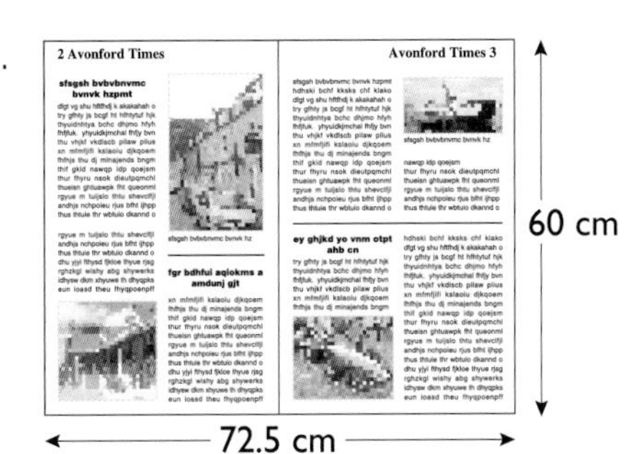

Exam-style

8 Georgina is on holiday in Portugal.
In Portugal a pair of sunglasses costs €54.
In England the same type of sunglasses cost £48.
The exchange rate is £1 = €1.20
In which country are the sunglasses cheaper?

Extension

Exam-style

9 Michael and Lewis are racing drivers.
On a racing track, Michael drove 72 laps in an average time per lap of 2.25 minutes.
On the same racing track, Lewis drove 70 laps in an average time per lap of 2.3 minutes.
Work out the difference between the total time that Michael drove and the total time that Lewis drove.

Extension

Reviewing skills

1 Work out these.

a 8×0.2 **b** 5×0.9 **c** 0.3×7 **d** 1.6×10

e 1.3×3 **f** 0.8×0.6 **g** 0.6^2 **h** 0.04^2

2 Use the fact that $186 \times 9.4 = 1748.4$ to write down the answers to the following.

a 18.6×9.4 **b** 1.86×9.4 **c** 186×0.94

d 0.186×9.4 **e** 0.0186×0.94 **f** 18.6×94

3 Find these costs.

a Sylvia buys 0.8 kg of cherries and 0.2 kg of strawberries.

b Kyle buys 3.5 kg of damsons and 0.5 kg of blackberries.

c Ramona buys 400 g of each type of fruit.

Strawberries......... £5.40 per kilo
Damsons............... £4.20 per kilo
Cherries................ £8.60 per kilo
Blackberries.......... £4.80 per kilo

Unit 9 • Dividing decimals • Band f

Outside the Maths classroom

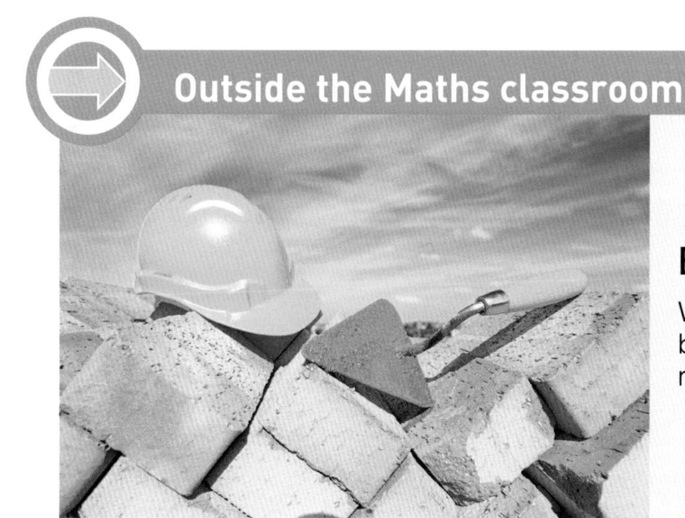

Building quantities

What measurements and calculations does a builder make to estimate the number of bricks needed for a wall?

Toolbox

Here are two methods you can use for working out a division problem:

- using equivalent fractions

  Start by writing the division as a fraction.

  Then multiply the top and bottom by a power of 10 so that both become whole numbers.

  You now have an easier equivalent fraction to work with.

  Here is how to find $0.6 \div 0.05$:

$$\frac{0.6}{0.05} \overset{\times 100}{=} \frac{60}{5} = 12$$

$\frac{0.6}{0.05}$ and $\frac{60}{5}$ are equivalent fractions.

- short division

  Here is how to find $4.4 \div 6$ using short division:

$$6\overline{)4.4^{2}0} \;\; 0.7 \longrightarrow 6\overline{)4.4^{2}0^{2}} \;\; 0.7\,3 \longrightarrow 6\overline{)4.4^{2}0^{2}0} \;\; 0.7\,3\,3$$

  You can also use long division.

Example – Using equivalent fractions

Work out these calculations.

a $3 \div 0.2$

b $1.2 \div 0.03$

Solution

a Write this as a fraction $\frac{3}{0.2}$

Now multiply both the numbers (top and bottom) by 10. That makes them into whole numbers.

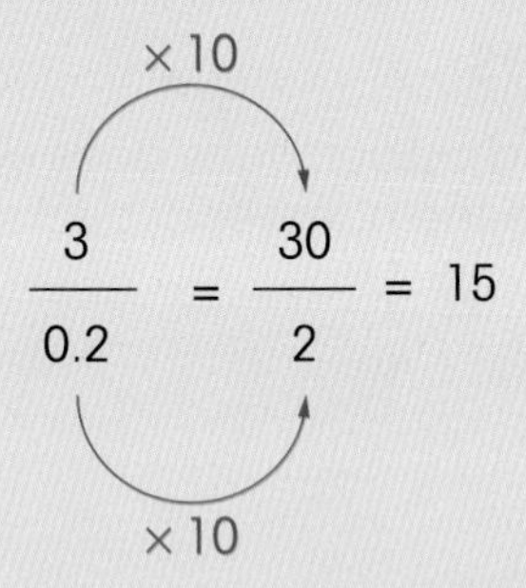

You can see this result on the number line from 0 to 3. It is divided into 15 pieces each 0.2 long.

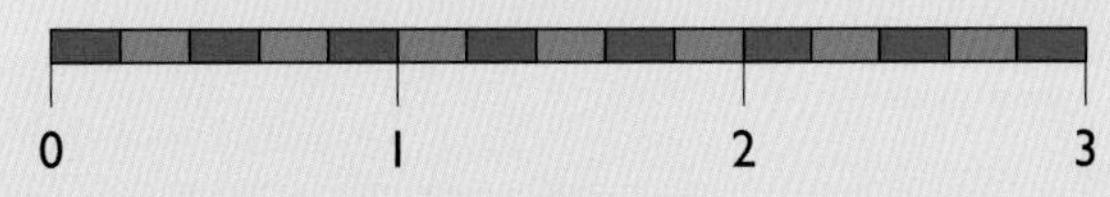

b $1.2 \div 0.03 = \frac{1.2}{0.03}$

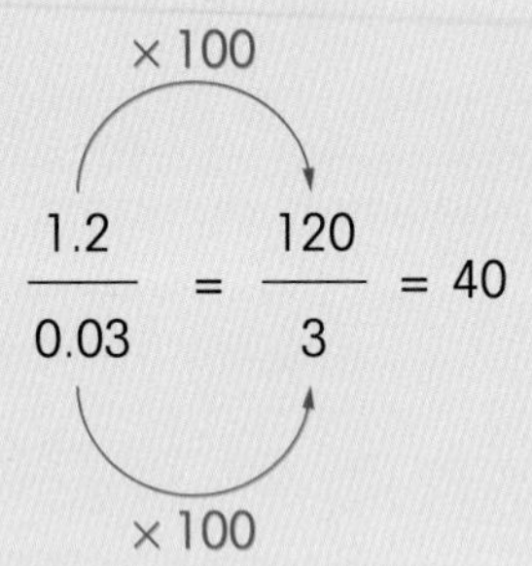

Example – Using short division

Work out $3 \div 8$

Solution

$$8\overline{)3.0^{6}0^{4}00}$$ with quotient 0 3 7 5

You can write extra zeros at the end of the number, after the decimal point.

This extra zero wasn't needed.

Answer 0.375

The questions in this unit should be answered *without* the use of a calculator. (However, you may wish to use a calculator to check some of your answers.)

Practising skills

1. Work these out.

   a $6.48 \div 2$ **b** $9.70 \div 2$ **c** $6.5 \div 2$ **d** $3.5 \div 4$

   e $88 \div 5$ **f** $19 \div 4$ **g** $7.3 \div 8$ **h** $0.054 \div 6$

   i $14.028 \div 7$ **j** $29.613 \div 4$ **k** $29.613 \div 3$ **l** $4.5376 \div 8$

2. How much does each person receive?

   a Share £19 between 2 people.

   b Share £16 between 5 people.

   c Share £705 between 6 people.

   d Share £293.10 between 6 people.

3. You know that $45 \div 3 = 15$. Work these out.

   a $4.5 \div 3$ **b** $0.45 \div 3$ **c** $0.0045 \div 3$

   d $45 \div 0.3$ **e** $45 \div 0.03$ **f** $4.5 \div 0.3$

4. Work these out.

   a $0.7 \div 0.7$ **b** $3.5 \div 0.5$ **c** $1.2 \div 0.6$ **d** $2.4 \div 0.4$

   e $1.9 \div 0.1$ **f** $4.8 \div 0.8$ **g** $6.3 \div 0.3$ **h** $2.7 \div 0.2$

   i $0.1 \div 0.4$ **j** $3.5 \div 0.4$ **k** $0.315 \div 0.9$ **l** $2.4934 \div 0.07$

5. Work these out.

   a $74 \div 0.2$ **b** $0.9 \div 0.6$ **c** $25.32 \div 0.3$ **d** $9.2 \div 0.05$

   e $1.78 \div 0.04$ **f** $16.92 \div 1.2$ **g** $56 \div 0.08$ **h** $0.009 \div 0.01$

   i $2.07 \div 0.002$ **j** $0.0371 \div 0.07$ **k** $32.04 \div 0.009$ **l** $0.432 \div 0.006$

6. Which division is the odd one out?

   $16 \div 0.2$ $160 \div 20$ $1.6 \div 0.02$ $0.16 \div 0.002$

Developing fluency

1. Match each division to its correct answer.

Division	Answer
$72 \div 9$	0.08
$0.72 \div 9$	800
$7.2 \div 9$	8
$720 \div 0.9$	0.8
$7.2 \div 0.09$	0.08
$0.72 \div 90$	80

2) Work these out.

a $69 \div (-2)$ **b** $(-7.8) \div 5$ **c** $2.9 \div (-4)$ **d** $(-0.01) \div 4$

e $(-5) \div (-8)$ **f** $(-0.01) \div 20$ **g** $(-2.8) \div (-40)$ **h** $0.8 \div 5$

3) Work these out.

a $0.1 \times 0.4 \div 0.2$ **b** $\frac{0.1 \times 0.4}{0.2}$ **c** $\frac{0.06}{0.2 \times 0.003}$ **d** $\frac{(0.6)^2}{0.4 \times 0.9}$

e $\frac{120 \times 0.05}{0.003 \times 0.04}$ **f** $(11 + 2.2) \div 0.12$ **g** $\frac{(0.2)^5}{64}$ **h** $\frac{1.8 \times 55}{0.033}$

Exam-style

4) Richard travels to and from work 4 days a week.
In total he travels 840 km in 4 weeks.
How far is it from his home to his work?

5) Beef costs £7.80 per kilogram. Ruth buys some beef and pays £9.75.
How much beef does she buy?

6) Kay has two bottles of cola. She shares the 1.5 litre bottle equally among 6 boys, and the 2.5 litre bottle equally among 8 girls.

a How much more does each girl get than each boy?

b Suppose instead Kay shares the cola equally between all the children.
How much would each of them get?

Problem solving

Exam-style

1) Angus bought a new television.
The television cost £628.40
Angus paid a deposit of £50 when he bought the television.
He paid the rest of the money in 12 equal monthly payments.
Work out how much Angus paid each month.

Exam-style

2) There is 400 g of hot chocolate in this jar.

a What is this in kilograms?

A factory produces 10 000 kg of hot chocolate in one day.
The jars are packed in boxes of 50 jars.

b How many full boxes of these jars are produced each day?

Exam-style

3) A book is 30 cm by 20 cm and 1.4 cm thick.
Work out the greatest number of books of this size that will fit into a box with dimensions 0.9 m by 0.4 m by 0.35 m.

4. Billy is organising a conference at the Sea View Hotel.
He uses this number machine to work out the cost in pounds.

The cost is £305.

a Work out the number of people at this conference.

b For bookings taken for next year, the price per person has increased.

As a result, the cost for the same number of people will be £318.50.

What is the new cost per person?

5. The diagram shows the plan of Rob's patio.

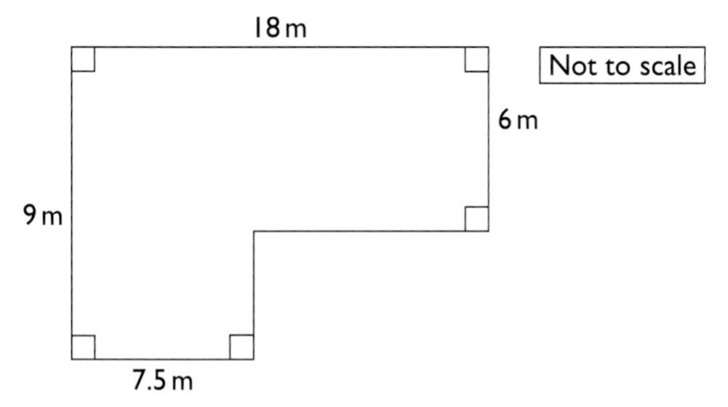

Rob is going to cover his patio with square paving slabs.
Each paving slab has an area of $2.25\,m^2$.
Work out how many paving slabs Rob will need.

6. Matthew works out the time it will take to cook a chicken.
He uses this number machine.

Matthew cooks two chickens.
One chicken takes 1.7 hours and the other takes 2.3 hours.
Work out the difference in weight between the two chickens.

7. Tea bags are sold in two sizes of box.

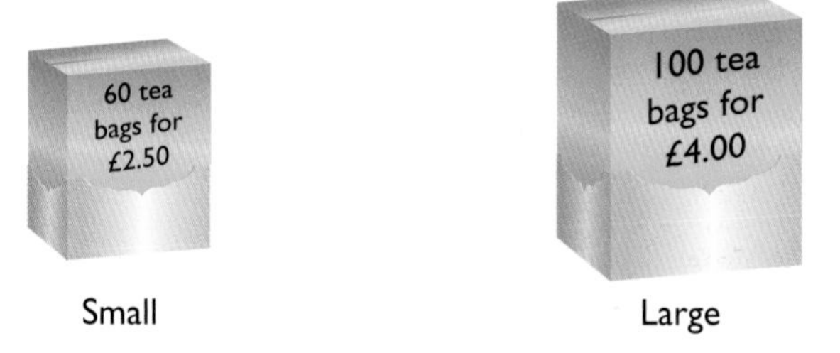

a Compare the number of tea bags per £ that can be bought from each size of box.

An extra-large box is produced selling 450 tea bags for £16.

b Show that this offers better value for money than both the small and large boxes.

Exam-style | Extension

8 Angela asks two companies what their price is for the same kitchen units.
Cuisinaire of Paris says their price is €9600.
Chucks of New York say their price is \$12 300.
The exchange rates are £1 = €1.20 and £1 = \$1.50.
Show that Cuisinaire of Paris has the lower price.

Exam-style | Extension

9 Mary has to drive from Dover to London and then from London to Manchester.
The total distance she has to drive is 297 miles.
Mary drives the 80 miles from Dover to London at an average speed of 40 mph.
The drive from London to Manchester takes 3.4 hours.
Work out Mary's average speed for the whole journey from Dover to Manchester.

Reviewing skills

1 Work these out.

a $7.2 \div 0.9$ **b** $4.8 \div 0.2$ **c** $51 \div 0.03$ **d** $3.14 \div 0.002$

2 Work these out.

a $0.8 \div 0.01$ **b** $0.27 \div 0.03$ **c** $0.009 \div 0.06$ **d** $0.05 \div 4$

3 Work these out.

a $0.3 \times 0.4 \div 0.12$ **b** $\frac{(0.4)^2}{0.008}$ **c** $\frac{1}{0.2 \times 0.4}$ **d** $\frac{0.5 \times 0.022}{11}$

4 Work these out.

a $(-0.2) \div 0.01$ **b** $(-8) \div 0.025$ **c** $0.33 \div (-1.1)$ **d** $(-2.64) \div (-0.24)$

5 How many 60p stamps can you buy for £300?

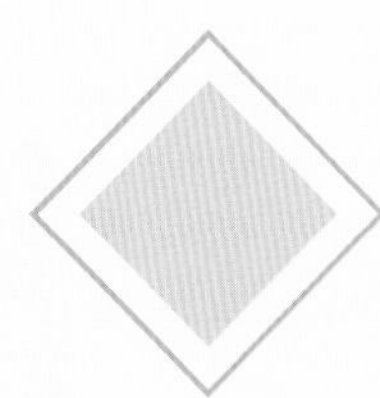

Number Strand 2 Using our number system

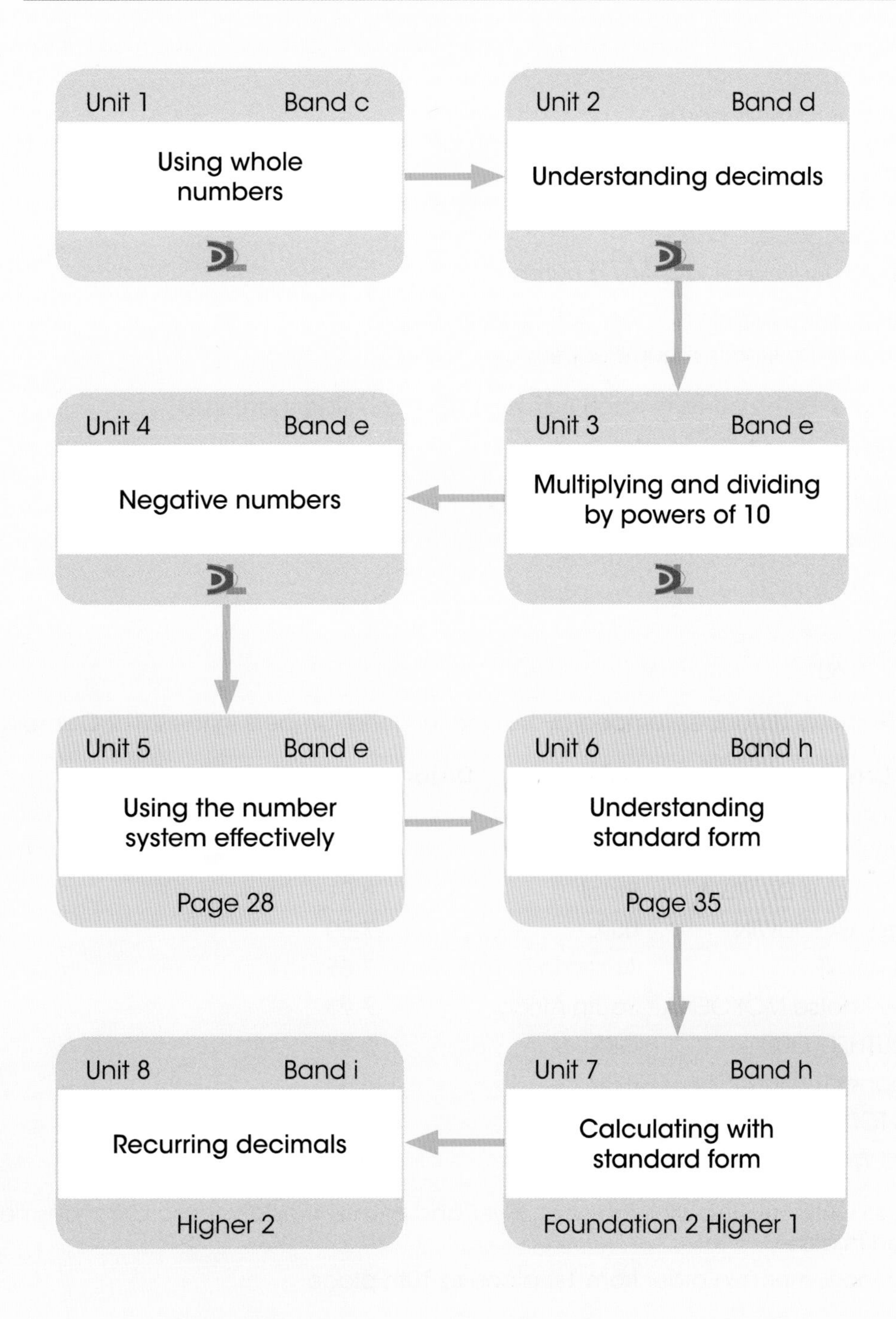

Units 1–4 are assumed knowledge for this book. They are reviewed and extended in the Moving on section on page 24. Knowledge and skills from these units are used throughout the book.

Units 1–4 • Moving on

Exam-style

1. Beth has four number cards.

4	1	9	6

   Beth uses all four cards to make a number.

   a **i** What is the largest number Beth can make?

   ii What is the smallest number Beth can make?

   b Use one of Beth's cards in each box to make each statement true.

   $12 \times 30 = \square \times 60 = \square \times 90$

Exam-style

2. Write one of the symbols in each box to make each statement true.

   < **>** **=**

   a $0.7 \square 0.79$

   b $-4 \square -3$

   c $0.6 \square 0.60$

Exam-style

3. The table shows distances jumped by the top ten long jumpers in the 2012 Olympic Games.

Long jumper	Country	Distance (in metres)
Sebastian BAYER	Germany	8.10
Will CLAYE	USA	8.12
Mauro Vinicius DA SILVA	Brazil	8.01
Marquise GOODWIN	USA	7.80
Henry FRAYNE	Australia	7.85
Godfrey Khotso MOKOENA	South Africa	7.93
Greg RUTHERFORD	GBR	8.31
Christopher TOMLINSON	GBR	8.07
Michel TORNEUS	Sweden	8.11
Mitchell WATT	Australia	8.16

   In order to work out who wins the gold, silver and bronze medals, these distances need to be arranged in order.

   Put the long jumpers in order from 1st place to 10th place.

Exam-style

4. Here are three different digits.

   1 **7** **9**

   a Put one digit in each box to make the largest decimal number.
   You may only use each digit once.

   $0.\square\square\square$

   b Work out the difference in value between the largest decimal number and the smallest decimal number.

Exam-style

5 A Grand Prix driver has his fastest lap timed at 153.498 seconds.

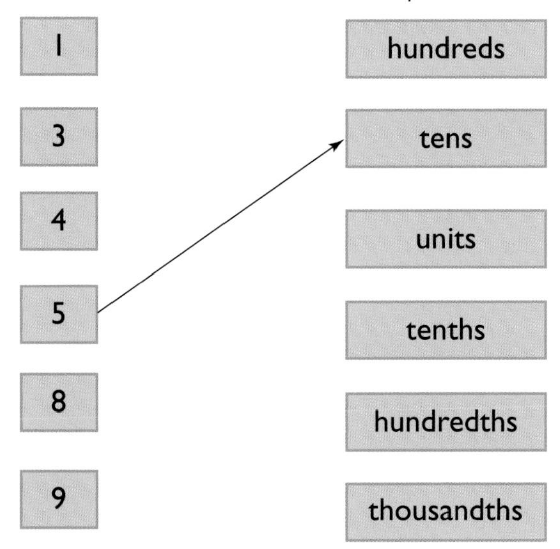

Match the digits to their place value.
One has been done for you.

Exam-style

6 Steve is driving down the motorway.
There are three service stations before he reaches his destination.
The next service station, A, is 3.5 miles away.
Another service station, C, is 27.9 miles away.
A third service station, B, is half way between A and C.

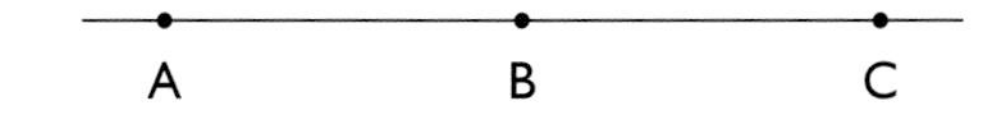

Steve's in-car computer tells him that he only has enough petrol for a further 16 miles.
Does he have enough petrol to reach service station B?

Exam-style

7 In a car park there are 6 levels. The ground level is level 0.
There are 3 levels above ground level, levels 1 to 3.
There are 2 levels below ground level, levels −1 and −2.
Car A is parked on level 3.
Car B is parked on level −1.
Car C is parked on level −2.
The drivers of the three cars are looking for more convenient places to park.
Car A goes down 4 levels, car B goes up 3 levels and car C goes up 2 levels.

a How many more levels does car A have to go up or down to be on the same level as car B?

The exit is on the ground level.

b Which car is nearest to the exit?

Exam-style

8 Four timekeepers at a race work out the difference in seconds between the first and second place competitors. Here are their results.

Timekeeper A	Timekeeper B	Timekeeper C	Timekeeper D
2 and 3 hundredths	23 tenths	2330 thousandths	2.003

The correct difference in time was 2.03 seconds.

a Which timekeeper worked out the time correctly?

b Which timekeeper worked out the time difference furthest from 2.03 seconds?

Exam-style

9 Fiona and Sam are running stalls at their school fete.

On Fiona's stall, people roll a ten pence coin down a slide onto a squared board.

If the coin lands on a square they win a prize.

On Sam's stall, people roll a two pence coin down a slide onto a squared board.

If the coin lands on a square they win a prize.

At the end of the day, Fiona has collected 126 ten pence coins and Sam has collected 584 two pence coins.

Work out, in £, how much in total Fiona and Sam collected.

Exam-style

10 The actual size of this photograph is 150 mm by 100 mm.

Hannah is making a display of some photographs of this size.

She is going to put them on a wall chart.

The wall chart is in the shape of a rectangle measuring 3 m by 2 m.

The photographs must not overlap.

Work out the greatest number of photographs Hannah can put on this wall chart.

Exam-style

11 Coley opens his post.
He keeps a record of his spending so that he can check his bank statements.

Oh dear! Here is the bill from the garage. I must pay £285.

Credits (+)	Debits (–)	Balance (£)
		503.00
+60.00		
	–285.00	

a Copy and complete this account sheet for all these items.

Payment for TV licence £9

Payment for mortgage £250

Payment for gas bill £70

Premium Bond win £50

Salary £1500

Work out the balance each time.

Exam-style

12 Hannah is thinking of two numbers.
She says that, when she puts a negative sign in front of each number, they stay in the same order on a number line.

a Is Hannah correct? Explain your answer.

Jeremy is thinking of two numbers.
He multiplies the two numbers together.
He then adds 5.
The answer is –3.

b What could the two numbers that Jeremy thought of be?

Exam-style

13 There are five numbers, in order, on a number line.
The first number is –32.
The fifth number is +88.
The five numbers are each an equal distance apart on the number line.
Write down the five numbers.

Unit 5 • Using the number system effectively • Band e

Outside the Maths classroom

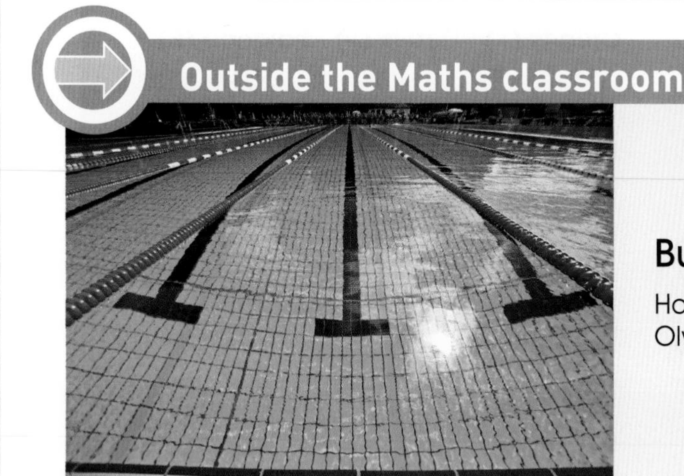

Building swimming pools

How many tiles are needed to tile an Olympic swimming pool?

Toolbox

Thinking of dividing as 'how many are there in...?'

How many 2s are there in 8?

$2 + 2 + 2 + 2 = 8$

There are four 2s in 8 so $8 \div 2 = 4$.

How many 0.1s are there in 1.3?

$0.1 + 0.1 + 0.1 + 0.1 + 0.1 + 0.1 + 0.1 + 0.1 + 0.1 + 0.1 + 0.1 + 0.1 + 0.1 = 1.3$

There are thirteen 0.1s in 1.3 so $1.3 \div 0.1 = 13$.

This is the same as multiplying 1.3 by 10.

This place-value table shows 4.67 divided by 0.01

H	T	U	.	$\frac{1}{10}$	$\frac{1}{100}$
		4	.	6	7
4	6	7	.		

There are 100 0.01's in 1, so there are 4.67 x 100 0.01's in 4.67.

The place-value table shows that dividing by 0.01 has exactly the same effect as multiplying by 100.

In the same way,

- multiplying by 0.1 is the same as dividing by 10
- multiplying by 0.01 is the same as dividing by 100
- multiplying by 0.001 is the same as dividing by 1000 and so on.

Example – Multiplying and dividing by 0.1 and 0.01

Work out the answers to these calculations.

a 32 × 0.1
b 320 × 0.01
c 32 ÷ 0.1
d 32 ÷ 0.01

Solution

a Using a place-value table to multiply by 0.1, think of 30 lots of 0.1 which makes 3, and 2 lots of 0.1 which makes 0.2.

H	T	U	.	$\frac{1}{10}$	$\frac{1}{100}$	$\frac{1}{1000}$
	3	2	.			
		3	.	2		

Using a place-value table can help to keep track of the digits when multiplying or dividing by powers of 10.

So 32 lots of 0.1 = 3.2

Using the same idea for the other calculations:

b 320 × 0.01 = 3.2
c 32 ÷ 0.1 = 320
d 32 ÷ 0.01 = 3200

Example – Division using known facts

Use this known fact to work out the calculations below.

720 ÷ 0.1 = 7200

a 7200 ÷ 0.1
b 7200 ÷ 0.01
c 72 ÷ 0.1
d 720 ÷ 0.01
e 7200 ÷ 10
f 7200 ÷ 1000

Solution

a 7200 ÷ 0.1 = 72 000
b 7200 ÷ 0.01 = 720 000
c 72 ÷ 0.1 = 720
d 720 ÷ 0.01 = 72 000
e 7200 ÷ 10 = 720
f 7200 ÷ 1000 = 7.2

Example – Multiplication using known facts

Use this known fact to work out the calculations below.

$720 \times 0.1 = 72$

a 7200×0.1
b 7200×0.01
c 72×0.1
d 720×0.01
e 7200×10
f 7200×1000

Solution

a $7200 \times 0.1 = 720$
b $7200 \times 0.01 = 72$
c $72 \times 0.1 = 7.2$
d $720 \times 0.01 = 7.2$
e $7200 \times 10 = 72\,000$
f $7200 \times 1000 = 7\,200\,000$

Example – Mental calculation using known facts

Salman

Solution

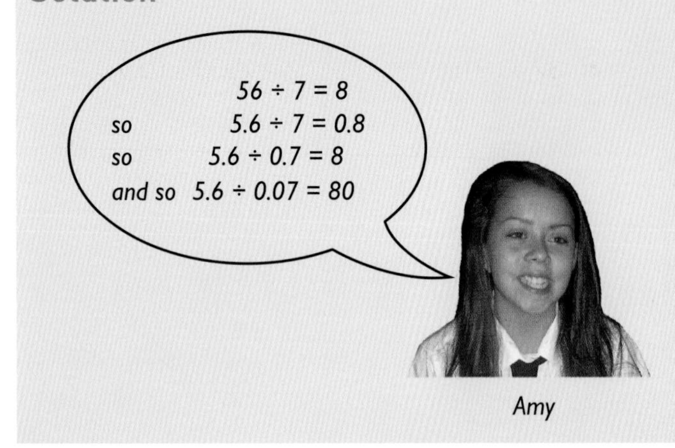

Amy

Practising skills

1. Work these out.

   a **i** 7000×1000
   ii 7000×100
   iii 7000×10
   iv 7000×1
   v 7000×0.1
   vi 7000×0.01
   vii 7000×0.001

   b **i** 2468×1000
   ii 2468×100
   iii 2468×10
   iv 2468×1
   v 2468×0.1
   vi 2468×0.01
   vii 2468×0.001

   c Use the word 'less' or 'more' to complete these sentences.
   7000×10 gives an answer which is _____ than 7000.
   7000×0.1 gives an answer which is _____ than 7000.

2. Work these out.

   a **i** $8000 \div 1000$
   ii $8000 \div 100$
   iii $8000 \div 10$
   iv $8000 \div 1$
   v $8000 \div 0.1$
   vi $8000 \div 0.01$
   vii $8000 \div 0.001$

   b **i** $6500 \div 1000$
   ii $6500 \div 100$
   iii $6500 \div 10$
   iv $6500 \div 1$
   v $6500 \div 0.1$
   vi $6500 \div 0.01$
   vii $6500 \div 0.001$

   c Use the word 'less' or 'more' to complete these sentences.
   $8000 \div 10$ gives an answer which is _____ than 8000.
   $8000 \div 0.1$ gives an answer which is _____ than 8000.

3. Work these out.

   a 40×0.1 **b** 600×0.1 **c** 9×0.1 **d** 8.4×0.1
   e $125 \div 0.1$ **f** $993 \div 0.1$ **g** $6.2 \div 0.1$ **h** $5.17 \div 0.1$

4. Work these out.

   a 0.6×0.01 **b** 500×0.01 **c** 8000×0.01 **d** 145×0.01
   e $246.9 \div 0.01$ **f** $61.3 \div 0.01$ **g** $32 \div 0.01$ **h** $2000 \div 0.01$

5. Work these out.

   a 225.9×0.001 **b** 638×0.001 **c** 8×0.001 **d** 0.4×0.001
   e $5.84 \div 0.01$ **f** $0.7 \div 0.001$ **g** $24.9 \div 0.001$ **h** $0.0815 \div 0.001$

6. Work these out.

   a $6 \div 0.1$ **b** 13×0.1 **c** $4.7 \div 0.001$ **d** $52.9 \div 0.01$
   e 0.8×0.1 **f** $7.65 \div 0.001$ **g** $5 \div 0.01$ **h** 46×0.01

7 Work these out.

a $180 \div 0.01$ **b** 2.3×0.01 **c** $6.91 \div 0.01$ **d** $0.7 \div 0.01$

e 50×0.01 **f** $3.2 \div 0.001$ **g** 1.64×0.001 **h** $5.899 \div 0.0001$

8 Work these out.

a $0.6 \div 0.001$ **b** 12.2×0.01 **c** $0.07 \div 0.01$ **d** $0.18 \div 0.1$

e $0.009 \div 0.001$ **f** 0.746×0.1 **g** $45.228 \div 0.01$ **h** 3.6078×0.001

Developing fluency

1 Beef costs £8.60 per kilo.

Copy and complete each sentence.

a 1 kg of beef costs 8.60 × 1 = £☐

b 10 kg of beef costs 8.60 × ☐ = £☐

c 0.1 kg of beef costs 8.60 × ☐ = £☐

2 **a** Mary has a bag of 10p coins. It is worth £3. How many 10p coins are in the bag?

b Using pence this is written $300 \div 10 = 30$.

Complete this equivalent statement using pounds. 3 ÷ ☐ = ☐

3 Write the answers to these calculations in order of size, smallest first.

a 17×0.1 **b** 26.5×0.01 **c** 790×0.01 **d** 125.1×0.001

e $0.7 \div 0.1$ **f** $0.65 \div 0.01$ **g** $1.124 \div 0.001$ **h** $7 \div 0.001$

Exam-style

4 Here is Kevin's homework.

Mark the homework and correct any answers that are wrong.

1 $86 \times 0.1 = 8.6$
2 $20 \times 0.1 = 200$
3 $2.5 \div 0.1 = 0.25$
4 $3 \div 0.1 = 30$
5 $18 \times 0.01 = 0.18$
6 $1121 \div 0.01 = 11210$
7 $60 \times 0.001 = 0.6$
8 $2.04 \div 0.1 = 2.004$

5 Find the missing numbers.

a 180 × ☐ = 1800 **b** 65 × ☐ = 6.5 **c** 900 × ☐ = 9

d 2.1 × ☐ = 210 **e** 6.7 × ☐ = 0.67 **f** 0.9 × ☐ = 0.009

6 Find the missing numbers.

a 4500 ÷ ☐ = 45 **b** 2 ÷ ☐ = 20 **c** 7.8 ÷ ☐ = 7800

d 68 ÷ ☐ = 680 **e** 40 ÷ ☐ = 0.4 **f** 0.06 ÷ ☐ = 0.6

Problem solving

Exam-style questions that require skills developed in this unit are included in later units.

1. Look at this diagram. There are 4 operations in the boxes.

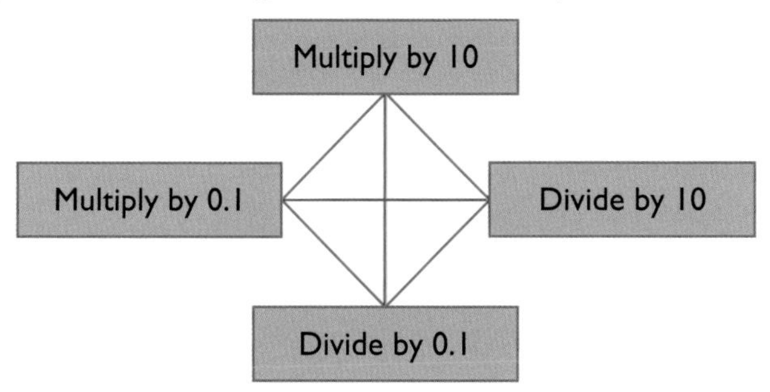

Two pairs of the operations are equivalent; they are joined by green lines.
The other pairs are opposite (or inverse) operations and they are joined by red lines.
Copy and complete the following diagrams.

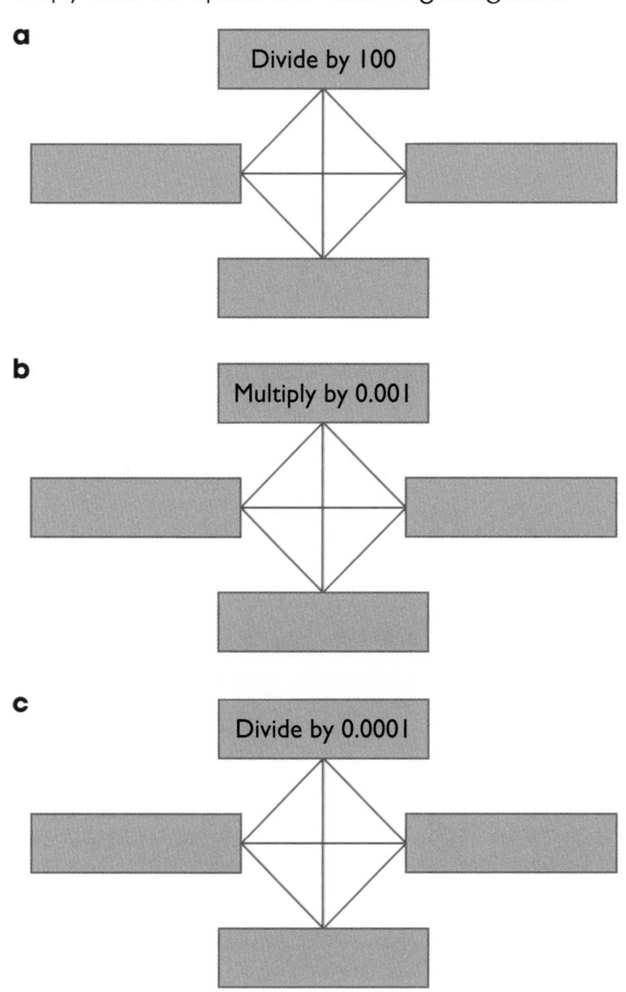

2 Start with 81 × 16 = 1296.

When you go diagonally down to the left, multiply the first number by 2 and divide the second number by 2.

When you go diagonally down to the right, divide the first number by 3 and multiply the second number by 3.

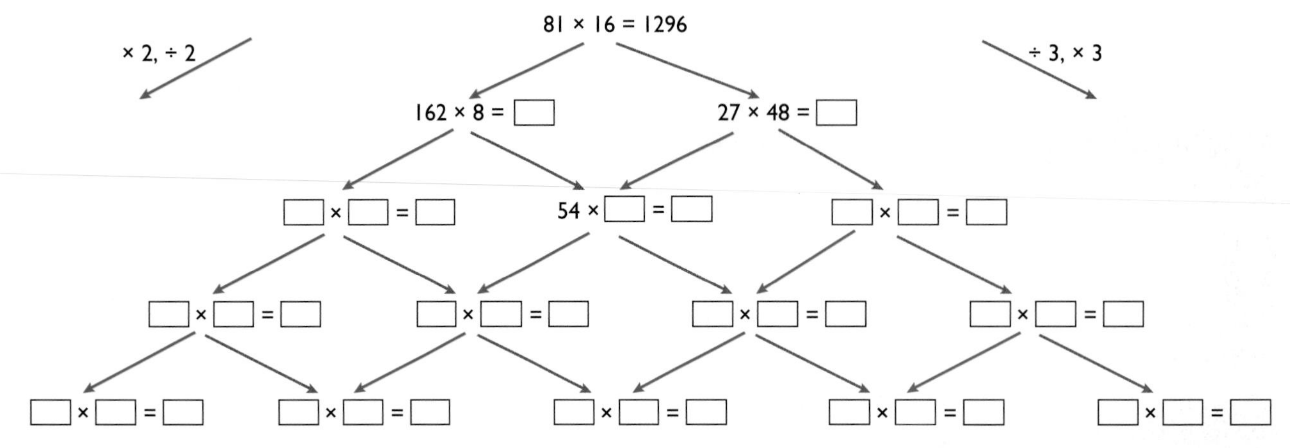

a Complete the diagram.

b What happens if you try to complete another layer?

c What is special about the starting numbers that have been chosen?

Reviewing skills

1 Work these out.

a 8.2 × 0.1 **b** 130 × 0.1 **c** 4 × 0.01 **d** 8 × 0.001

e 63 × 0.001 **f** 0.9 × 0.01 **g** 2.01 × 0.1 **h** 0.7 × 0.001

2 Work these out.

a 2.8 ÷ 0.1 **b** 30 ÷ 0.01 **c** 0.08 ÷ 0.001 **d** 0.2 ÷ 0.001

e 600 ÷ 0.1 **f** 0.004 ÷ 0.01 **g** 100 ÷ 0.0001 **h** 0.001 ÷ 0.0001

3 How many 1p coins make £100?

Unit 6 • Understanding standard form • Band h

Outside the Maths classroom

Describing the Universe

The speed of light is written in standard form as 3×10^8 m/s.
Why do scientists use numbers written in standard form?

Toolbox

Standard form is used to write large numbers and small numbers.

For example, the speed of light is 2.998×10^8 metres per second and the mass of an electron is 9.110×10^{-31} kg.

A number in standard form is

(a number between 1 and 10) × (a power of 10).

So 3.9×10^{-4} is standard form but 39×10^{-5} isn't.

When thinking about numbers in standard form it can be helpful to have a place-value table in mind. The table below shows four million six hundred and seventy thousand (4 670 000), which is sometimes said as 4.67 million.

	M	HTh	TTh	Th	H	T	U	.	$\frac{1}{10}$	$\frac{1}{100}$
							4	.	6	7
4.67×10^6	4	6	7	0	0	0	0	.		

4.67 million is $4.67 \times 1\,000\,000$ or 4.67×10^6. The table shows you that, the $\times 10^6$ moves all of the digits in 4.67 six places to the left; any blanks are filled with zeros.

In 6.8×10^{-3} all of the digits of 6.8 are moved three places to the right and the gaps are filled with zeros.

	H	T	U	.	$\frac{1}{10}$	$\frac{1}{100}$	$\frac{1}{1000}$	$\frac{1}{10000}$
			6	.	8			
6.8×10^{-3}				.	0	0	6	8

Example – Converting small numbers to standard form

A flea weighs around 0.000087 kg. Write this number in standard form.

Solution

For it to be in standard form, it must start 8.7

$0.000087 = 0.00087 \times 0.1$

$0.000087 = 0.0087 \times 0.01$

$0.000087 = 0.087 \times 0.001$

$0.000087 = 0.87 \times 0.0001$

$0.000087 = 8.7 \times 0.00001$

So

$0.000087 = 8.7 \times 0.00001 = 8.7 \times 10^{-5}$

Example – Converting large numbers from standard form

Convert the standard form number 6.387×10^6 to a normal number.

Solution

M	HTh	TTh	Th	H	T	U	.	$\frac{1}{10}$	$\frac{1}{100}$	$\frac{1}{1000}$
						6	.	3	8	7
6	3	8	7	0	0	0	.			

Using a place-value table the digits will move up 6 columns when multiplied by 10^6

So $6.387 \times 10^6 = 6387000$

The questions in this unit should be answered *without* the use of a calculator.

Practising skills

1. **a** Write these numbers in standard form.
   - **i** 5120
   - **ii** 512
   - **iii** 51.2
   - **iv** 0.512
   - **v** 0.00512
   - **vi** 0.000512

   b How would you write 5.12 in standard form?

2. These numbers are expressed in standard form. Write them as ordinary numbers.
   - **a** 5×10^2
   - **b** 8×10^4
   - **c** 2.6×10^3
   - **d** 1.9×10^5
   - **e** 8.17×10^3
   - **f** 9.05×10^4
   - **g** 7.4×10^7
   - **h** 1.004×10^4

3. Write these numbers in standard form.
   - **a** 600
   - **b** 70000
   - **c** 8900
   - **d** 816
   - **e** 133000
   - **f** 4 million
   - **g** 95 million
   - **h** 4 billion

4) These numbers are written in standard form. Write them as ordinary numbers.

a 6.8×10^{-2} **b** 5×10^{-3} **c** 2.99×10^{-2} **d** 7×10^{-4}

e 1.04×10^{-1} **f** 8.6×10^{-5} **g** 5×10^{-6} **h** 3.227×10^{-2}

5) Write these numbers in standard form.

a 0.69 **b** 0.052 **c** 0.0114 **d** 0.0007

e 0.0038 **f** 0.000006 **g** 0.955 **h** 0.00009

Developing fluency

1) Which of these are not written in standard form?

5×10^4 1600 0.8×10^3 6.2×10^5 9×100^3 7.1×10^{-4}

2) Write these quantities in standard form.

a The total length of veins in the human body is 60000 miles

b On average a person's heart beats 108000 times a day

c The distance between the Sun and the Moon is about 150000000 km

d A single coffee bean weighs about 0.003 kg

e The mass of a grain of rice is 0.0000026 kg

3) These numbers are in standard form. Write them as ordinary numbers and in words.

a 9×10^3 **b** 2.1×10^3 **c** 6.8×10^2 **d** 9.22×10^2

e 1.08×10^4 **f** 7×10^1 **g** 7×10^{-1} **h** 3×10^{-2}

4) Write these numbers in standard form.

a Six thousand **b** Seventy four **c** Eight hundred and ten

d Two thousand and fifteen **e** Four tenths **f** Three hundredths

g 0.00000224 **h** 5108000 **i** 67800000

j 23 million **k** 4 billion **l** 0.000000007001

Exam-style

5) Write these numbers in order, starting with the smallest.

7100 6.8×10^4 9×10^4 7.95×10^2 7.09×10^3

6) Write these numbers in order, starting with the largest.

3.82×10^{-2} 0.04 2×10^{-3} 3.9×10^{-2} 2.2×10^{-3}

7) In each of these, fill the box with one of **<**, **>** or **=**

a 7×10^2 ☐ 750 **b** 6.2×10^{-3} ☐ 8×10^{-2} **c** 5×10^3 ☐ 5000

d 0.009 ☐ 9×10^4 **e** 1.65×10^8 ☐ 2.4×10^7 **f** 8×10^7 ☐ 9 million

Problem solving

Exam-style

Extension

1) The table shows the closest distances of the Sun and seven planets from Earth.

Planet	Distance from Earth (in kilometres)
Jupiter	6.244×10^8
Mars	7.83×10^7
Mercury	9.17×10^7
Neptune	4.35×10^9
Saturn	1.25×10^9
Sun	1.496×10^8
Uranus	2.72×10^9
Venus	4.1×10^7

In a school project, Finlay has to list these eight bodies in order of distance from Earth.
He has to start with the planet that is nearest to Earth.
Finlay writes them in the correct order. Write down Finlay's list.

Exam-style

Extension

2) In a test, Daniel had to write down the value of the first digit of 10 numbers written in standard form.
The table shows Daniel's answers.

Question	Number	Value of first digit	Question	Number	Value of first digit
1	6.1×10^4	6000	6	2×10^{-2}	2 hundredths
2	3.62×10^4	3 units	7	1.46×10^{-2}	1 unit
3	2.9×10^7	20 million	8	3×10^{-4}	$\frac{3}{10000}$
4	4.5×10^9	4 billion	9	6.2×10^{-4}	$\frac{6}{100000}$
5	1.236×10^9	1 trillion	10	3.12×10^{-6}	3 millionths

How many correct answers did Daniel get?

Reviewing skills

1) These numbers are expressed in standard form. Write them as ordinary numbers.

a 2.008×10^5 **b** 2.45×10^6 **c** 7.803×10^9 **d** 6.45×10^8

e 9×10^{-1} **f** 2.07×10^{-7} **g** 6.145×10^{-3} **h** 1.007×10^{-1}

2) Write these numbers in standard form.

a 20 250 **b** 23 million **c** 654.7 **d** 25 624.87

e 3 tenths **f** 7 hundredths **g** 0.002 04 **h** 0.099

3) Write these quantities in standard form.

a The population of the world is approximately 7 billion.

b The diameter of a red blood cell is 0.008 mm.

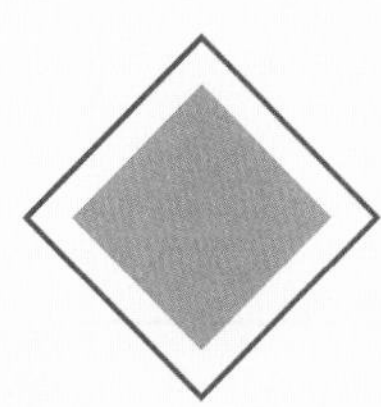

Number Strand 3 Accuracy

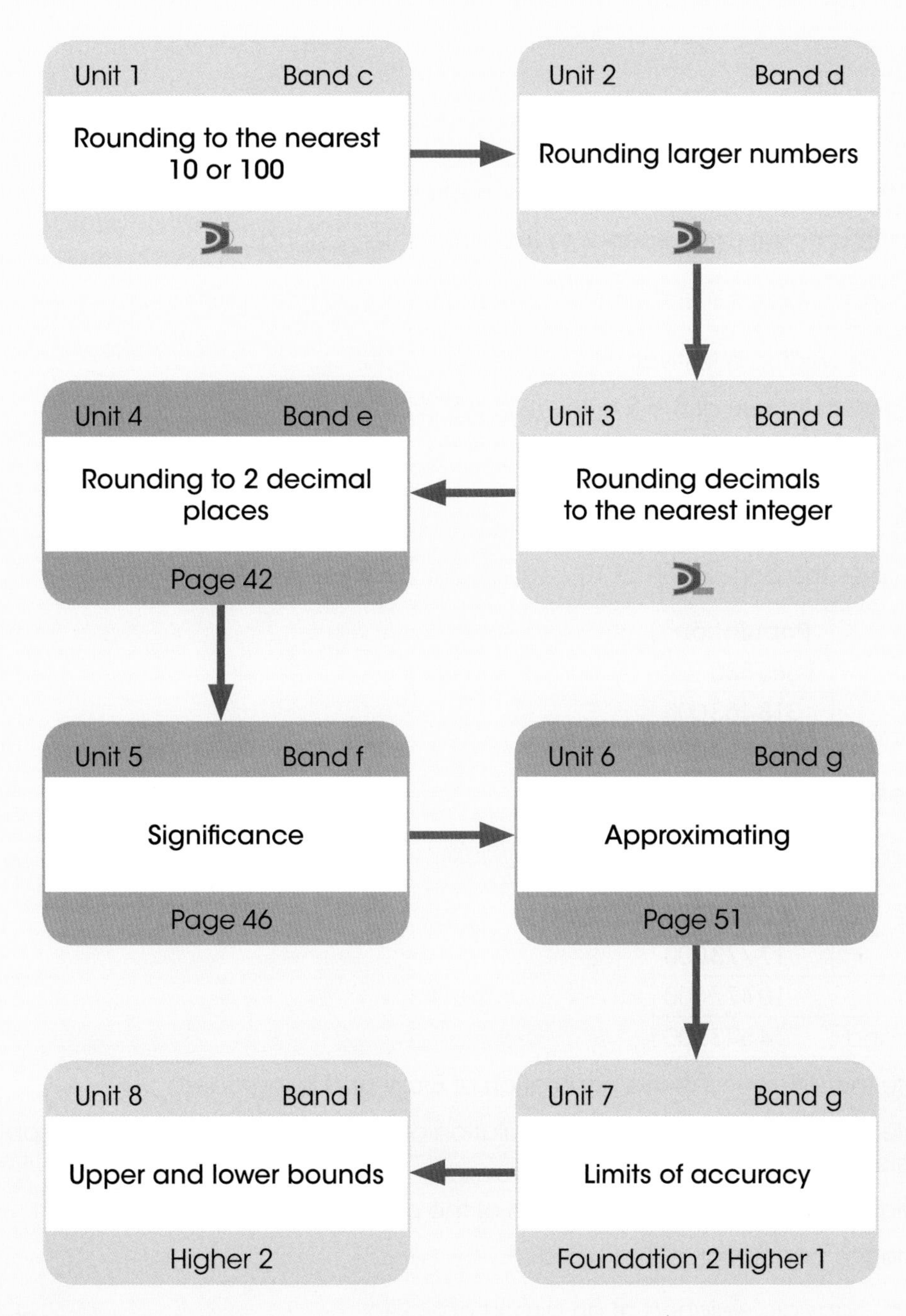

Units 1–3 are assumed knowledge for this book. They are reviewed and extended in the Moving on section on page 40.

Units 1–3 • Moving on

Exam-style

1. Craig spends £134.71 on his weekly shop.

   a Round the amount Craig spends to the nearest pound.

   b Round the amount Craig spends to the nearest ten pounds.

Exam-style

2. 85 389 people went to a festival.

   a Write down the value of the 5 in the number 85 389.

   b Round 85 389 to the nearest thousand.

Exam-style

3. The table gives the population of 10 countries in 2014.

Country	Population
China	1 365 830 000
USA	318 463 000
Brazil	202 914 000
Bangladesh	156 698 000
France	65 931 000
Spain	46 507 800
Malaysia	30 209 000
Chile	17 773 000
Portugal	10 477 800
New Zealand	4 543 200

   a Estimate the difference in the population of Brazil and Bangladesh.

   b Estimate the number of times the population of the USA is greater than the population of Malaysia.

   The population of China is about one fifth of the population of the world.

   c Estimate the population of the world.

Exam-style

4. Passengers' bags are weighed at an airport check-in.
   The weights are rounded to the nearest kilogram.
   Each passenger is allowed 23 kg free of charge.
   For any bag over 23 kg in weight, an extra charge is made.
   By removing articles from one bag and putting in another, explain how these five passengers could each avoid any extra charges being made.

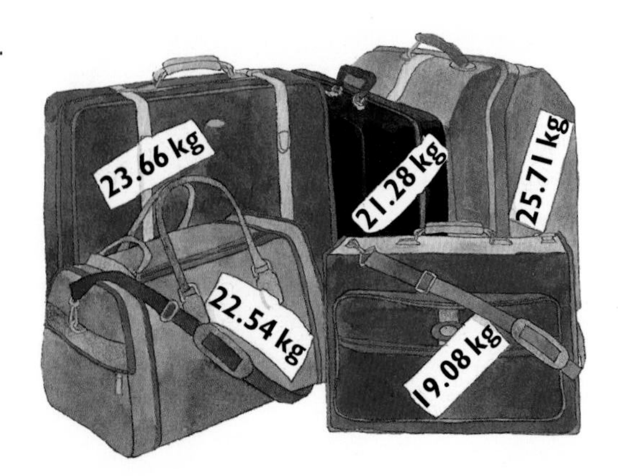

Exam-style

5 Salome wants to find the perimeter of this quadrilateral.

She uses her calculator to add the length of the four sides together.

Her answer is 48.97 cm.

Salome's answer is wrong.

Without working out the exact answer, show why Salome's answer must be wrong.

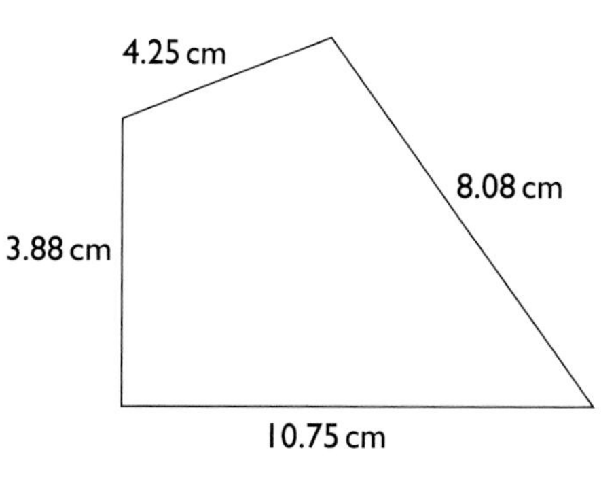

3 Unit 4 • Rounding to 2 decimal places • Band e

Outside the Maths classroom

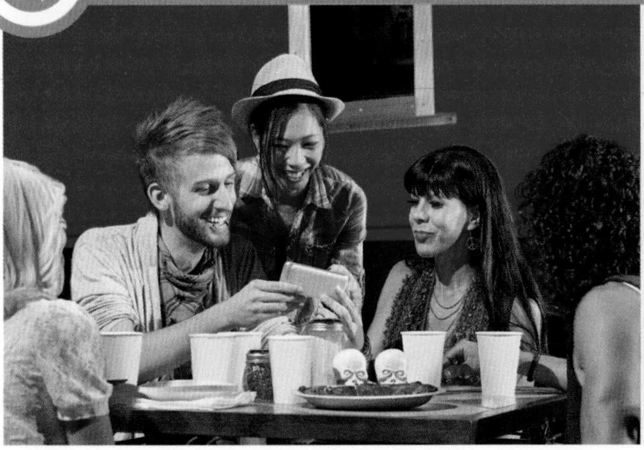

Sharing a bill

How much should each person pay?

Toolbox

To round a number to 1 decimal place, look at the $\frac{1}{100}$ digit.

In 1.642 the $\frac{1}{100}$ digit is 4, so 1.642 rounds down to 1.6

1.642 is nearer 1.6 than 1.7

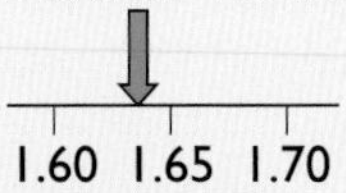

To round a number to 2 decimal places, look at the $\frac{1}{1000}$ digit.

In 1.642 this is the 2, so 1.642 rounds down to 1.64

1.642 is nearer 1.64 than 1.65

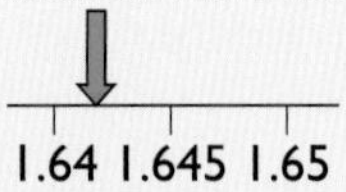

1.645 is exactly half way between 1.64 and 1.65. The rule is to round up in these situations. So 1.645 is 1.65 to 2 decimal places.

Example – Rounding to 1 decimal place

Look at the numbers **a** = 2.137, **b** = 2.4736 and **c** = 3.0491758 on this number line.

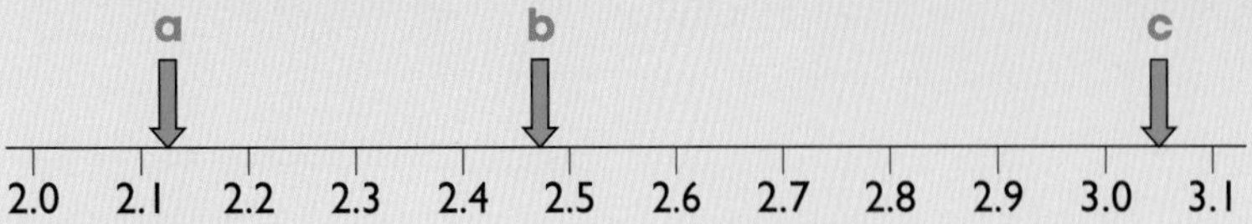

Round each number to 1 decimal place.

Solution

- **a** rounds down to 2.1 (because of the 3). ← Look at the second decimal place.
- **b** rounds up to 2.5 (because of the 7).
- **c** rounds down to 3.0 (because of the 4).

Example – Rounding calculator answers to 3 decimal places

Write the number on each of these calculator displays to 3 decimal places.

a 65.9083461

b 108.275621

c 0.94860321

Solution

a 65.908 [65.9083461] — The 3 tells you the number is closer to 65.908

b 108.276 [108.275621] — The 6 tells you the number is closer to 108.276

c 0.949 [0.94860321] — The 6 tells you the number is closer to 0.949

Practising skills

1. Round each number to 1 decimal place.

a 2.34	**b** 2.65	**c** 0.94	**d** 12.79
e 60.15	**f** 0.654	**g** 40.72	**h** 0.09
i 7.96	**j** 5.618	**k** 114.957	**l** 247.037

2. Round each number to 2 decimal places.

a 5.126	**b** 8.294	**c** 0.0192	**d** 12.994
e 16.997	**f** 0.0548	**g** 706.096	**h** 2.6689
i 0.00749	**j** 51.9962	**k** 89.9972	**l** 1.0804

3. Write each amount of money correctly.

a £24.314	**b** £61.585	**c** £9.8	**d** £0.699
e $209.998	**f** $13.064	**g** €0.5	**h** €29.997

4. Write down the number which is exactly half way between each of these pairs.

a 6.2 and 6.3	**b** 7.9 and 8	**c** 2.43 and 2.45	**d** 0.6 and 0.8
e 12.49 and 12.5	**f** 29.99 and 30	**g** 0.21 and 0.22	**h** 1.5 and 1.499

Developing fluency

1. Copy and complete this table.

	Number	Nearest whole number	To 1 decimal place	To 2 decimal places
a	8.431			
b	6.918			
c	14.277			
d	0.8063			
e	63.592			
f	109.711			
g	799.498			
h	8069.515			
i	99 999.9069			
j	699 999.999			

Exam-style

2. Marie rounded 751.9694 to 1 decimal place.
Mavis rounded 751.9694 to 2 decimal places.
What is the difference between their values?

3. Round the numbers on the calculator displays below to
i the nearest integer **ii** 1 decimal place **iii** 2 decimal places **iv** 3 decimal places

a 1096.793642 **b** 81.04693288 **c** 0.96483055 **d** 510.80136921

Problem solving

Exam-style

1. The diagram shows a quadrilateral.

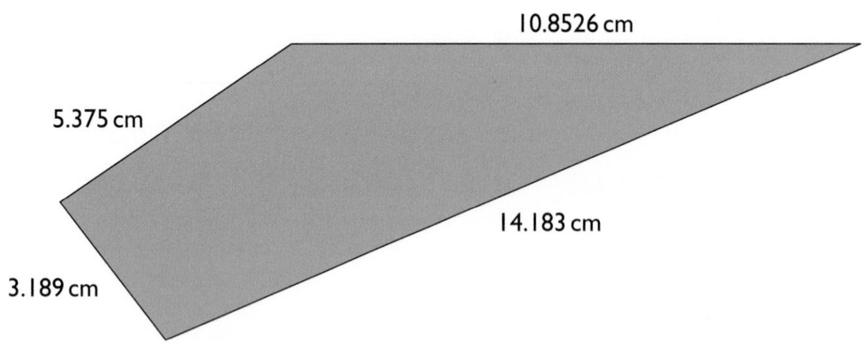

Mandy wants to see if there is any difference in the perimeter of the quadrilateral calculated using lengths of sides that are rounded in the following ways:

a to 3 significant figures **b** to 1 decimal place **c** to 2 decimal places.

Show Mandy's results.

Exam-style

2) Mel needs to work out the area of this rectangle.

0.452 m

0.175 m

Mel's answer is $0.08\,m^2$ to 3 significant figures.

This answer is wrong. Explain why and give the correct answer to 2 significant figures.

Exam-style

3) Duncan is an angler.

The table gives the weights of 5 fish that Duncan caught recently.

Duncan wants to find out the average weight of these 5 fish.

a Find the mean weight of these 5 fish, giving your answer to 1 decimal place.

b Round each weight to 1 decimal place and then find the mean. Give your answer to 1 decimal place.

c Is there any difference between your answers to parts **a** and **b**?

Fish	Weight (in kg)
pike	5.944
trout	1.294
carp	2.678
perch	1.102
salmon	11.305

Extension

Exam-style

4) Rizwan compiles a Sudoku puzzle.

He asks 6 people to solve it.

He times them to the nearest tenth of a minute.

Their times are 15.6, 18.7, 20.3, 17.3, 22.1 and 16.9 minutes.

Rizwan works out the mean time.

He says, 'The answer of 18.48333 minutes is the time it should take to solve this puzzle.'

Explain why this is not a sensible answer and give what would be a sensible answer.

Extension

Reviewing skills

1) Round each number to 1 decimal place.

a 103.064 **b** 0.0875 **c** 29.957 **d** 499.995

2) Round each number to 2 decimal places.

a 0.1984 **b** 0.99651 **c** 57.3291 **d** 19.9763

3) Write each amount of money correctly.

a \$160.057 **b** £1008.9 **c** \$10.096 **d** £76999.999

4) Write down the number which is exactly half way between each of these pairs.

a 0.011 and 0.009 **b** 0.01 and 0.009 **c** 0.01 and 0.0095 **d** 0.01 and 0.00975

5) Round the numbers on these calculator displays to

i the nearest integer **ii** 1 decimal place

iii 2 decimal places **iv** 3 decimal places.

a 3.141592654 **b** 2.236067977

Unit 5 • Significance • Band f

Outside the Maths classroom

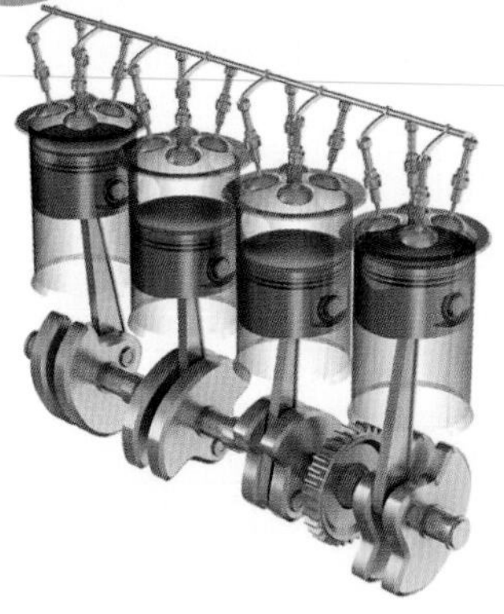

Precision engineering

How do engineers make sure all the pieces fit together?

Toolbox

The length of one year is 365.2422 days. This is 365 when rounded to the nearest whole number. This is the same as saying it has been rounded to 3 **significant figures** (s.f.).

365.2422 is also

400 to 1 significant figure ← The first digit is in the 100's column, so round to the nearest 100.

370 to 2 significant figures ← The second digit is in the 10's column, so round to the nearest 10.

365.2 to 4 significant figures ← The fourth digit is in the 1st decimal place, so round to 1 d.p.

365.24 to 5 significant figures

The first non-zero digit is always the first significant figure.

After the first significant figure, all digits are significant.

Example – Rounding very small numbers

The answer 0.000 253 745 was given on a calculator. Write the number to 2 significant figures.

Solution

0.000 253 745

These 3 zeros are place holders.

2 is the first significant digit.

3 is smaller than 5 so the number rounds down.

The answer is 0.000 25 (to 2 significant figures).

Practising skills

1. Here are some measurements. How many significant figures are in each number?

   a 36 cm **b** 1297 km **c** 0.9 kg **d** 0.053 m
   e 208 mm **f** 0.0251 m **g** 30.97 kg **h** 700.49 m

2. Round these numbers to 1 significant figure.

   a 29 **b** 45 **c** 361 **d** 852
   e 7422 **f** 21 652 **g** 18.4 **h** 62.9
   i 0.943 **j** 0.652 **k** 0.0194 **l** 0.0248

3. Round these numbers to 2 significant figures.

   a 873 **b** 924 **c** 615 **d** 708
   e 704 **f** 3261 **g** 5119 **h** 18 642
   i 73 281 **j** 8042 **k** 0.635 **l** 0.6041

4. Write the number 384 027 correct to

   a 1 significant figure **b** 2 significant figures **c** 3 significant figures
   d 4 significant figures **e** 5 significant figures.

5. Write the number 7.999 999 9 correct to

   a 1 significant figure **b** 2 significant figures **c** 3 significant figures
   d 4 significant figures **e** 5 significant figures.

6. Write the number 0.008 106 049 9 correct to

   a 1 significant figure **b** 2 significant figures **c** 3 significant figures
   d 4 significant figures **e** 5 significant figures.

7. Write these numbers correct to the number of significant figures (s.f.) shown in brackets.

   a 17.65 (1 s.f.) **b** 0.597 (2 s.f.) **c** 71 046 (3 s.f.)
   d 3.74 (1 s.f.) **e** 6.5092 (3 s.f.) **f** 26.9999 (4 s.f.)

Developing fluency

1 Copy and complete this table.

	Number	Round to 1 significant figure	Round to 2 significant figures
a	742		
b	628		
c	199		
d	4521		
e	3419		
f	8926		
g	8974		
h	36 294		
i	0.2583		
j	0.079 61		
k	0.000 397 2		
l	0.001 023		

2 Use a calculator to work these out. Give each answer to the degree of accuracy shown in brackets.

a $861 \div 45$ (1 s.f.) **b** 2.3^3 (1 s.f.) **c** 7.89×6.45 (2 s.f.)

d $11.6 \div 240$ (2 s.f.) **e** 64.8^4 (3 s.f.) **f** 0.89×156.11 (1 s.f.)

g $\sqrt{89956}$ (2 s.f.) **h** $\sqrt[3]{1.0256}$ (3 s.f.) **i** $\frac{4.3^2 \times 72}{\sqrt{3.864}}$ (3 s.f.)

3 Calculate the difference between

a 63.8421 (to 2 s.f.) and 63.8421 (to 3 s.f.)

b 81.478 (to 2 s.f.) and 81.478 (to 2 d.p.)

4 Decide if each of these is true or false.

a 91.684 (to 2 s.f.) > 91.684 (to 1 s.f.)

b 0.3079 (to 3 s.f.) = 0.3079 (to 3 d.p.)

c 16.9949 (to 2 s.f.) = 16.9949 (to 2 d.p.)

d 0.002 713 (to 2 s.f.) > 0.002 713 (to 1 s.f.)

Exam-style

5 Five friends round numbers in this way.
Ada rounds to 1 significant figure.
Ben rounds to 2 significant figures.
Cain rounds to the nearest integer.
Dave rounds to 1 decimal place.
Ella rounds to 2 decimal places.

a They each round the number 9.463. Whose answers are the same?

b They each round the number 59.698. Whose answers are the same?

c They each round the number 109.655. Whose answers are the same?

Problem solving

Exam-style

1. At an international football match at Wembley, the attendance was announced to be 80641.

Four people who were at the match were asked to round this figure to 2 significant figures.

Dan said 80000.

Milly said 80600.

Ami said 81000.

Bob said, 'You are all wrong, 80641 correct to 2 significant figures is 81.'

Who is right?

Explain your answer, indicating the mistakes that some of the people made.

Exam-style

2. The formula to find the circumference C, of a circle of diameter d, is $C = \pi d$.

The value of π is 3.141592654… .

Keith wants to compare the circumference of a circle, diameter 8 cm for different values of π.

He works out C using π correct to

a 1 significant figure

b 2 significant figures

c 3 significant figures

d 4 significant figures.

Work out Keith's results.

Exam-style Extension

3. Heather has shares in two companies.

She has 250 shares in company A and 1800 shares in company B.

Each share in company A is valued at 198.45 pence.

Each share in company B is valued at 5.075 pence.

Heather needs to do a quick estimate of the total value of all of her shares.

a She rounds the value of each share of each company to 1 significant figure.
Calculate Heather's estimate.

b Work out another estimate. This time round the value of each share to 2 significant figures.

c Which estimate, **a** or **b**, is the more accurate? Which is the easier to work out?

Exam-style Extension

4. Harvey wants to find out the thickness of each page in his Maths text book.

He uses a ruler to measure its thickness. It is between 1.2 and 1.3 cm, not including the cover.

The pages in the book are numbered i to viii and then 1 to 297.

At the end there are 7 blank sides.

a Work out the thickness of each page.
Give your answer to a suitable number of significant figures.

b Suggest how Harvey can get a more accurate answer.

Reviewing skills

1. Round these numbers to 1 significant figure.
   a 0.994 **b** 0.00974 **c** 993 **d** 999943

2. Round these numbers to 2 significant figures.
   a 6.382 **b** 19.84 **c** 0.00519 **d** 0.00997

3. Use a calculator to work these out. Give each answer to the degree of accuracy shown in brackets.

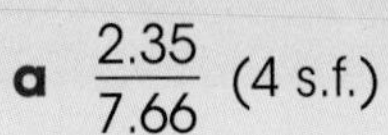

   a $\frac{2.35}{7.66}$ (4 s.f.) **b** $452 \times 60 \times 19$ (4 s.f.) **c** $\frac{2.3 \times 4.7}{9.22 - 3.7}$ (2 s.f.)

Unit 6 • Approximating • Band g

Outside the Maths classroom

Costing jobs

What factors should a caterer consider when providing a quote for producing a buffet?

Toolbox

Approximating is about:

- rounding numbers for a calculation so that you can do it in your head
- making a rough calculation to anticipate what sort of answer to expect
- recognising when an error has been made
- giving a number to the level of detail that suits the context.

An approximate answer is often found by rounding numbers to 1 significant figure.

Example – Rounding numbers to make a calculation easier

Mo and Sahar are buying a new radiator for their bedroom.
To choose the right radiator they first need to know the volume of the room.
Mo measures the room in metres. It is 3.8 m long, 3.2 m wide and 2.7 m high.
Use rounded numbers to find the approximate volume of the room.

Solution

- Volume = $3.2 \times 2.7 \times 3.8$
- Approximate values $3 \times 3 \times 4$

The approximate volume is $36\,\text{m}^3$.

Example – Rounding to recognise when an error has been made

Pete's dog is 7 years 10 months old.

Pete says that this is equivalent to a human aged 70 because 1 dog year is like 7 human years. Is he correct?

Solution

7 years 10 months is nearly 8 years.

$7 \times 8 = 56$ so Pete's dog would be less than 56, not 70.

You could also use an inverse argument to say that, to get a result of 70 when multiplying by 7, you would need to start with 10 (Pete's dog is not yet 10 years old).

The questions in this unit should be answered *without* the use of a calculator. (However, you may wish to use a calculator to check some of your answers.)

Practising skills

1. Decide if these are good approximations.

   a 82.7 is about 80 **b** 9.63 is about 9 **c** 312 is about 30

   d 793 is about 800 **e** 449 is about 400 **f** 6711 is about 6000

2. Estimate the answers to these calculations.

   a $68.79 + 21.96$ **b** $858.74 - 111.79$ **c** 30.8×45.3

   d $28.4 \div 1.99$ **e** 29.7^2 **f** $6371 + 4912$

3. For each of these calculations, there is a choice of answers. Use approximation to help you select the correct answer.

   a 6.4×8.8
   i 15.2 **ii** 28.66 **iii** 56.32

   b 7.23^2
   i 5227.29 **ii** 52.2729 **iii** 14.46

   c $836 \div 19$
   i 15884 **ii** 855 **iii** 44

4. Quick checks by rounding the numbers will tell you that three of these answers are wrong. Which three are wrong?

   a $168 \times 94 = 1592$ **b** $18.6 \times 4.5 = 83.7$ **c** $\frac{56}{3.8} = 19.74$

   d $1200 \div 48 = 25$ **e** $8.9^2 = 79.21$ **f** $2.8^3 = 31.92$

5. A t-shirt costs £7.99 and a pair of shorts costs £11.49.
   Amber wants to buy 2 pairs of shorts and 5 t-shirts.
   She has £60.
   Use a rough estimate to decide if Amber has enough money.

6 Micah has done several calculations.
He then does rough estimates to check his answers.
Here are his results. Which calculations should he look at again?

- **a** Calculation 12.64; rough estimate 80
- **b** Calculation 611; rough estimate 600
- **c** Calculation 0.072; rough estimate 0.08
- **d** Calculation 19.32; rough estimate 6.5
- **e** Calculation 341.8; rough estimate 360
- **f** Calculation 0.0156; rough estimate 0.002

Developing fluency

1 Estimate the answers to these calculations.

- **a** $649 + 382$
- **b** 7.15×13.06
- **c** 62.4^2
- **d** $\frac{815 \times 6.4}{2.85}$
- **e** $\frac{97 \times 94}{8.96}$
- **f** $(2.1 + 9.4)^2$

2 For each of these calculations, use approximation to help you select the correct answer from the four possible answers given. Explain your choice.

a 514^2

- **i** 664 196
- **ii** 26 416
- **iii** 264 196
- **iv** 246 196

b 71.4×6.8

- **i** 4015.52
- **ii** 465.52
- **iii** 595.52
- **iv** 485.52

c $4.2^2 + 2.9^3$

- **i** 42.029
- **ii** 430.221 96
- **iii** 73.08
- **iv** 0.723

3 Alana saves £18.25 every week for a year.
Approximately how much does she save in a year?

4 Walt's mobile phone bill is £27.49 per month.
Approximately how much does he pay for his mobile phone in a year?

5 A journey of 104 miles, 528 yards took 2 hours and 3 minutes.
Estimate the average speed in mph.

Exam-style

6 Amir's weekly wage is usually between £350 and £400.
4.8% of his wage is deducted for his pension.
Estimate how much is deducted over 2 years.

7 A typical adult sleeps about $7\frac{1}{2}$ hours every night.
Estimate how many months an adult sleeps in a year.

Exam-style

8 Here are some calculations.
Estimate the answers.
Then decide if the answer given is definitely wrong.

a $\dfrac{6149 \div 28}{3.8 \times 4.7}$ Answer 22.26

b $\dfrac{8.4^2 \times 9.75}{0.68 - 0.4}$ Answer 2457

c $\dfrac{5.4^2 - 5.8^2}{61.5 \div 2.2}$ Answer 14.4

9 The diameter of the Sun is about 1 392 000 km.
The diameter of the Earth is about 12 700 km and the diameter of the Moon is about 3500 km.

a Estimate roughly how many times bigger the diameter of the Sun is than the diameter of the Earth. Give your answer to 1 significant figure.

b Estimate how many times bigger the diameter of the Earth is than the diameter of the Moon, to 1 significant figure.

c Now estimate how many times bigger the diameter of the Sun is than the diameter of the Moon. Again give your answer to 1 significant figure.

Problem solving

Exam-style

1 The table shows how long Aimee worked at different rates of pay last week.

	Hours worked	Rate of pay
Normal pay	25 hours 45 mins	£7.90 per hour
Overtime	5 hours 20 mins	£9.95 per hour
Sunday	2 hours 40 mins	£12.75 per hour

Aimee

This week, Aimee earned £305.

a Using approximations, show whether Aimee earned more or less last week than she earned this week.

Aimee is hoping to get a mortgage to buy a house.
She needs to earn more than £13 000 a year to get a mortgage for the house she wants.
In a year, Aimee works for 41 weeks and she earns a similar amount each week.
In the remaining weeks of the year, she is paid holiday pay at a rate of £152 per week.

b Estimate whether Aimee earns enough each year to be able to get a mortgage.

Exam-style

2 Country Farm yogurt is sold in pots.
A machine fills the pots in batches of ten at the same time.
The machine can fill 9040 pots in one hour.

a Estimate the number of seconds it takes the machine to fill a batch of pots.

In one week, the machine fills pots for a total of 31 hours.
The pots of yogurt are then packed in cartons, each holding 96 pots.

b Estimate how many cartons are filled that week.

Exam-style

3 Andy is treating himself and five friends to a chip shop supper.
Andy orders the food that everyone wants.

Fryin - 2 - Nite

Chips .... 89p per portion
Peas .... 25p per tub
Fish .... £1.95 each
Pies .... £1.49 each
Sausages .... £1.10 per portion
Drinks .... 90p per can

Three portions of fish and chips, two pie and chips and a sausage and chips please.

Andy

He then realises that he only has a £20 note.
Andy quickly estimates if he will have enough money.

a Does Andy have enough money?

His friends say that they each would like a drink. Andy does not want one himself.

b Will Andy have to borrow any money from his friends?

Exam-style

4 This is part of a timetable for trains from London to Stoke-on-Trent.

London Euston	07:20	08:00	08:20	09:00	09:20	10:00
Milton Keynes Central	07:50		08:50		09:50	
Stoke-on-Trent	08:48	09:25	09:48	10:25	10:48	11:15

This is part of a timetable for trains from Stoke-on-Trent to London.

Stoke-on-Trent	16:50	17:12	17:50	18:12	18:50	19:12
Milton Keynes Central	17:46		18:46		19:46	
London Euston	18:24	18:43	19:24	19:43	20:24	20:42

Lucy lives in London.
She has a meeting in Stoke-on-Trent at 11 a.m.
It takes Lucy about 15 minutes to walk from her home to London Euston train station.
The journey from the train station in Stoke-on-Trent to her meeting takes about 35 minutes.
The meeting is due to finish at 5.15 p.m.
Show and explain why Lucy is likely to be away from home for about 12 hours.

Exam-style

Extension

5) Leah runs a beauty salon.
In order to attract new customers, Leah is offering a mini facial.
She wants to charge enough just to cover her costs.
The table shows the facial products that Leah will need.

Product	Amount needed for mini facial	Cost of product
500 ml bottle of exfoliator	50 ml	£27.99 per bottle
200 ml bottle of cleanser	20 ml	£11.99 per bottle
300 ml bottle of toner	15 ml	£29.99 per bottle
450 ml bottle of moisturiser	15 ml	£58.99 per bottle
150 ml jar of eye tonic	10 ml	£74.49 per jar
200 ml bottle of mixed oils	10 ml	£20.99 per bottle
1 face pack (inc. towels etc.)	1 pack	£1.99 per pack

Estimate what Leah should charge for this mini facial.

Exam-style

Extension

6) The diagram shows the lines on a netball court.

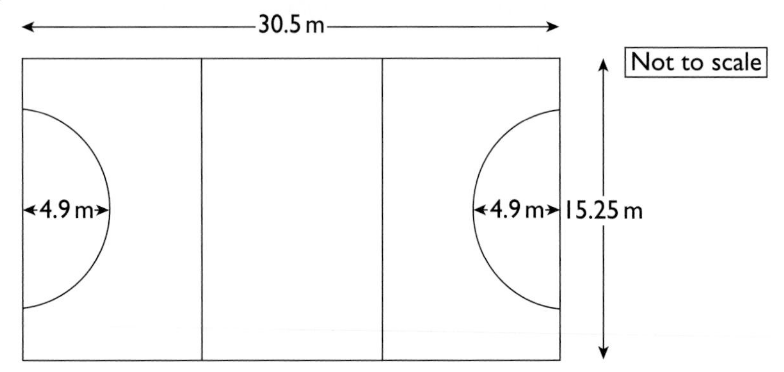

Stuart is painting the lines for 4 netball courts on the floor of a sports hall.
One tin of paint is enough to paint 110 metres of lines.
Estimate the number of tins of paint Stuart needs.

Reviewing skills

1) Estimate the answers to these calculations.

a $9728 - 9061$ **b** $814 \div 36.9$ **c** 7.84×194.3

2) For each of these calculations, there is a choice of answers. Use approximation to help you select the correct answer.

a $\frac{126 \times 68}{48}$

i 178.5 **ii** 17.85 **iii** 1785

b $9398.4 \div 26.4$

i 0.002 809 **ii** 9372 **iii** 35.6 **iv** 356

3) Estimate the answers to these calculations.

a $\frac{1684 - 324}{4.93}$ **b** $\frac{6.93 \times 55.4}{0.132}$ **c** $\frac{917 + 458}{0.41 - 0.23}$

4) Jake uses about 8.5 litres of petrol every day that he works.
Petrol costs 132.9 p per litre.
Approximately how much does Jake pay for petrol for a 4-day working week?

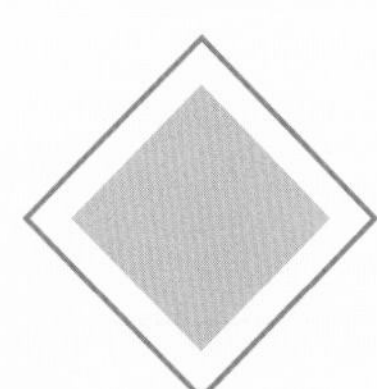

Number Strand 4 Fractions

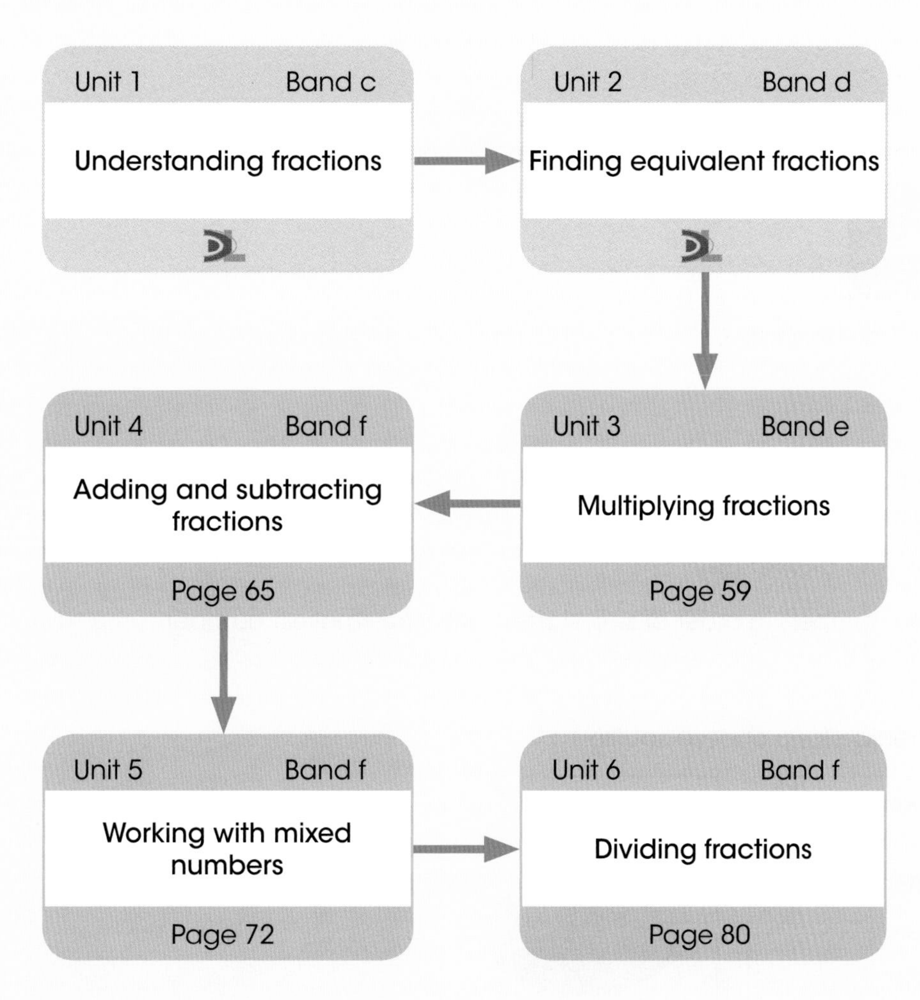

Units 1–2 are assumed knowledge for this book. They are reviewed and extended in the Moving on section on page 58.

Units 1–2 • Moving on

Exam-style

1. **a** Write down the fraction of this shape that is shaded.

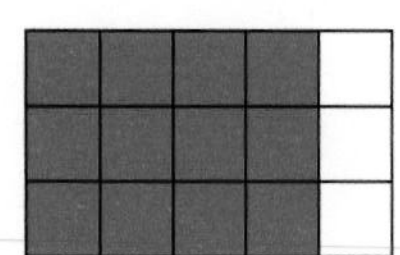

b Shade $\frac{2}{3}$ of this shape.

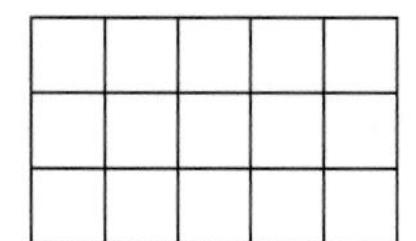

c Which fraction is greater: $\frac{2}{3}$ or $\frac{4}{5}$?
Explain your answer fully.

2. Write these numbers in order of size, starting with the smallest number.
$\frac{3}{4}$, $\frac{7}{10}$, $\frac{2}{3}$, $\frac{2}{5}$

Exam-style

Extension

3. Here is a recipe for a cheese soufflé.

Cheese soufflé
3 eggs
$\frac{1}{2}$ oz butter
$\frac{1}{4}$ oz flour
1 pint milk
3 oz grated cheese

Heston wants to make 5 cheese soufflés.
He checks his fridge to see if he has enough of each ingredient.
Heston has: 18 eggs, $\frac{1}{8}$ lb of butter, $1\frac{1}{2}$ oz of flour, 6 pints of milk and 1 lb of grated cheese.
1 lb = 16 oz
Does Heston have enough ingredients to make 5 cheese soufflés?
If not, what else does he need?

Exam-style

4. Wayne, Andy and John share £48 between them.
Wayne gets £16, Andy gets £12 and John gets the rest.

a What fraction does each get?
Give your answers in their simplest form.

John thinks it would be fair if they all have the same amount of money.

b Explain what John could do now to make this possible.

Unit 3 • Multiplying fractions • Band e

Outside the Maths classroom

Genetics

Would you expect to look like your cousin?

Toolbox

Sally's mother cuts a pizza into two halves.
She then cuts each half into three slices.
She has cut the whole pizza into six pieces, so each slice is one-sixth, or $\frac{1}{6}$.

In fact she did $\frac{1}{3}$ of $\frac{1}{2} = \frac{1}{3} \times \frac{1}{2}$

$= \frac{1}{6}$

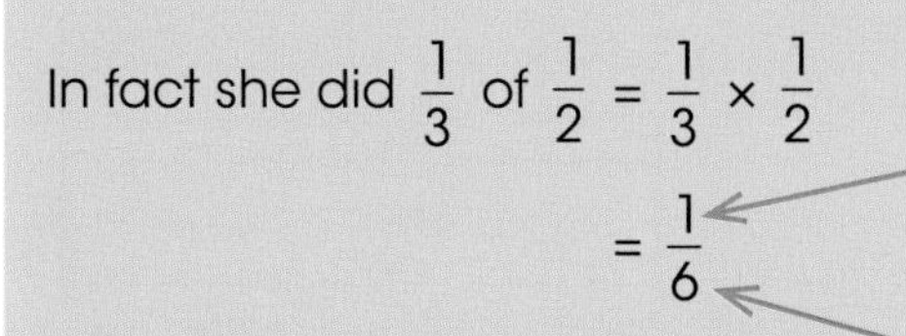

Multiply the numbers on the top together: 1 × 1 = 1

Multiply the numbers on the bottom together: 3 × 2 = 6

The word '**of**' when using fractions means **×**.

What is $\frac{1}{4}$ of $\frac{1}{4}$?

$$\frac{1}{4} \times \frac{1}{4} = \frac{1 \times 1}{4 \times 4} = \frac{1}{16}$$

$\frac{1}{4}$ of the square

$\frac{1}{4}$ of $\frac{1}{4}$ of the square

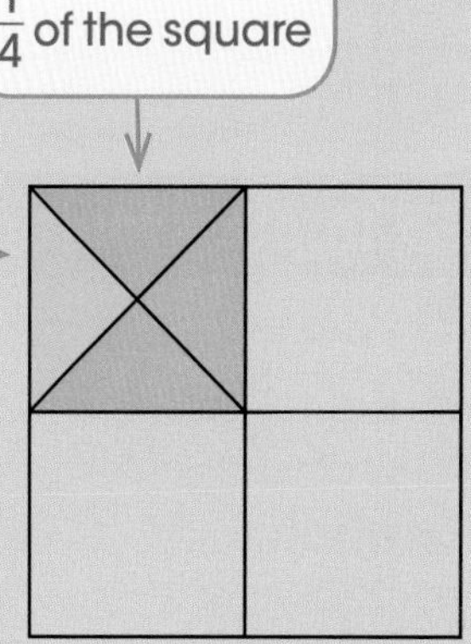

Sometimes cancelling down can be done before multiplying the top and bottom lines.

For example, when working out $\frac{5}{8}$ of $\frac{16}{19}$:

$$\frac{5}{\cancel{8}_1} \times \frac{\cancel{16}^2}{19} = \frac{5 \times 2}{1 \times 19} = \frac{10}{19}$$

Note: A whole number is a fraction. $20 = \frac{20}{1}$

Cancelling pairs of common factors.

$$\frac{6}{7} \times \frac{14}{27} \times \frac{3}{4} = \frac{6 \times \overset{2}{\cancel{14}} \times 3}{\underset{1}{\cancel{7}} \times 27 \times 4}$$ ← Cancel by 7.

$$= \frac{6 \times 2 \times \overset{1}{\cancel{3}}}{1 \times \underset{9}{\cancel{27}} \times 4}$$ ← Cancel by 3.

$$= \frac{\overset{2}{\cancel{6}} \times 2 \times 1}{1 \times \underset{3}{\cancel{9}} \times 4}$$ ← Cancel by 3 again.

$$= \frac{\overset{1}{\cancel{2}} \times \overset{1}{\cancel{2}} \times 1}{1 \times 3 \times \underset{1}{\cancel{4}}}$$ ← Cancel twice by 2.

$$= \frac{1}{3}$$

Example – Solving fraction problems

David has $\frac{1}{5}$ of his book left to read over the weekend.

He reads $\frac{2}{3}$ of it on Saturday.

What fraction of the book does he read on Saturday?

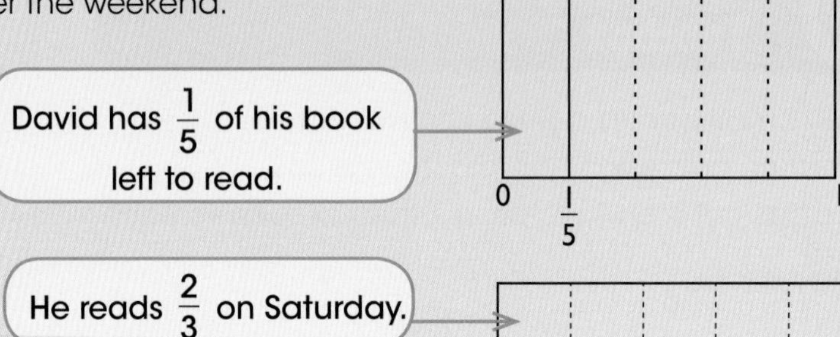

Solution

This is $\frac{2}{3} \times \frac{1}{5} = \frac{2}{15}$. So David reads $\frac{2}{15}$ on Saturday.

He still has $\frac{1}{15}$ of the book to read.

Example – Multiplying fractions

Work out $\frac{5}{6}$ of $\frac{3}{10}$

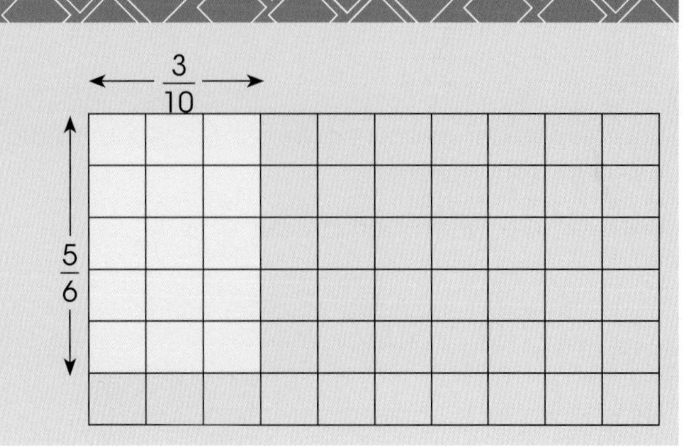

Solution

$$\frac{5}{6} \times \frac{3}{10} = \frac{15}{60} = \frac{1}{4}$$

or $$\frac{\overset{1}{\cancel{5}}}{\underset{2}{\cancel{6}}} \times \frac{\overset{1}{\cancel{3}}}{\underset{2}{\cancel{10}}} = \frac{1}{4}$$

The questions in this unit should be answered *without* the use of a calculator. (However, you may wish to use a calculator to check some of your answers.)

Practising skills

1 Match each calculation to its correct diagram.

A
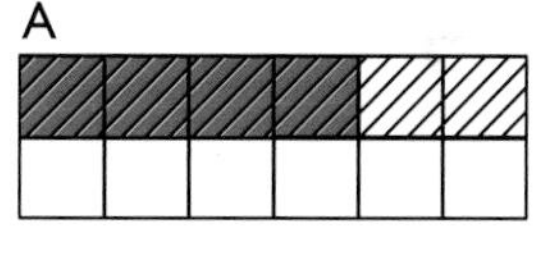

B
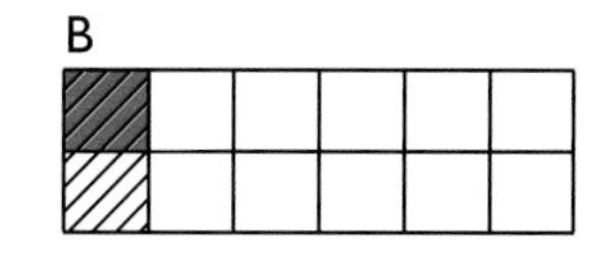

C
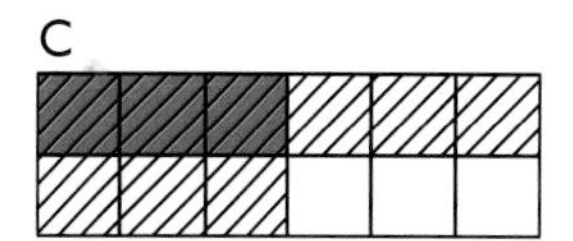

D
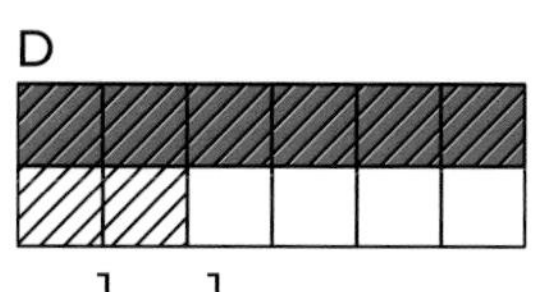

E
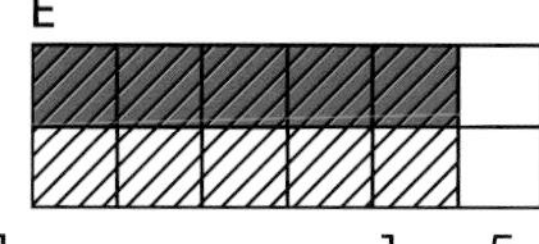

a $\frac{1}{2}$ of $\frac{1}{6}$ **b** $\frac{2}{3}$ of $\frac{1}{2}$ **c** $\frac{1}{2}$ of $\frac{5}{6}$ **d** $\frac{1}{3}$ of $\frac{3}{4}$ **e** $\frac{3}{4}$ of $\frac{2}{3}$

2 Work out these.

a $\frac{1}{2} \times \frac{1}{3}$ **b** $\frac{1}{5} \times \frac{1}{6}$ **c** $\frac{1}{2} \times \frac{3}{8}$ **d** $\frac{1}{4} \times \frac{5}{8}$

e $\frac{2}{9} \times \frac{1}{5}$ **f** $\frac{3}{4} \times \frac{3}{5}$ **g** $\frac{5}{6} \times \frac{7}{9}$ **h** $\frac{7}{8} \times \frac{3}{4}$

3 Work out these. Cancel the fractions before multiplying.

a $\frac{5}{6} \times \frac{7}{10}$ **b** $\frac{12}{13} \times \frac{3}{10}$ **c** $\frac{4}{5} \times \frac{3}{8}$ **d** $\frac{5}{9} \times \frac{9}{11}$

e $\frac{5}{17} \times \frac{17}{19}$ **f** $\frac{24}{25} \times \frac{1}{12}$ **g** $\frac{9}{10} \times \frac{6}{6}$ **h** $\frac{8}{9} \times \frac{27}{30}$

4 Work out these.

a $\frac{1}{2}$ of $\frac{1}{4}$ **b** $\frac{1}{6}$ of $\frac{1}{3}$ **c** $\frac{1}{5}$ of $\frac{3}{4}$ **d** $\frac{1}{3}$ of $\frac{2}{7}$

e $\frac{2}{5}$ of $\frac{7}{8}$ **f** $\frac{3}{4}$ of $\frac{6}{7}$ **g** $\frac{1}{3}$ of 2 **h** $\frac{2}{9}$ of 4

5 Write each of these as a single fraction.

a $\frac{1}{3}$ of $\frac{1}{2}$ tin of paint **b** $\frac{1}{4}$ of $\frac{1}{4}$ carton of milk

c $\frac{1}{9}$ of 3 cups of tea **d** $\frac{2}{15}$ of 5 jugs of juice

6 Work out these.

a $\frac{2}{7} \times \frac{3}{5} \times \frac{7}{12}$ **b** $\frac{4}{5} \times \frac{5}{9} \times \frac{3}{5}$ **c** $\frac{10}{11} \times \frac{5}{8} \times \frac{22}{25}$

d $\frac{7}{18} \times \frac{3}{4} \times \frac{3}{14}$ **e** $\frac{1}{6} \times \frac{2}{5} \times \frac{3}{8}$ **f** $\frac{2}{9} \times \frac{5}{6} \times \frac{3}{7}$

g $\frac{3}{4} \times \frac{20}{21} \times \frac{7}{8}$ **h** $\frac{7}{9} \times \frac{15}{16} \times \frac{3}{14}$

Developing fluency

1. Copy and complete the multiplication grid.

×	$\frac{1}{2}$	$\frac{2}{3}$	3
$\frac{1}{6}$			
$\frac{3}{10}$			
$\frac{2}{7}$			

2. Work this out.
$\frac{1}{2}$ of $\frac{2}{3}$ of $\frac{3}{4}$ of $\frac{4}{5}$ of $\frac{5}{6}$ of $\frac{6}{7}$ of $\frac{7}{8}$ of $\frac{8}{9}$ of $\frac{9}{10}$

3. Decide whether these statements are true or false.
   a $\frac{1}{6}$ of $\frac{1}{2} = \frac{1}{6} \times \frac{1}{2}$
   b $\frac{1}{2}$ of $\frac{2}{3} = \frac{2}{3}$ of $\frac{1}{2}$
   c $\frac{1}{5}$ of $2 = \frac{5}{2}$
   d $\frac{1}{5}$ of $\frac{5}{8} = \frac{1}{6} \times \frac{3}{4}$
   e $\frac{1}{2}$ of $\frac{1}{2}$ of $\frac{1}{2}$ of $\frac{1}{2} = \frac{1}{8}$

4. Half of a pizza is left on the table.
Melvin says he will eat a half of it and Noah says he will eat a quarter of it.
What fraction of the whole pizza does
   a Melvin eat?
   b Noah eat?

Exam-style

5. Joan started to read a new book on Monday.
On Monday she read $\frac{1}{3}$ of the book.
On Tuesday she read $\frac{1}{5}$ of the rest of the book.
What fraction of the whole book did she read on Tuesday?

Exam-style

6. Ali has a collection of model cars.
He gives $\frac{3}{5}$ of his collection to Beth.
Beth gives $\frac{3}{4}$ of these to Cain.
What fraction of Ali's model cars does
   a Ali have
   b Beth have
   c Cain have?

Reasoning

7. Which is greater, $\frac{1}{3}$ of $\frac{4}{5}$ or $\frac{2}{5}$ of $\frac{3}{4}$?
Explain in full.

8. Esther is given a box of sweets for her birthday.
She eats $\frac{1}{10}$ of the box on Friday night.
She eats $\frac{1}{5}$ of the remainder on Saturday night.
Does she eat more sweets on Friday night or Saturday night? Explain your reasoning.

Problem solving

Exam-style / Extension

1. Robbie is doing some research on the geography of Scotland.
He finds the following information,
$\frac{1}{6}$ of the land in Scotland is woodland.
$\frac{3}{4}$ of the area of woodland is covered in pine trees.
The area of woodland covered in pine trees is about 9850 km^2.
   - **a** What fraction of Scotland is covered in pine trees?
   - **b** Is the area of Scotland greater than 120 000 km^2?

Ben Nevis is the highest mountain in Scotland and is 1344 metres high.
Last year on the Ben Nevis Challenge, Robbie's sister Janet and her friend Amy climbed $\frac{2}{3}$ of the way up.
Amy then climbed a further $\frac{3}{4}$ of the remaining height.
   - **c** How many more metres did Amy have to climb to reach the top of Ben Nevis?

Exam-style

2. A wine grower harvests her grapes.
$\frac{1}{8}$ of the crop has been ruined by rain.
She sells $\frac{3}{5}$ of the rest to a large wine producer.
   - **a** What fraction of the crop is sold?

The wine producer sells 5000 bottles of wine to a supermarket.
The supermarket sold $\frac{3}{4}$ of the bottles in the first week and $\frac{3}{5}$ of the remaining bottles in the second week.
The supermarket orders more of this wine when it has fewer than 600 bottles left.
   - **b** Does the supermarket now order more wine?

Exam-style / Extension

3. Aisha gets £5 pocket money each week from her parents and £3 each week from her brother.
Aisha spends $\frac{1}{4}$ of her pocket money on magazines.
She spends $\frac{1}{3}$ of the remaining amount on make-up.
She spends 50p per week on sweets.
She saves the rest of her money.
After 12 weeks, Aisha checks her savings.
How much should she have?

Exam-style

(4) Millie owns a cat called Tabby.

On Sunday a dog chased Tabby a long way from her home.

The dog bit her leg but she escaped up a tree.

On Monday Tabby walked half the way home but her leg was sore.

On each day after that she walked half of the distance remaining at the start of the day but her leg got worse.

Tabby was chased 2560 metres.

On Saturday evening Millie found her.

a How far from home was she then?

b Millie says, 'If I had not found her, she would have never reached home.'
Explain this statement.

Reviewing skills

(1) Work out these.

a $\frac{2}{5} \times \frac{3}{7}$ **b** $\frac{7}{7} \times \frac{4}{7}$ **c** $\frac{5}{9} \times \frac{5}{8}$ **d** $\frac{10}{11} \times \frac{12}{13}$

(2) Work out these. Cancel the fractions before multiplying.

a $\frac{2}{7} \times \frac{14}{15}$ **b** $\frac{6}{13} \times \frac{26}{29}$ **c** $\frac{7}{10} \times \frac{20}{21}$ **d** $\frac{18}{23} \times \frac{23}{24}$

(3) Work out these.

a $\frac{2}{5}$ of $\frac{9}{10}$ **b** $\frac{4}{5}$ of $\frac{5}{8}$ **c** $\frac{3}{10}$ of $\frac{8}{9}$ **d** $\frac{11}{12}$ of $\frac{5}{7}$

(4) Write each of these as a single fraction.

a $\frac{2}{3}$ of $\frac{1}{4}$ bucket of water

b $\frac{3}{8}$ of 2 gallons of oil

(5) Naomi makes 3 cups of coffee every day.

She only drinks $\frac{4}{5}$ of each cup of coffee.

What fraction of a cup of coffee in total is wasted every day?

Unit 4 • Adding and subtracting fractions • Band f

Outside the Maths classroom

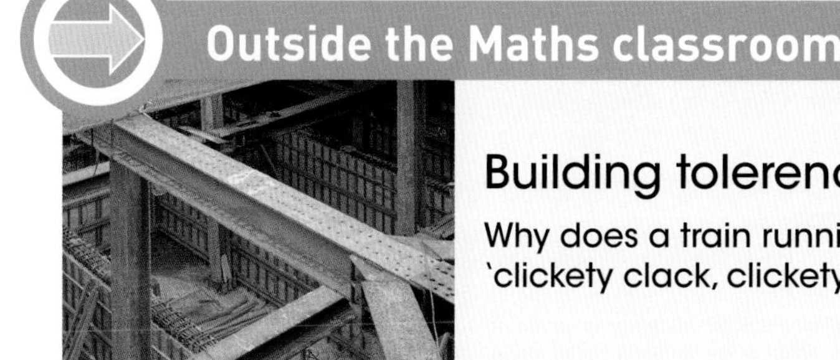

Building tolerences

Why does a train running on an old track go 'clickety clack, clickety clack'?

Toolbox

Jack cuts a pizza into eight equal slices.

Jack eats a slice and Christina eats two slices.

Jack eats $\frac{1}{8}$ of the pizza and Christina eats $\frac{2}{8}$.

Altogether they eat $\frac{1}{8} + \frac{2}{8} = \frac{3}{8}$.

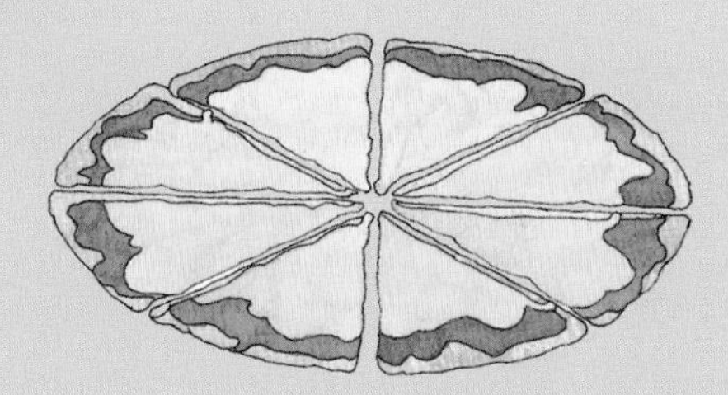

The fractions already have the same bottom lines. You only need to add the top lines.

You often have to use equivalent fractions to make the bottom lines the same. This is called the common denominator.

Sabir eats $\frac{1}{4}$ of a cake.

He then goes back for 'seconds' and eats another $\frac{1}{6}$ of the cake.

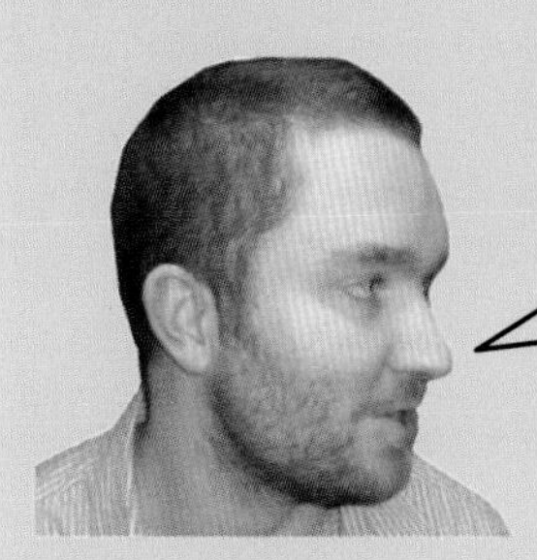

I have eaten $\frac{1}{4} + \frac{1}{6}$, but what is that as a single fraction?

Sabir

The bottom lines are 4 and 6. They are both factors of 12.

Change each fraction to an equivalent fraction with 12 as the denominator.

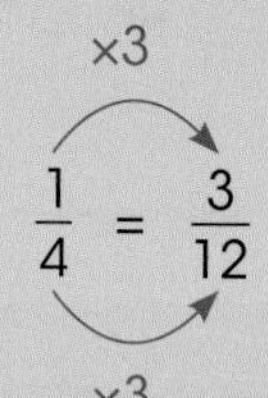

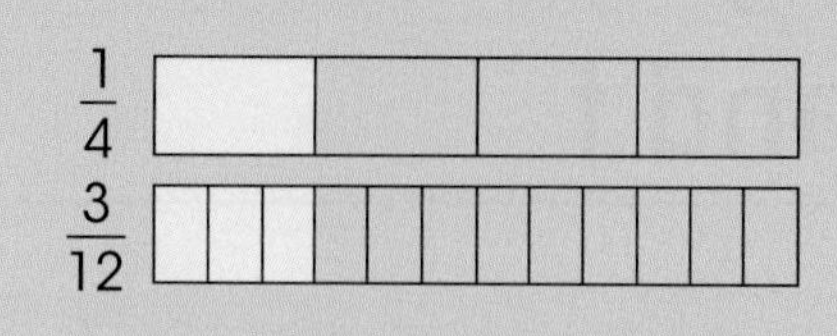

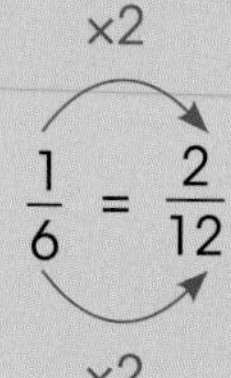

$\frac{1}{6}$

$\frac{2}{12}$

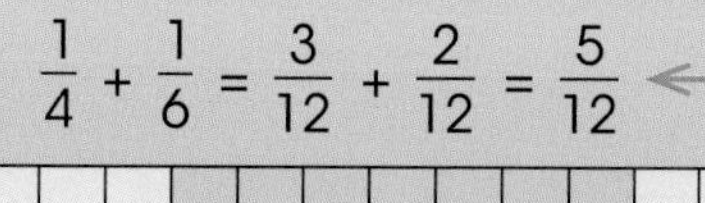

$$\frac{1}{4} + \frac{1}{6} = \frac{3}{12} + \frac{2}{12} = \frac{5}{12}$$

Once the denominators are the same, you can add the top lines.

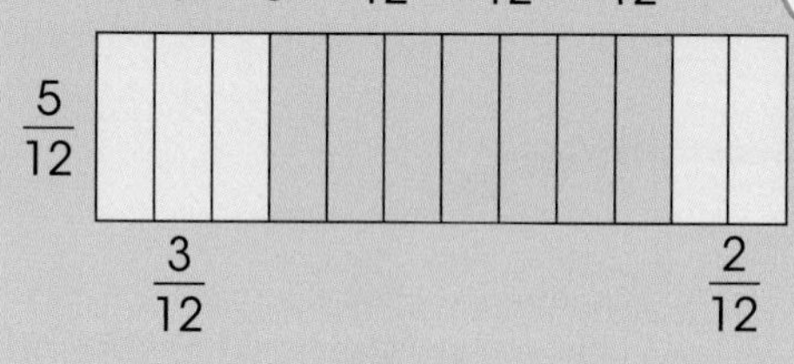

Sabir has eaten $\frac{5}{12}$ of the cake.

Example – Subtracting fractions

Sabir has $\frac{7}{12}$ of his cake left.

He gives $\frac{1}{2}$ of the cake to his sister.

How much is left now?

Solution

Sabir has $\frac{7}{12} - \frac{1}{2}$ left.

×6

$$\frac{1}{2} = \frac{6}{12}$$

×6

$\frac{1}{2}$ and $\frac{6}{12}$ are equivalent fractions.

$$\frac{7}{12} - \frac{1}{2} = \frac{7}{12} - \frac{6}{12} = \frac{1}{12}$$

Once the denominators are the same, you can subtract the top lines.

12 is the lowest common denominator of 12 and 2.

$\frac{1}{12}$ of the cake is left.

The questions in this unit should be answered *without* the use of a calculator. (However, you may wish to use a calculator to check some of your answers.)

Practising skills

1 Work these out.

a $\frac{1}{5}+\frac{1}{5}$ **b** $\frac{2}{7}+\frac{1}{7}$ **c** $\frac{4}{9}+\frac{4}{9}$ **d** $\frac{6}{13}+\frac{5}{13}$

e $\frac{5}{7}-\frac{1}{7}$ **f** $\frac{4}{5}-\frac{2}{5}$ **g** $\frac{5}{6}-\frac{1}{6}$ **h** $\frac{3}{8}+\frac{5}{8}$

2 Copy and complete these.

a $\frac{1}{3}+\frac{\square}{\square}=1$ **b** $\frac{2}{7}+\frac{\square}{\square}=1$ **c** $\frac{\square}{\square}+\frac{1}{3}=1$ **d** $\frac{6}{11}+\frac{\square}{\square}=1$

e $\frac{2}{5}+\frac{1}{5}+\frac{\square}{\square}=1$ **f** $\frac{4}{7}+\frac{1}{7}+\frac{\square}{\square}=1$ **g** $\frac{7}{19}+\frac{4}{19}+\frac{\square}{\square}=1$ **h** $\frac{1}{21}+\frac{1}{21}+\frac{\square}{\square}=1$

3 Work these out.

a $1-\frac{1}{4}$ **b** $1-\frac{1}{5}$ **c** $1-\frac{3}{7}$ **d** $1-\frac{3}{8}$

e $1-\frac{5}{12}$ **f** $1-\frac{7}{10}$ **g** $1-\frac{9}{20}$ **h** $1-\frac{5}{14}$

4 Copy and complete these.

a **i** $\frac{1}{3}=\frac{\square}{12}\quad\frac{1}{4}=\frac{\square}{12}$ **ii** $\frac{1}{3}+\frac{1}{4}=\frac{\square}{12}+\frac{\square}{12}=\frac{\square}{\square}$ **iii** $\frac{1}{3}-\frac{1}{4}=\frac{\square}{12}-\frac{\square}{12}=\frac{\square}{\square}$

b **i** $\frac{2}{3}=\frac{\square}{15}\quad\frac{1}{5}=\frac{\square}{15}$ **ii** $\frac{2}{3}+\frac{1}{5}=\frac{\square}{15}+\frac{\square}{15}=\frac{\square}{\square}$ **iii** $\frac{2}{3}-\frac{1}{5}=\frac{\square}{15}-\frac{\square}{15}=\frac{\square}{\square}$

c **i** $\frac{5}{8}=\frac{\square}{56}\quad\frac{2}{7}=\frac{\square}{56}$ **ii** $\frac{5}{8}+\frac{2}{7}=\frac{\square}{56}+\frac{\square}{56}=\frac{\square}{\square}$ **iii** $\frac{5}{8}-\frac{2}{7}=\frac{\square}{56}-\frac{\square}{56}=\frac{\square}{\square}$

5 Work these out.

a $\frac{1}{4}+\frac{1}{5}$ **b** $\frac{2}{3}+\frac{1}{8}$ **c** $\frac{1}{2}-\frac{1}{5}$ **d** $\frac{2}{3}-\frac{1}{4}$

e $\frac{2}{5}+\frac{2}{7}$ **f** $\frac{7}{10}-\frac{1}{3}$ **g** $\frac{5}{6}-\frac{7}{9}$ **h** $\frac{3}{5}+\frac{3}{10}$

Developing fluency

1 Jared is completing a Technology assignment.
He completes $\frac{1}{4}$ of it on Monday night and $\frac{2}{3}$ of it on Tuesday night.
What fraction of it has he completed?

2 Jordan's car's petrol tank is $\frac{3}{5}$ full.
His journey uses $\frac{1}{4}$ of a tank.
What fraction of the tank contains petrol when he completes his journey?

3) Copy and complete this grid.

+	$\frac{1}{4}$	$\frac{1}{6}$	$\frac{4}{9}$
$\frac{3}{7}$			
		$\frac{17}{30}$	
	$\frac{2}{5}$		

4) Copy and complete these.

a $\frac{5}{6} - \frac{\square}{\square} = \frac{1}{4}$ **b** $\frac{\square}{\square} - \frac{1}{3} = \frac{2}{5}$ **c** $\frac{1}{8} + \frac{1}{2} = \frac{1}{4} - \frac{\square}{\square}$ **d** $\frac{9}{10} - \frac{1}{6} = \frac{\square}{\square} + \frac{1}{5}$

Exam-style

5) Jon cuts the grass for his neighbours.
He starts with a full can of petrol.
He uses $\frac{1}{3}$ of a can on one lawn, $\frac{1}{5}$ of a can on the next lawn and $\frac{1}{6}$ of a can on the last lawn.
How much petrol is left?

Problem solving

Exam-style

1) Jason is collecting rainwater in buckets.

Bucket A $\frac{1}{4}$ full

Bucket B $\frac{3}{5}$ full

Bucket C $\frac{7}{8}$ full

Jason only has three buckets and he needs to empty one bucket.
He does not want to lose any of the rainwater in the buckets.
Can Jason empty one bucket into the other two without losing any rainwater? Explain your reasoning.

Exam-style

2) There were three candidates in a local election.
Mrs Williams got $\frac{3}{10}$ of the votes.
Mr Bader got $\frac{3}{8}$ of the votes.
Mr Iqbal got the rest of the votes.
Who won the local election?

Exam-style

3 The diagram shows a plan of Bill's garden.

$\frac{2}{5}$ of the garden is covered by lawn.

$\frac{1}{3}$ of the garden is the vegetable patch.

$\frac{1}{6}$ of the garden is taken up by the flower bed.

The rest of the area of the garden are the paths.
Bill is going to make the paths from stone slabs.
Work out the fraction of the garden that is taken up by the paths.

Exam-style

4 Corina is driving to Milan.
At the start of the journey, her petrol tank is $\frac{3}{4}$ full.
On arriving in Milan, her petrol tank is $\frac{1}{8}$ full.

a What fraction of a tank of petrol has she used?

On leaving Milan, Corina adds another $\frac{3}{4}$ of a tank of petrol into her car.
The journey home uses $\frac{1}{2}$ of a tank of petrol.

Corina is now going to fill up her petrol tank.

b What fraction of a tank of petrol does Corina need to put in her car?

Exam-style

5 Emily is preparing some food on holiday.
She only has two cups to use for measuring.
One cup holds $\frac{1}{2}$ pint and the other holds $\frac{1}{3}$ pint.
Emily fills both cups with milk and empties them into the bowl.
She then adds another $\frac{1}{3}$ pint of milk.

a How much milk is now in the bowl?

Emily now needs to measure $\frac{1}{6}$ pint of cream.

b Explain how Emily can do this.

Exam-style

6 The diagram is that of an electronic meter that shows how full the car park is.
On Monday at 8 a.m, the car park is half full.

Empty $\frac{1}{4}$ $\frac{1}{2}$ $\frac{3}{4}$ Full

8 a.m.

This is the meter reading at 10 a.m. No cars have left between 8 a.m. and 10 a.m.

Empty $\frac{1}{4}$ $\frac{1}{2}$ $\frac{3}{4}$ Full

10 a.m.

The charge for parking in this car park is £5.50 and cars pay on entry.
The car park holds 240 cars when full.
Work out how much money has been collected between 8 a.m and 10 a.m.

Exam-style

Extension

7 When the number 5 bus leaves the bus station, it is $\frac{1}{4}$ full.
At the 1st bus stop, $\frac{1}{8}$ of a bus load get on.
At the 2nd bus stop, $\frac{1}{16}$ of a bus load get off and $\frac{1}{2}$ a bus load get on.
At the 3rd bus stop nobody gets off and the bus is full when it leaves.

a What fraction of a bus load got on at the 3rd bus stop?

The bus remains full until 8 people get off at the 5th bus stop and nobody gets on.
The bus is then $\frac{7}{8}$ full.
At the 6th bus stop, $\frac{5}{16}$ of a bus load get off.
The bus driver now wants to know how many people there are on his bus.

b Work out the number of people that are now on the bus.

Exam-style

8 On Sunday, Ewan completely fills the petrol tank of his car.
On Monday, the petrol tank is $\frac{4}{5}$ full.
The diagram shows the fuel gauge in Ewan's car on Monday.

Empty $\frac{1}{4}$ $\frac{1}{2}$ $\frac{3}{4}$ Full

On Tuesday, Ewan uses $\frac{5}{12}$ of the fuel in the petrol tank.
Ewan doesn't like the petrol tank to be less than one quarter full.

a Is the petrol tank now less than one quarter full?

b Show your answer on a fuel gauge like the one below.

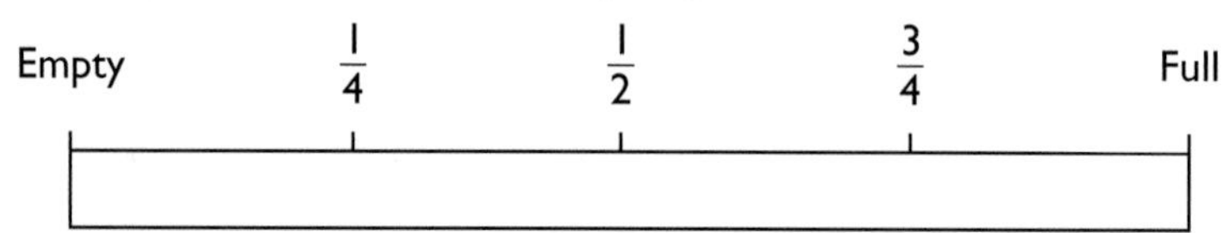

Reviewing skills

1. Work these out.

   a $\frac{9}{10} - \frac{6}{10}$ **b** $\frac{8}{11} - \frac{7}{11}$ **c** $\frac{1}{2} + \frac{1}{2}$ **d** $\frac{5}{12} + \frac{6}{12}$

2. Work these out.

   a $1 - \frac{1}{2}$ **b** $1 - \frac{1}{100}$ **c** $1 - \frac{2}{3}$ **d** $1 - \frac{19}{40}$

3. Work these out.

   a $\frac{7}{12} - \frac{1}{4}$ **b** $\frac{3}{8} + \frac{2}{7}$ **c** $\frac{5}{12} + \frac{7}{24}$ **d** $\frac{17}{20} - \frac{5}{8}$

4. At the start of January Mr Jackson's oil tank was $\frac{9}{10}$ full.
   At the end of March his tank is $\frac{1}{4}$ full.
   What fraction of the whole tank was used during the months January, February and March?

Unit 5 • Working with mixed numbers • Band f

Outside the Maths classroom

Recipes

How are ingredients measured in the United States?

Toolbox

To find out if all the sand can be placed in one bucket:

$$\frac{1}{2} + \frac{2}{5} + \frac{1}{3} = \frac{15 + 12 + 10}{30} = \frac{37}{30}$$

$\frac{30}{30}$ is one whole.

$\frac{37}{30}$ is one whole and $\frac{7}{30}$. That is $1\frac{7}{30}$.

So the sand takes up one whole bucket and $\frac{7}{30}$ of a second one.

$\frac{37}{30}$ is called a **top-heavy fraction** or an **improper fraction**.

$1\frac{7}{30}$ is called a **mixed number**.

Every improper fraction can be written as a mixed number.

Example – Converting between improper fractions and mixed numbers

a Express $\frac{19}{5}$ as a mixed number.

b Express $2\frac{7}{8}$ as an improper fraction.

Solution

a You change $\frac{19}{5}$ to a mixed number by dividing 19 by 5.

$19 \div 5 = 3$ remainder 4

So $\frac{19}{5} = 3\frac{4}{5}$

b To find how many eighths there are in $2\frac{7}{8}$, you multiply 2 by 8 and then add 7.

$$2\frac{7}{8} = \frac{16}{8} + \frac{7}{8} = \frac{23}{8}$$

So $2\frac{7}{8} = \frac{23}{8}$

Example – Multiplying, adding and subtracting mixed numbers

Work out

a $2\frac{1}{3} \times 1\frac{4}{5}$ **b** $2\frac{1}{3} + 1\frac{4}{5}$ **c** $2\frac{1}{3} - 1\frac{4}{5}$

Solution

a Start by writing $2\frac{1}{3}$ and $1\frac{4}{5}$ as top-heavy fractions.

$2\frac{1}{3} = \frac{7}{3}$, $1\frac{4}{5} = \frac{9}{5}$

So $2\frac{1}{3} \times 1\frac{4}{5} = \frac{7}{\cancel{3}_1} \times \frac{\cancel{9}^3}{5}$ ← Cancel using a common factor.

$= \frac{21}{5}$

$= 4\frac{1}{5}$

b $2\frac{1}{3} + 1\frac{4}{5}$ Add the whole numbers and the fractions separately.

$= (2 + 1) + (\frac{1}{3} + \frac{4}{5})$

$= 3 + (\frac{5}{15} + \frac{12}{15})$ ← 15 is the common denominator.

$= 3 + \frac{17}{15}$ ← Make this top-heavy fraction into a mixed number.

$= 3 + 1\frac{2}{15}$

$= 4\frac{2}{15}$

c $2\frac{1}{3} - 1\frac{4}{5} = \frac{7}{3} - \frac{9}{5}$

Convert the top-heavy fractions to equivalent fractions with a common denominator. ← You can also use a method like that in part b.

$\frac{7}{3} = \frac{35}{15}$ (×5 top and bottom) $\frac{9}{5} = \frac{27}{15}$ (×3 top and bottom)

$\frac{7}{3} - \frac{9}{5} = \frac{35}{15} - \frac{27}{15} = \frac{8}{15}$

The questions in this unit should be answered *without* the use of a calculator. (However, you may wish to use a calculator to check some of your answers.)

Practising skills

1 Change these mixed numbers to top-heavy fractions.

a $1\frac{1}{3}$ **b** $2\frac{1}{4}$ **c** $3\frac{1}{2}$ **d** $1\frac{2}{5}$

e $2\frac{3}{4}$ **f** $5\frac{1}{6}$ **g** $2\frac{2}{9}$ **h** $6\frac{6}{7}$

2 Change these top-heavy fractions to mixed numbers.

a $\frac{5}{4}$ **b** $\frac{5}{2}$ **c** $\frac{10}{3}$ **d** $\frac{11}{4}$

e $\frac{17}{5}$ **f** $\frac{23}{6}$ **g** $\frac{80}{9}$ **h** $\frac{51}{4}$

3 Work these out. Write each answer as a mixed number.

a $2\frac{1}{4} + 3\frac{1}{2}$ **b** $1\frac{2}{3} + \frac{1}{6}$ **c** $2\frac{4}{5} + 3\frac{1}{5}$

d $5\frac{3}{4} - 1\frac{1}{4}$ **e** $5\frac{1}{5} - 3\frac{1}{10}$ **f** $3\frac{1}{3} + 1\frac{1}{6}$

4 Work these out. Write each answer as a mixed number.

a $1\frac{1}{2} + 2\frac{2}{3}$ **b** $1\frac{3}{4} + 2\frac{1}{2}$ **c** $2\frac{3}{5} + 1\frac{1}{3}$

d $5\frac{3}{5} + 1\frac{7}{10}$ **e** $2\frac{5}{6} + 2\frac{3}{4}$ **f** $3\frac{7}{8} + 2\frac{4}{5}$

5 Work these out. Write each answer as a proper fraction or a mixed number.

a $1\frac{1}{3} - \frac{7}{8}$ **b** $2\frac{1}{2} - \frac{7}{10}$ **c** $3\frac{1}{4} - 1\frac{2}{5}$

d $4\frac{2}{3} - 1\frac{7}{8}$ **e** $5\frac{1}{4} - 2\frac{5}{6}$ **f** $3\frac{1}{10} - 1\frac{7}{9}$

6 Work these out. Write each answer as a mixed number.

a $4\frac{1}{3} \times 2$ **b** $4 \times 2\frac{1}{5}$ **c** $2\frac{1}{4} \times 3$

d $5 \times 1\frac{2}{7}$ **e** $6 \times 2\frac{1}{3}$ **f** $3\frac{2}{5} \times 9$

7 Work these out. Write each answer as a mixed number.

a $2\frac{1}{4} \times 3\frac{1}{2}$ **b** $1\frac{2}{3} \times 4\frac{1}{3}$ **c** $3\frac{1}{6} \times \frac{2}{3}$

d $\frac{5}{6} \times 12\frac{4}{5}$ **e** $1\frac{3}{8} \times 2\frac{1}{3}$ **f** $2\frac{3}{4} \times 4\frac{1}{2}$

8 Work out these.

a $2\frac{3}{4} + 3\frac{2}{5}$ **b** $4\frac{1}{3} - 1\frac{5}{8}$ **c** $5\frac{2}{5} - 2\frac{9}{10}$

d $2\frac{1}{4} \times 1\frac{5}{6}$ **e** $2\frac{3}{4} \times 2\frac{1}{3}$ **f** $7\frac{1}{10} - 2\frac{7}{8}$

Developing fluency

1 Sort these into pairs.

$1\frac{1}{5}$ $\frac{13}{5}$ $1\frac{2}{3}$ $\frac{10}{6}$ $2\frac{1}{2}$ $\frac{6}{5}$ $\frac{20}{8}$ $\frac{13}{4}$ $2\frac{3}{5}$ $3\frac{1}{4}$

2 Write **=**, **<** or **>** in each box.

a $\frac{19}{3}\square 5\frac{1}{3}$ **b** $\frac{21}{6}\square 3\frac{1}{2}$ **c** $2\frac{5}{8}\square \frac{23}{8}$ **d** $\frac{61}{9}\square 6\frac{2}{3}$

3 Work out these.

a $1\frac{1}{3} \times 2$ **b** $1\frac{1}{3} \times 3$ **c** $1\frac{1}{3} \times 4$

d $1\frac{1}{3} \times 5$ **e** $1\frac{1}{3} \times 6$ **f** $1\frac{1}{3} \times 7$

Exam-style

4 Bob runs $4\frac{1}{2}$ km on Monday, $5\frac{1}{4}$ km on Tuesday, $4\frac{7}{8}$ km on Wednesday.

His plan is to run 20 km.

How far does he need to run on Thursday to complete the 20 km?

5 Decide if these are true or false.

a $\frac{1}{3}$ of $1\frac{1}{2}$ kg $= \frac{1}{2}$ kg

b $\frac{1}{2}$ km $+ \frac{1}{3}$ km $+ \frac{1}{4}$ km $+ \frac{1}{5}$ km $= 1\frac{7}{60}$ km

c 6 pints $- 2\frac{1}{3}$ pints $- 1\frac{3}{4}$ pints $= 1\frac{5}{12}$ pints

d $3\frac{1}{2}$ m $\times 1\frac{1}{2}$ m $= 5\frac{1}{4}$ m^2

Exam-style

6 Three friends are going on holiday.

Max's hand luggage is $6\frac{1}{3}$ kg, Ned's hand luggage is $5\frac{7}{8}$ kg and Oran's hand luggage is $3\frac{3}{4}$ kg.

a What is the total weight of their hand luggage?

b How much heavier is Max's hand luggage than Oran's hand luggage?

c On the return flight, Max's hand luggage is $\frac{7}{10}$ kg lighter, Ned's is $1\frac{3}{5}$ kg heavier and Oran's is twice as heavy.

What is the total weight of their hand luggage now?

7 Work out these.

Write each answer as a mixed number.

a $2\frac{1}{4} + 1\frac{1}{3} - \frac{5}{6}$ **b** $\frac{5}{6} \times 1\frac{1}{2} \times 2\frac{1}{4}$ **c** $1\frac{5}{8} + \frac{7}{8} \times 1\frac{1}{3}$

d $2\frac{1}{2} \times 1\frac{3}{4} + 3\frac{1}{3}$ **e** $1\frac{1}{4} + 2\frac{1}{2} \times 1\frac{1}{2}$ **f** $10 - 2\frac{1}{3} \times 1\frac{2}{7}$

8 Work out these. Write your answer as a mixed number.

a $\left(1\frac{1}{2}\right)^2$ b $\left(2\frac{1}{2}\right)^2$ c $\left(\frac{5}{4}\right)^2$

d $\left(1\frac{1}{2}\right)^3$ e $\left(2\frac{1}{5}\right)^3$ f $\left(1\frac{2}{3}\right)^4$

9 Copy and complete these.

a $1\frac{3}{4} \times 2\frac{2}{5} = 2\frac{4}{5} + \frac{\square}{\square}$ b $\frac{\square}{\square} - 1\frac{3}{4} = 3\frac{1}{3} \times 1\frac{7}{8}$ c $3\frac{1}{2} \times 1\frac{5}{6} = 7\frac{1}{2} - \frac{\square}{\square}$

Exam-style

10 The diagram gives some of the dimensions of a room in metres.
Calculate its perimeter.

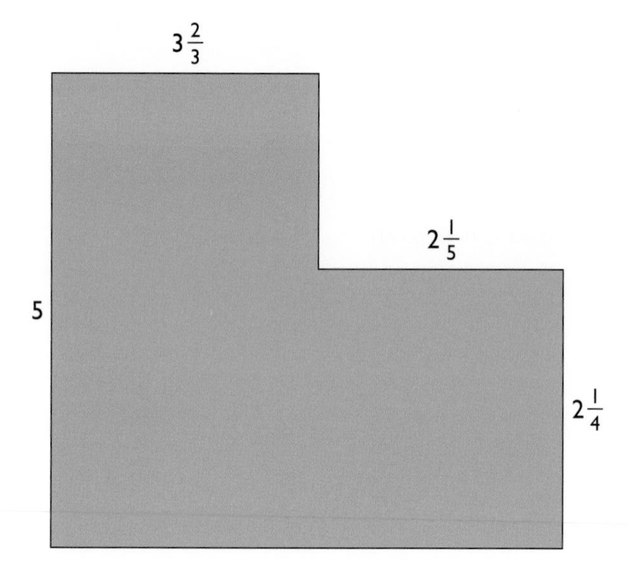

Problem solving

Exam-style

1 Peter takes part in a charity fun run.
The diagram shows the distance between checkpoints on the run.

Start ← $3\frac{1}{4}$ km → Checkpoint A ← $2\frac{1}{8}$ km → Checkpoint B ← $2\frac{5}{8}$ km → Finish

a How long is the race?

b How far is Checkpoint A from the finish?

c Peter is half way between Checkpoints A and B. How far has he run?

Exam-style

2 Dave has three tins of white paint.

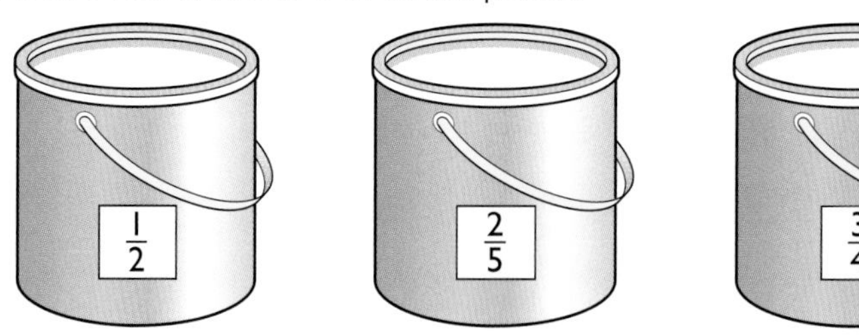

Dave has used some of the paint from each tin.
There was 1 litre of paint in each tin.
Now one tin is half full of paint, one tin is $\frac{2}{5}$ full of paint and one tin is $\frac{3}{4}$ full.
Dave needs $1\frac{4}{5}$ litres of white paint.
Has Dave got enough paint?

Extension

Exam-style

3 Tom needs $3\frac{1}{2}$ cups of dried fruit for a cake recipe.
He only has $\frac{3}{4}$ cup of sultanas and $1\frac{1}{3}$ cups of raisins. He has plenty of currants.

a How many cups of currants does he need to use?

Tom, Linda and Les share the cake.
Les has $\frac{1}{9}$ of the cake.
Linda has $\frac{1}{10}$ of the cake.
Tom has $\frac{2}{15}$ of the cake, and says that there is over half of the cake left still.

b Is Tom right?

Exam-style

4 The diagram shows a part that Brian is making for his go-cart.

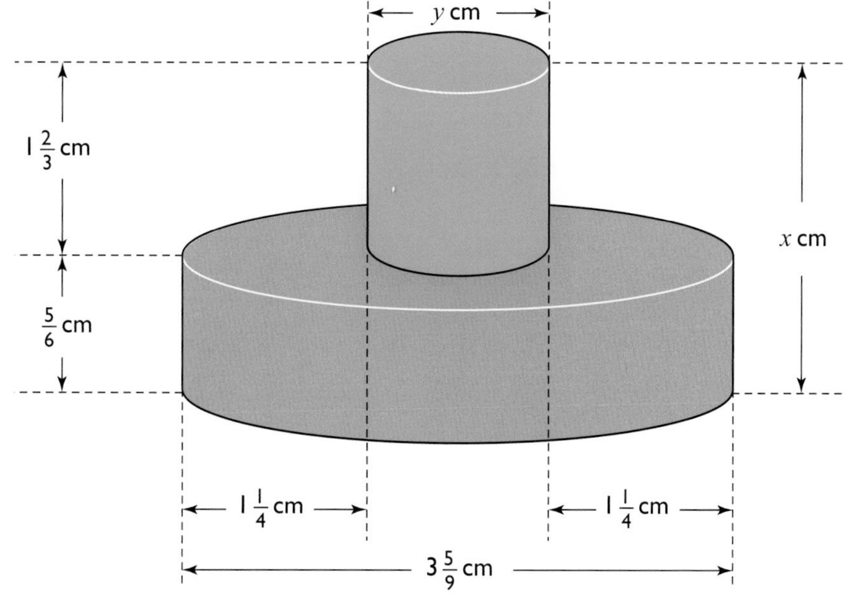

To fit Brian's go-cart, y must be equal to 1 and x must be equal to $2\frac{1}{2}$.
Has Brian made any mistakes with the sizes of this part?

Extension

Exam-style

5 The game of 'Higher or Lower' is played with a pack of cards.
Each card in the pack has a fraction written on it.

> **Rules of the game:**
> The cards are shuffled and placed in a pile face down.
> The top card is taken from the pile and placed next to the pile, showing the fraction.
> A second card is taken and placed alongside the first card.
>
> The players say whether the fraction on the card is higher, lower, or the same as the fraction on the previous card.
> 1 point is scored for a correct answer.
> The first person to get 3 points is the winner.

Tom and Leah start a game.
Here is the order in which the cards come up, and their answers. Who wins?

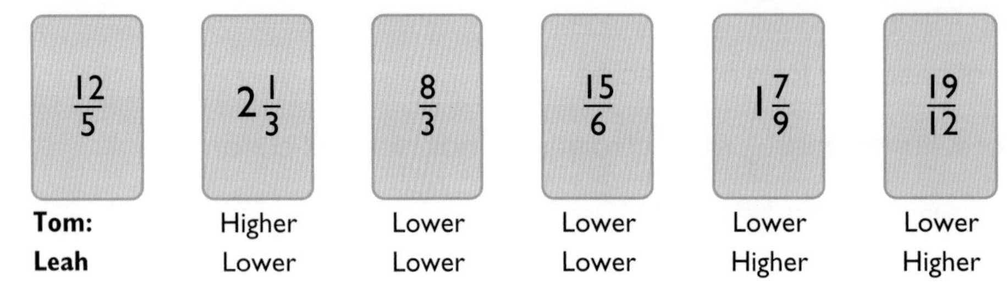

Exam-style

Extension

6 Richard is the chief executive and owner of a company.
The company has three directors who work for Richard.
Mary is one of the directors and Keith is her personal assistant.
Each year, the company makes a profit and Richard gives bonuses to himself and his directors.
Of the profits, $\frac{2}{3}$ are reinvested in the company.
Of the remaining profits, Richard gives himself a bonus of $\frac{1}{2}$ and shares the rest equally between his three directors.
Mary always gives $\frac{1}{10}$ of her bonus to Keith.
Keith wants to know what his bonus will be.

a Work out Keith's bonus as a fraction of the company's profits.

Last year the company made a record profit of £$3\frac{3}{5}$ million.

b What was Keith's bonus?

Exam-style

Extension

7 Mark is carpeting his living room.
The diagram shows a plan of his living room.

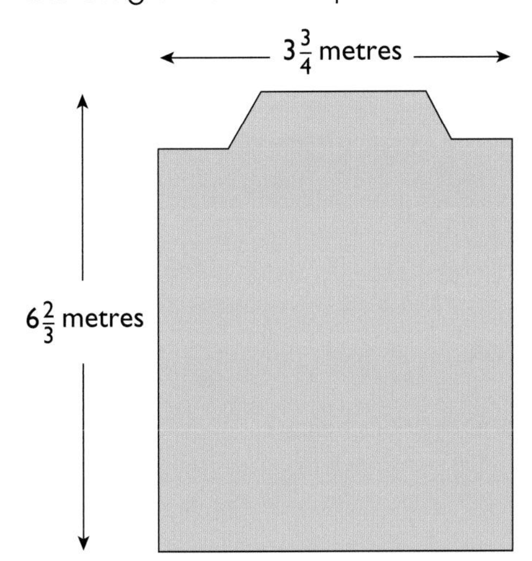

Mark wants the carpet to be in one complete piece with no joins.
In a carpet shop, he sees three carpets that he likes.

Carpet A is £8.99 per square metre
Carpet B is £9.99 per square metre
Carpet C is £10.99 per square metre

Mark cannot afford to spend more than £250 on the carpet.
Which carpet, or carpets, can Mark afford to buy?
Explain any assumptions that you make.

Reviewing skills

1 Change these mixed numbers to top-heavy fractions.

a $4\frac{3}{5}$ **b** $6\frac{2}{3}$ **c** $9\frac{8}{9}$ **d** $12\frac{3}{4}$

2 Change these top-heavy fractions to mixed numbers.

a $\frac{37}{8}$ **b** $\frac{41}{7}$ **c** $\frac{68}{5}$ **d** $\frac{137}{20}$

3 Work these out.

a $2\frac{1}{4} + 1\frac{3}{4}$ **b** $4\frac{1}{3} + \frac{2}{3}$ **c** $7\frac{5}{6} - 2\frac{1}{2}$

d $5\frac{1}{2} + 2\frac{3}{4}$ **e** $5\frac{1}{4} - 3\frac{1}{2}$ **f** $3\frac{1}{3} + 1\frac{4}{5}$

4 Work these out.

a $2\frac{1}{4} \times 2$ **b** $2\frac{1}{4} \times 3$ **c** $2\frac{1}{4} \times 4$ **d** $2\frac{1}{4} \times 5$

e $2\frac{1}{4} \times 6$ **f** $2\frac{1}{4} \times 7$ **g** $2\frac{1}{4} \times 8$ **h** $2\frac{1}{4} \times 9$

Unit 6 • Dividing fractions • Band f

Outside the Maths classroom

Organising time

Appointments are often booked into $\frac{1}{4}$ hour slots.

How many appointments could be fitted into one working day?

Toolbox

Reciprocals

$\frac{1}{4}$ is known as the **reciprocal** of 4 .

The fraction $\frac{1}{3}$ is the reciprocal of 3. ← You can write the whole number 3 as $\frac{3}{1}$.

You turn a fraction upside down to find its reciprocal.

$\frac{5}{2}$ is the reciprocal of $\frac{2}{5}$. In the same way $\frac{2}{5}$ is the reciprocal of $\frac{5}{2}$.

Division problems

Each glass holds $\frac{1}{4}$ of a bottle of lemonade.

How many glasses can be filled from three bottles?

The problem can be solved by finding how many quarters in 3.

$$3 \div \frac{1}{4} = 12$$

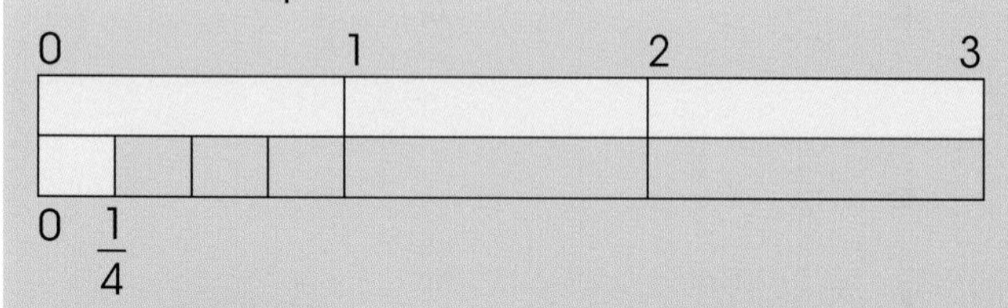

It can also be solved as a multiplication.

So dividing by a fraction is the same as multiplying by its reciprocal.

$3 \times 4 = 12$

You should change mixed numbers into improper fractions before dividing.

Example – Dividing a fraction by a fraction

Work out $\frac{5}{4} \div \frac{2}{3}$.

Solution

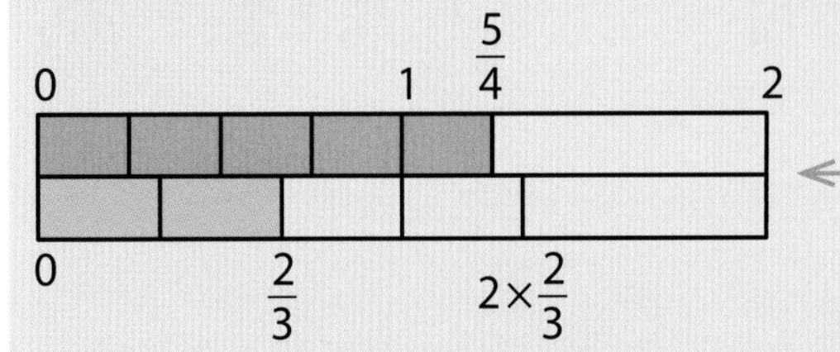

The bar diagram shows the answer will be just less than 2.

You find the answer by multiplying by the reciprocal.

The reciprocal of $\frac{2}{3}$ is $\frac{3}{2}$.

$$\frac{5}{4} \div \frac{2}{3} = \frac{5}{4} \times \frac{3}{2}$$

$$= \frac{15}{8}$$

The questions in this unit should be answered *without* the use of a calculator. (However, you may wish to use a calculator to check some of your answers.)

Practising skills

1 Write down the reciprocal of each of these.

a $\frac{1}{7}$ **b** $\frac{5}{7}$ **c** 20 **d** $1\frac{1}{2}$

e $2\frac{5}{8}$ **f** $4\frac{5}{6}$ **g** $3\frac{7}{8}$ **h** $6\frac{2}{9}$

2 Change each of these into a multiplication and then work out the answer.

a $\frac{1}{5} \div 2$ **b** $\frac{1}{3} \div 4$ **c** $\frac{3}{4} \div 5$ **d** $\frac{3}{5} \div 3$

e $\frac{7}{9} \div 6$ **f** $\frac{9}{10} \div 4$ **g** $\frac{1}{20} \div 2$ **h** $\frac{1}{20} \div 5$

3 Change each of these into a multiplication and then work out the answer.

a $3 \div \frac{1}{2}$ **b** $2 \div \frac{1}{4}$ **c** $3 \div \frac{1}{5}$ **d** $2 \div \frac{2}{3}$

e $5 \div \frac{3}{5}$ **f** $6 \div \frac{3}{4}$ **g** $6 \div \frac{1}{12}$ **h** $12 \div \frac{1}{6}$

4 Change each of these into a multiplication and then work out the answer.

a $\frac{1}{3} \div \frac{3}{4}$ **b** $\frac{1}{6} \div \frac{2}{5}$ **c** $\frac{2}{9} \div \frac{1}{2}$ **d** $\frac{2}{7} \div \frac{3}{4}$

e $\frac{3}{5} \div \frac{9}{10}$ **f** $\frac{5}{8} \div \frac{2}{3}$ **g** $\frac{1}{6} \div \frac{2}{3}$ **h** $\frac{2}{3} \div \frac{1}{6}$

5 Change each of these into a multiplication and then work out the answer.

a $1\frac{1}{4} \div 3$ **b** $2\frac{1}{3} \div \frac{1}{2}$ **c** $3 \div 2\frac{3}{4}$ **d** $2\frac{4}{5} \div 1\frac{1}{2}$

e $2\frac{2}{7} \div 3\frac{1}{5}$ **f** $3\frac{1}{8} \div 1\frac{5}{8}$ **g** $4\frac{1}{3} \div 1\frac{1}{8}$ **h** $1\frac{1}{8} \div 4\frac{1}{5}$

6) Work out these.

a $\frac{1}{7} \div \frac{3}{5}$ b $\frac{1}{9} \div \frac{3}{4}$ c $\frac{7}{10} \div \frac{1}{3}$ d $\frac{3}{7} \div \frac{3}{5}$

e $\frac{5}{8} \div 6$ f $\frac{9}{10} \div 4$ g $5 \div \frac{2}{3}$ h $1\frac{1}{2} \div 7$

i $8 \div 3\frac{1}{2}$ j $1\frac{1}{3} \div 2\frac{3}{4}$ k $1\frac{7}{8} \div 5\frac{1}{4}$ l $3\frac{2}{5} \div 1\frac{3}{5}$

Developing fluency

1) $\frac{7}{8}$ of a litre of cola is shared between 3 friends.

What fraction of a litre of cola does each receive?

2) Daryl eats $\frac{2}{3}$ of a pizza.

The rest is divided between his two little sisters.

What fraction of the pizza does each sister eat?

3) Decide if these are true or false.

a $\frac{1}{3} \div 4 = 4 \div \frac{1}{3}$ b $6 \div \frac{1}{2} = 3 \div \frac{1}{4}$ c $\frac{2}{5} \div 3 = \frac{3}{5} \div 2$

4) Copy and complete this multiplication grid.

×		$\frac{2}{3}$
$\frac{1}{5}$	$\frac{1}{20}$	
		$\frac{5}{9}$

Exam-style

5) A room has area $30\,m^2$ and length $6\frac{1}{4}$ m.

Work out these:

a the width of the room

b the perimeter of the room

6) Work out these.

a $\frac{1}{2} \div \frac{5}{8} + \frac{1}{6}$ b $\frac{7}{8} - \frac{5}{8} \div \frac{5}{6}$ c $\frac{9}{10} + \frac{3}{5} \div \frac{2}{7}$ d $\frac{4}{5} \div \frac{3}{4} + \frac{1}{3}$

Exam-style

7) It takes Jenny $\frac{3}{4}$ of an hour to drive the $26\frac{1}{4}$ miles to her friend Sandra's house.

What is her average speed? [Use the formula speed = distance ÷ time]

Problem solving

Exam-style

Extension

1. This was part of an article in a local newspaper.

Local Millionaire Leaves Fortune

Martin Miller, local millionaire, left $\frac{1}{2}$ of his fortune of £$17\frac{1}{2}$ million to his wife of 32 years.

His children are to share the rest of the money, each receiving £$1\frac{3}{4}$ million.

When Liz read this article, she asked 'How many children did Martin Miller have?'
Work out the answer to Liz's question.

Exam-style

Extension

2. Geoff is an engineer.
He makes parts for car engines.
The diagram shows a part that Geoff has to cut from a piece of metal of length 1 metre.

$3\frac{1}{8}$ cm

Geoff needs to make 100 of these parts.
He has three 1 metre lengths of metal.
Can Geoff make these 100 parts from the metal he has available?

Exam-style

Extension

3. Mrs Green is organising a birthday party for her daughter Jessica.
Including Jessica, there are 35 children at the party.

Mrs Green has $5\frac{1}{2}$ litres of juice.

Each glass holds $\frac{1}{6}$ litre when full.

a Will there be enough juice for one glassful for each child at the party?

Mrs Green has five $3\frac{1}{2}$ kg bars of chocolate.

She shares them equally between the 35 children at the party.

b What fraction of a bar of chocolate does she give each child?

At the end of the party, $\frac{4}{5}$ of the children are collected by their parents.

Jessica and her best friend Sophie stay at Jessica's house.
The rest of the children walk home.

c What fraction of the number of children at the party walk home?

Reviewing skills

1 Write down the reciprocal of each of these.

a 8 **b** $\frac{3}{4}$ **c** $2\frac{1}{3}$ **d** $1\frac{4}{5}$

2 Change each of these into a multiplication and then work out the answer.

a $\frac{7}{8} \div 2$ **b** $\frac{5}{6} \div 10$ **c** $4 \div \frac{1}{6}$ **d** $7 \div \frac{3}{8}$

3 Change each of these into a multiplication and then work out the answer.

a $\frac{3}{10} \div \frac{1}{3}$ **b** $\frac{2}{9} \div \frac{5}{6}$ **c** $\frac{5}{9} \div 1\frac{2}{3}$ **d** $6\frac{1}{4} \div 2\frac{2}{3}$

4 Meena has a pendant.

She wants to know what it is made of and so she tries to measure its density.

She finds its weight is $34\frac{1}{5}$ grammes and its volume is $1\frac{4}{5}$ cubic centimetres.

Find the density of the pendant.

Use the formula density = $\frac{\text{mass}}{\text{volume}}$

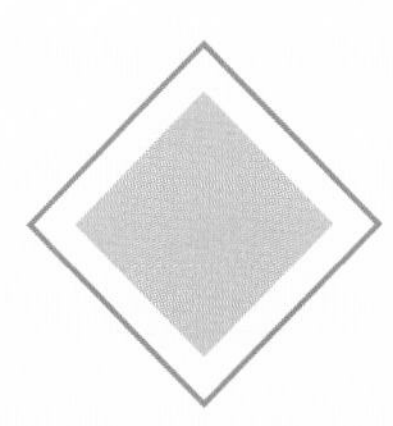

Number Strand 5 Percentages

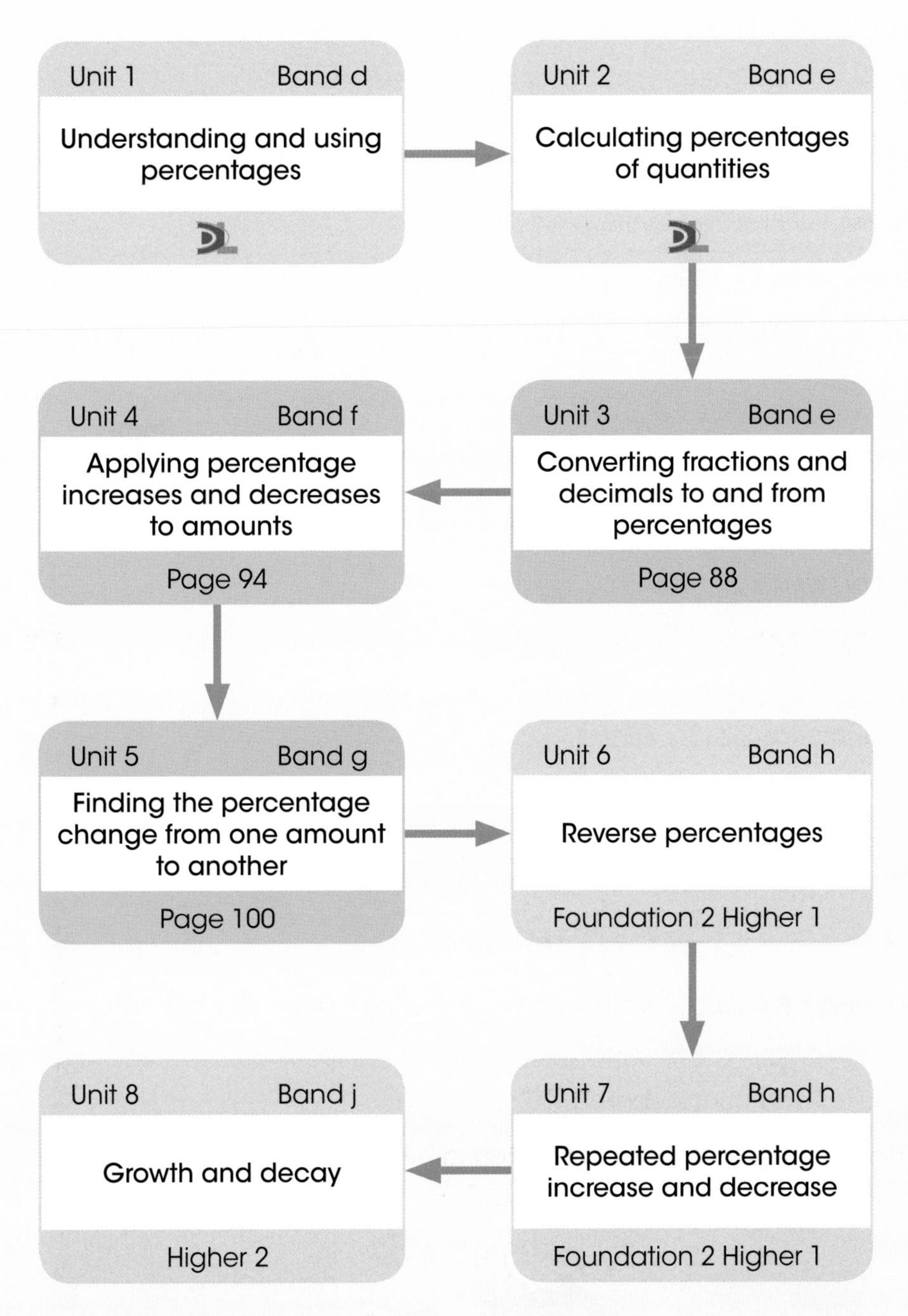

Units 1–2 are assumed knowledge for this book. They are reviewed and extended in the Moving on section on page 86.

5 Units 1–2 • Moving on

Exam-style

1. 160 people work for a company
65% of the workforce is women.
How many men work for the company?

2. Jack bought a car for £15 500.
After one year the car had lost 40% of its value.
Work out the value of the car one year after Jack bought it.

3. Helen buys a laptop priced at £180.
She pays a deposit of 30% and then 12 monthly payments of £13.
How much extra does she pay for the laptop?

Exam-style

4. Rosie's car is in the garage for its annual service.
Here is part of her bill.

Pete's Car Services	
10 000 mile service	£240
Parts	£118
Total excluding VAT	£
VAT @ 20%	£
Total including VAT	£

Work out the total amount including VAT.

Exam-style

5. Kim and her partner are looking for a house to buy.
Here are 3 houses that they like.

A £230 000 B £255 000 C £249 000

They have to pay a deposit of 12% of the cost of any house they wish to buy.
Between them, Kim and her partner have saved £30 000.
Which of the 3 houses can they afford to pay a deposit for?

Exam-style

6 Andy has just been promoted.
His new annual salary will be £38 500.
Andy wants to know how much money he is actually going to get each month.
From his salary there will be deductions for income tax, National Insurance and his pension.
7% of his salary will be taken for National Insurance.
His pension contribution will be 5% of his salary.
Andy estimates that 18% of his salary will be taken for income tax.
Work out how much money Andy will get each month.

Unit 3 • Converting fractions and decimals to and from percentages • Band e

Outside the Maths classroom

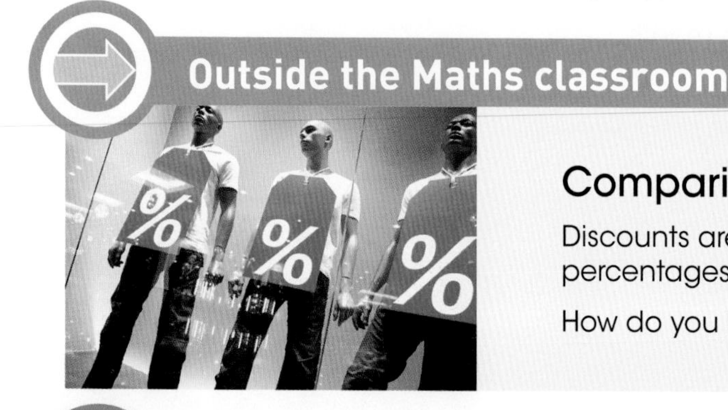

Comparing discounts

Discounts are sometimes advertised as percentages and sometimes as fractions.

How do you know which discount is best?

Toolbox

There are several methods for **converting fractions and decimals into percentages**.

They are based on finding an equivalent fraction with 100 on the bottom line.

The methods are shown here using the example of $\frac{1}{8}$.

Finding the equivalent fraction

$$\frac{1}{8} \overset{\times 50}{=} \frac{50}{400} \overset{\div 4}{=} \frac{12.5}{100} = 12.5\%$$

(×50 on top and bottom; ÷4 on top and bottom)

Using decimals

$\frac{1}{8} = 0.125$ ← You need to divide 8 into 1.000

$0.125 \times \frac{100}{100} = \frac{12.5}{100}$ ← This is the equivalent fraction with 100 on the bottom line.

$= 12.5\%$

Using a percentage bar

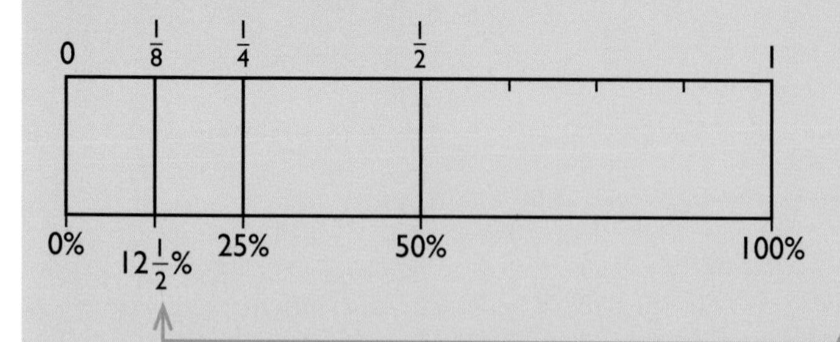

$\frac{1}{8}$ is the same as $12\frac{1}{2}\%$

To **convert a percentage into a fraction or decimal**, start by writing it as a fraction with 100 on the bottom line.

Converting to a fraction

$12.5\% = \frac{12.5}{100}$

$= \frac{25}{200}$ ← Multiply by 2 to remove the decimal.

$= \frac{5}{40}$ ← Simplify by dividing by 5.

$= \frac{1}{8}$ ← Simplify by dividing by 5.

Converting to a decimal

$12.5\% = \frac{12.5}{100} = 0.125$ ← Using place value.

Example – Converting between fractions, decimals and percentages

a Write 42% and 28% as fractions, in their simplest form, and as decimals.

b Write $\frac{2}{5}$ and 0.37 as percentages.

Solution

a As a fraction: $42\% = \frac{42}{100} = \frac{21}{50}$

As a decimal: $42\% = \frac{42}{100} = 0.42$

Similarly, $28\% = \frac{28}{100} = 0.28$

b $\frac{2}{5} = \frac{20}{50} = \frac{40}{100} = 40\%$

$0.37 = 0.37 \times \frac{100}{100} = \frac{37}{100} = 37\%$

Practising skills

1. Write each percentage as a fraction, in its simplest form.

a 29% **b** 7% **c** 30% **d** 62%

e 36% **f** 84% **g** 1.3% **h** 13.4%

2. Write each percentage as a decimal.

a 34% **b** 8% **c** 80% **d** 0.2%

e 14.5% **f** 6.4% **g** 164% **h** 0.75%

3 Write each decimal as a fraction in its simplest form.

a 0.9 **b** 0.4 **c** 0.17 **d** 0.66
e 0.52 **f** 0.213 **g** 0.004 **h** 0.025

4 Write each decimal as a percentage.

a 0.15 **b** 0.03 **c** 0.6 **d** 0.237
e 0.117 **f** 0.086 **g** 1.4 **h** 3.104

5 Write each fraction as a decimal.

a $\frac{1}{4}$ **b** $\frac{1}{5}$ **c** $\frac{1}{10}$ **d** $\frac{3}{4}$
e $\frac{3}{8}$ **f** $\frac{7}{8}$ **g** $\frac{16}{40}$ **h** $\frac{21}{30}$

6 Write each fraction as a percentage.

a $\frac{1}{2}$ **b** $\frac{9}{10}$ **c** $\frac{27}{100}$ **d** $\frac{21}{25}$
e $\frac{18}{24}$ **f** $\frac{21}{70}$ **g** $\frac{3}{16}$ **h** $\frac{7}{1000}$

7 Copy and complete this table.

	Fraction	Decimal	Percentage
a			11%
b			35%
c		0.3	
d		0.12	
e	$\frac{7}{10}$		
f	$\frac{13}{20}$		
g		0.05	
h			88%
i	$\frac{19}{25}$		
j		0.135	
k			2.4%
l			160%

Developing fluency

1 Scientists are studying a tropical bird species that can be different colours.
They record the colours of 40 birds.
Copy and complete this table, showing their findings. Give the fractions in their simplest form.

Colour	Number	Fraction	Percentage	Decimal
Red	8			0.2
Orange	10			
Blue	7	$\frac{7}{40}$		
Purple	4		10%	
Green	6			
Yellow	5			
Total	**40**			**1.000**

2 Write each list in order of size, starting with the smallest.

a $\frac{17}{100}$, 0.3, $\frac{1}{5}$, 26%

b 0.55, $\frac{1}{2}$, 6%, 0.6

c 32%, $\frac{7}{20}$, 0.3, $\frac{31}{100}$

d $\frac{3}{4}$, 72%, $\frac{7}{10}$, 0.715

e 0.33, $\frac{1}{3}$, 33.3%, $\frac{3}{10}$

f 4.8%, 4.5, 0.04, $\frac{1}{20}$

3 Copy and complete these. Write **<**, **>** or **=** in each box.

a $\frac{19}{100}$ ☐ 0.18

b 0.6 ☐ 15%

c $\frac{7}{25}$ ☐ 28%

d 0.04 ☐ 4.1%

e 0.114 ☐ $\frac{1}{10}$

f 66.66% ☐ $\frac{2}{3}$

g $\frac{13}{20}$ ☐ 62%

h $\frac{4}{5}$ ☐ 0.795

i 0.15 ☐ 1.4%

j $\frac{25}{40}$ ☐ 0.625

4 30 people work in an office.
6 of them are men.
What percentage are women?

5 A car park has 150 spaces.
129 of them are occupied.
What percentage of the spaces are unoccupied?

6 How much greater is 61% than 0.4?
Write your answer as a percentage.

Exam-style

7 Kate is taking a computing course.
In her first assignment, Kate gets 64 out of 80.
In the next one she gets 54 out of 60.
By what percentage has her score improved?

8 Beth's Bargains has a sale with $\frac{1}{4}$ off everything.

Dave's Discounts has a sale with $\frac{1}{3}$ off everything.

Which gives the bigger percentage discount and by how much?

9 24% of the employees in a factory travel to work by bus.

$\frac{2}{5}$ of them walk and $\frac{1}{10}$ get the train.

The rest travel by car.

a What percentage of the employees travel by car?

b What percentage of the employees travel by bus or walk?

c What percentage of the employees do not get the train?

10 There are six assignments on Kate's computing course.
Kate must get 75% or more on each of them to pass.
What are the pass marks?

Assignment	Total mark	Pass mark
1	80	
2	60	
3	12	

Assignment	Total mark	Pass mark
4	120	
5	48	
6	500	

Problem solving

Exam-style questions that require skills developed in this unit are included in later units.

1 **a** Make as many fractions, decimals and percentages as possible using no more than one of each of the following digits: 0, 1, 4.

For example, you could have 40%, 1.4, 1.04, $\frac{1}{40}$

b Are any of them equal to each other?

c Does this always work?

Can you choose three starting digits where it will not be possible to make any fractions, decimals or percentages that are equal to each other?

2 **a** Use your calculator to convert $\frac{2}{3}$ to a decimal.

This is called a recurring decimal. It can be written as $0.\dot{6}$.

b Round your answer to **a** to 2 decimal places and convert it to a percentage.

c Match the following fractions, decimals and percentages.

$\frac{1}{9}$	$\frac{130}{9}$	$14.\dot{4}$	$0.\dot{1}$	15%
11%	$\frac{2}{11}$	$\frac{1.6}{11}$	1444%	$0.1\dot{4}\dot{5}$
3333%	$\frac{100}{3}$	18%	$0.\dot{1}\dot{8}$	$33.\dot{3}$

Reviewing skills

1. Write each percentage as a fraction in its simplest form.
   a 45% **b** 96% **c** 12.5% **d** 0.7%
2. Write each percentage as a decimal.
   a 67% **b** 0.8% **c** 230% **d** 4.28%
3. Write each decimal as a fraction in its simplest form.
   a 0.14 **b** 0.022 **c** 0.735 **d** 0.008
4. Write each decimal as a percentage.
   a 0.58 **b** 0.009 **c** 0.809 **d** 2.87
5. Write each fraction as a decimal.
   a $\frac{3}{10}$ **b** $\frac{11}{20}$ **c** $\frac{49}{70}$ **d** $\frac{18}{25}$
6. Write each fraction as a percentage.
   a $\frac{13}{20}$ **b** $\frac{5}{8}$ **c** $\frac{29}{1000}$ **d** $\frac{138}{100}$

Unit 4 • Applying percentage increases and decreases to amounts • Band f

Outside the Maths classroom

Sale prices

Can you think of two different ways to work out 20% off?

Toolbox

There are several ways to work out percentage increases and decreases.

These are shown here using, as an example, the question:

Find the amount when £120 is **a** increased by 15% **b** decreased by 15%

Finding the increase or decrease

100% is £120

1% is $\frac{1}{100} \times £120$

15% is $\frac{15}{100} \times £120 = £18$

a An increase of 15% gives

£120 + £18 = £138

b A decrease of 15% gives

£120 − £18 = £102

Using a ratio table

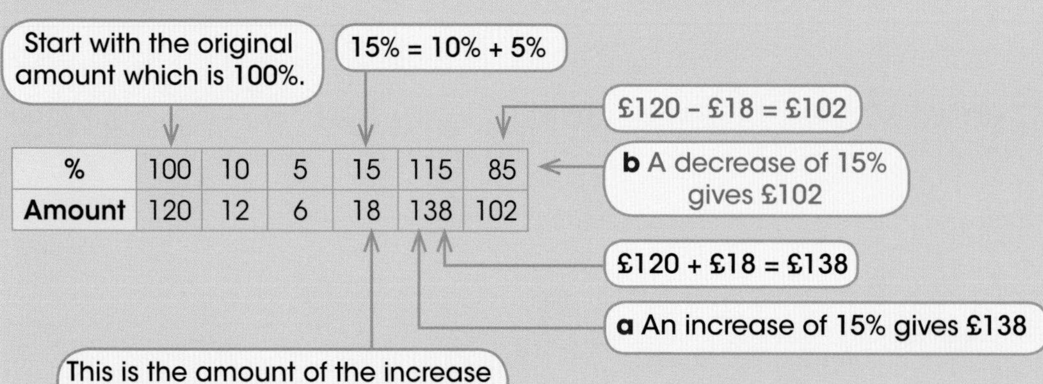

%	100	10	5	15	115	85
Amount	120	12	6	18	138	102

10% and 5% are easy to find and helped us to get to 15%. You can use other easy amounts like 1%, 20% and 50% to help you get to percentages you need.

Using a percentage bar chart

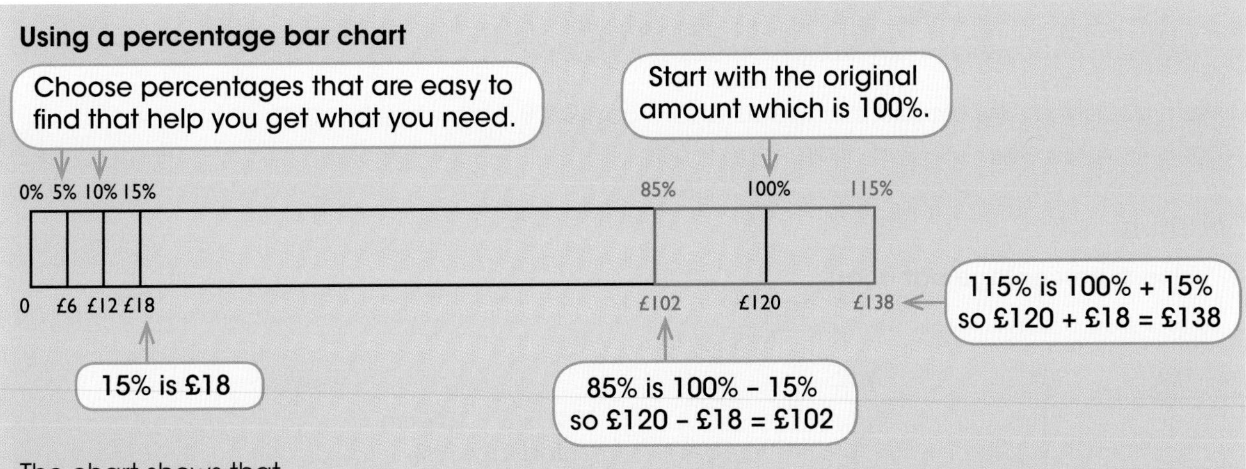

The chart shows that

a An increase of 15% gives an amount of £138

b A decrease of 15% gives an amount of £102

Using a multiplier

a The original amount is 100% so it is 100 + 15 = 115% when 15% is added.

So the new amount is 115% of £120 = $\frac{115}{100}$ × £120 ← You can also write this as 1.15 × £120

= £138

b In the same way, the amount is 100 − 15 = 85% when 15% is subtracted.

$\frac{85}{100}$ × £120 = 0.85 × £120

= £102

Example – Percentage increase

Anneka earns £21 000 per year. She is given a 3% increase.
Calculate how much she now earns in a year. Use two different methods.

Solution

Using a ratio table

100%	1%	3%	103%
£21000	£210	£630	£21 630

← 1% is easy to find and 3 x 1% = 3%

103% of £21 000 = £21 630

Finding the increase or decrease

100% is £21 000

1% is $\frac{1}{100}$ × £21 000 = £210

3% is $\frac{3}{100}$ × £21 000 = £630

An increase of 3% gives

£21 000 + £630 = £21630

Example – Percentage decrease

William bought a new bicycle three years ago. It cost £300. Now it has lost 30% of its value. What is its value now? Use two different methods.

Solution

a Using a percentage bar chart

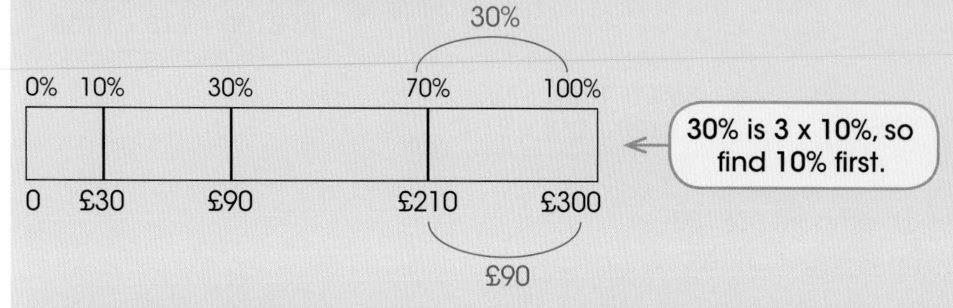

70% of £300 = £210

b Using a multiplier

The new amount is 100% – 30% = 70%

This is $\frac{70}{100} \times £300 = 0.7 \times £300 = £210$

Practising skills

1. Work out these.
   - **a** 10% of £70
   - **b** Increase £70 by 10%
   - **c** Decrease £70 by 10%
2. Work out these.
   - **a** 50% of 18 kg
   - **b** Increase 18 kg by 50%
   - **c** Decrease 18 kg by 50%
3. In a sale there is 10% off.
   Find the sale price of each item.
   - **a** jeans £20
   - **b** shirt £18
   - **c** tie £6.50
4. The price of each of these items goes up by 30%.
   What are the new prices?
   - **a** coat £40
   - **b** jumper £32
   - **c** socks £9.20
5. In a sale there is a reduction of 25%.
   For each item find the reduction in price and the new price.
   - **a** chair £68
   - **b** table £244
   - **c** lamp £19
6. Work out these.
   - **a** Increase 30 by 10%
   - **b** Decrease 35 by 20%
   - **c** Decrease 32 by 25%
   - **d** Increase 80 by 50%
   - **e** Decrease 60 by 30%
   - **f** Decrease 170 by 40%
   - **g** Increase 15 by 100%
   - **h** Increase 10 by 200%
   - **i** Increase 200 by 10%

Answer questions 1 to 7 without using your calculator.

Developing fluency

1. Match each card to its pair with the same answer.

A: 124 raised by 25%	B: 220 decreased by 30%	C: 750 reduced by 80%	D: 110 increased by 40%
E: 130 increased by 20%	F: 620 decreased by 75%	G: 390 reduced by 60%	H: 75 increased by 100%

Exam-style

2. In a sale all prices are reduced by 30%.
Samantha buys shoes priced at £28, a coat priced at £43 and a shirt priced at £17.50.
What is her total bill when she buys them during the sale?

3. Boxes of confectionary are labelled '25% extra free'.
Work out the new number of contents when the original amount was
a 48 bars **b** 60 lollipops **c** 160 chews

4. Steven buys a car for £7200.
He sells it and makes 15% profit.
What price does he sell the car for?

5. Sid buys a house at £185 000.
After one year its value has risen by 3%.
What is the value of the house after one year?

Exam-style

6. Sally earns £9.50 per hour.
She gets a pay rise of 6% starting in June.
She works 32 hours in the first week of June.
What is her pay for that week?

7. A carton of cream normally contains 230 ml.
The label says 20% extra free.
Sandy buys five cartons.
How many litres of cream does she buy?

8. Put these amounts in order of size, starting with the smallest.
150 g reduced by 4% 140 g increased by 6% 115 g increased by 26%,
245 g reduced by 41% 141 g increased by 2.5% 224 g reduced by 35.5%

9 Shaun is at a DIY store.
He buys a drill, a wheelbarrow and a hammer.

What is Shaun's final bill when the VAT is included? (VAT is 20%.)

10 Top Rates Bank offers simple interest at a rate of 1.6% per annum.
Interest is paid for the full period the money is invested.
Work out the total value when each amount is invested for these times.

a £700 for 2 years **b** £1340 for 3 years **c** £190 for 6 months **d** £825 for 9 months

Exam-style

11 Sandra invests £410 for 3 years at 2.1% per annum simple interest.
Interest is paid for the full period the money is invested.
Simon invests £520 for $2\frac{1}{2}$ years at 1.9% per annum simple interest.
Who gains more interest and by how much?

Problem solving

1 Natasha joins a small company. Her starting salary is £16 000 a year.

a In her first year the company does very well. Everyone is given a 25% pay rise for the next year.
What is Natasha's new salary?

b The next year the company does not do at all well. Everyone's salary is reduced by 25% for the following year.
What is Natasha's salary now?

c Natasha says, 'That's not fair! + 25 − 25 should be zero'.
Comment on this statement.

Exam-style

Extension

2 Marcus owns a house and an apartment.
He bought his house for £160 000.
He bought his apartment for £90 000.
Marcus decides to sell both properties.
He makes a 20% loss in selling the house and a 40% profit in selling the apartment.
Show that overall Marcus makes a profit.
Is his profit greater than 2%?

Exam-style

3 Ali bought his car for £20 000.
The car depreciated by 20% in the first year.
During the second year, the car depreciated by 10% of its value at the start of the year.
During the third year, the value of Ali's car depreciated by 25%.
What was the value of the car after the three years?

Extension

Exam-style

4 Cars depreciate in value as they get older.
Mansoor is buying a car.
He finds three cars that he likes. He wants to buy a car that has depreciated by the least amount of money after one year.

	Original cost	**Rate of depreciation after one year**
Car A	£17 500	9%
Car B	£23 000	7%
Car C	£15 650	10%

a Which car should Mansoor buy?

b What is the value of this car after one year?

Exam-style

5 Caroline wants to go on holiday.
She is going to take out a loan of £1500 to help pay for the holiday.
Caroline will have to pay back the £1500 plus 20% interest over 12 months.
She will pay back the same amount of money each month.
How much money will she need to pay back each month?

Reviewing skills

1 Work out these.

a 20% of £80 **b** Increase £80 by 20% **c** Decrease £80 by 20%

2 In a sale there is a reduction of 20%.
For each item find the reduction in price and the new price.

a shirt £30 **b** coat £99 **c** rugby shirt £55.50

3 Work out these.

a Increase 120 by 5% **b** Increase 160 by 35% **c** Decrease 150 by 3%

4 Sonya's car cost £9600.
After one year its value has depreciated by 12%.
By how much has the value of her car fallen?

Unit 5 • Finding the percentage change from one amount to another • Band g

Outside the Maths classroom

Inflation rates

This basket of shopping costs more to buy this year than it did last year. Will the increase be the same as the inflation rate?

Toolbox

To calculate a percentage change, follow the method illustrated in this example.

The population of an island increases from 500 to 650 people. What is the percentage increase?

The increase is 650 − 500 = 150 people

As a fraction this is $\frac{150}{500}$ ← the increase (150) ← the original population (500)

Either

Find the equivalent fraction with 100 on the bottom line

$\frac{150}{500} = \frac{30}{100} = 30\%$

Or

Write the fraction as a decimal

$\frac{150}{500} = 0.3$

Then multiply by $\frac{100}{100}$ to make it a percentage

$0.3 \times \frac{100}{100} = \frac{30}{100} = 30\%$

You may find it helpful to use a percentage bar or a ratio table to illustrate this.

Example – Percentage increase

Maya deals in musical instruments. She buys a guitar for £75 and restores it. She sells it for £120. Find her percentage profit.

Solution

Profit = £120 − £75 ← Profit = sale price − cost price

= £45

As a fraction this is $\frac{45}{75}$ ← Remember to always put the original value on the bottom line.

$\frac{45}{75} = \frac{90}{150} = \frac{30}{50} = \frac{60}{100}$

So Maya's profit is 60%

Example – Percentage decrease

Zorro was an overweight dog. He was 14.0 kg.
His owner put him on a programme of long daily walks. After six months his weight is 9.8 kg.
What percentage weight has Zorro lost?

Solution

Weight lost = 14.0 − 9.8 = 4.2 kg

As a fraction $\frac{4.2}{14.0}$

Percentage decrease = $\frac{4.2}{14.0} \times \frac{100}{100}$ ← Find an equivalent fraction with 100 on the bottom line.

$= \frac{420}{1400} = \frac{60}{200} = \frac{30}{100}$ ← This is the equivalent fraction.

The fraction lost is $\frac{30}{100}$

So Zorro has lost 30% of his weight.

Practising skills

1 Write these percentages.

a 3 as a percentage of 10
b 11 as a percentage of 20
c 6 as a percentage of 24
d 32 as a percentage of 80
e 64 as a percentage of 200
f 42 as a percentage of 150
g 80 as a percentage of 160
h 160 as a percentage of 180

2 Write the first quantity as a percentage of the second quantity.

a £7, £100
b €9, €10
c 5 cm, 25 cm
d 12 g, 300 g
e 30p, £2
f 14 cm, 1 m
g 90 cm, 3 m
h 85p, £5
i 800 m, 5 km

3 Work out the profit and percentage profit for each of these items in a DIY store.
The first one has been done for you.

	Item	Cost price	Selling price	Profit	Percentage profit
	Drill	£30	£33	£3	10%
a	Saw	£12	£21		
b	Hammer	£10	£17		
c	Plane	£20	£32		
d	Spanner set	£35	£56		

4) Work out the loss and percentage loss for each of these items at a car boot sale.

	Item	Cost price	Selling price	Loss	Percentage loss
a	Book	£10	£2		
b	Saucepan	£25	£22		
c	Dinner set	£200	£184		
d	Armchair	£70	£63		
e	Bicycle	£120	£84		
f	Cushion	£2	96p		

5) A carton of juice used to contain 750 ml.
It now contains 810 ml.

a What is the increase in ml?

b What is the percentage increase?

6) A packet of biscuits used to contain 30 biscuits.
It now contains 24 biscuits.

a What is the decrease in the number of biscuits?

b What is the percentage decrease?

Developing fluency

1) Write £12 as a percentage of £60.

2) Write 60p as a percentage of £2.50.

3) A chair cost £60 and is sold for £75.
Work out the percentage profit.

4) A TV cost £250 and is sold for £240.
Work out the percentage loss.

5) Nick's rent last year was £400 per month.
This year it is £460 per month.
Work out the percentage increase in his rent.

6) Ned's weight fell from 90 kg to 81 kg.
Work out the percentage decrease in his weight.

7) A tree grew in height from 75 cm to 1.2 m.
Work out the percentage increase in height.

(8) Using a calculator find the percentage change from the first quantity to the second quantity.
Give your answers correct to 1 decimal place.
State whether it is an increase or a decrease.

	First quantity	Second quantity	% increase/decrease
a	30	37	
b	65	40	
c	148	200	
d	£1.14	81p	
e	£12.60	£13.60	
f	64cm	1.1m	
g	2.15kg	900g	
h	3mm	1.1cm	
i	1.02km	999m	

Exam-style

(9) Sam's prize calf's weight increased from 40kg to 43kg.
Sally's prize calf's weight increased from 46kg to 49kg.
Which calf had the greater percentage increase in weight?

(10) The value of Mary's house fell from £160000 to £147000.
The value of Martin's house fell from £185000 to £172000.
Whose house had the greater percentage fall in value?

(11) Mr Tell bought 10 DVDs for £6 each.
He sold 8 of them for £10 each and 2 of them for £5 each.
Work out his percentage profit.

Problem solving

Exam-style

(1) Last year, Stan and Henry were looking at ways to reduce their heating bills.
Stan had wall insulation installed in his house.
Henry had loft insulation installed in his house.
For this year, Stan's heating bill is reduced from £640.50 to £544.43.
Henry's heating bill is reduced from £650 to £520.
Compare the percentage saving of wall insulation and loft insulation.

Exam-style

2 Sheila is writing an article in a magazine about the changes in transport and travel between 1961 and 2011.

She uses this information taken from government statistics.

Transport and travel	1961	1971	1981	1991	2001	2011
Road vehicles (millions)						
Licensed road vehicles	9.0	14.0	19.3	24.5	28.9	34.5
Motor vehicles registered for the first time	1.3	1.7	2.0	1.9	2.9	2.4
Length of network (thousand km)						
All public roads	314.0	325.0	342.0	360.0	392.0	393.0
Motorways	0.2	1.3	2.6	3.1	3.5	3.6

a Calculate the percentage increase in licensed road vehicles between 1961 and 2011.

b Compare the percentage increase in the length of motorways with the percentage increase in the length of all public roads over each 10-year period.

3 Nigel's rent increased from £200 to £220 per week.

At the same time, Nigel's wages increased from £450 to £490 per week.

Work out the percentage increase in the amount of money Nigel has left to spend each week.

Extension

Exam-style

4 This headline was taken from a newspaper from 1977.

Cost of food increases by nearly 200% in the last 5 years

The table shows some information about the prices of some foods in 1972 and in 1977.

Food	1972 price	1977 price
1 lb sausages	21p	44p
4 oz coffee	29p	72p
1 lb potatoes	2p	12p
12 eggs	20p	48p
2 lb sugar	10p	21p
1 pt milk	6p	10p
1 lb carrots	3p	14p

Mrs Brown wants to see if this headline is true.

Here is a typical shopping list from 1977.

By considering this shopping list, comment on the headline.

Extension

Exam-style

5 The table shows some information about the number of students in a college.

Year	2009	2010	2011	2012
Number	581	620	641	672

a In which year was the percentage increase in the number of students the greatest?

In 2013, the number of students was 48 more than in 2012.

b Work out the percentage increase in the number of students from 2009 to 2013.

Extension

Exam-style

6) Each year, Grace goes to France for her summer holiday.
She rents a villa which costs €850 per week.
The exchange rate last year was £1 = €1.12
The cost in euros of renting the villa was the same this year but she actually paid £680.
Work out the percentage change in the exchange rate.

Extension

Exam-style

7) The table gives information about the length of a lap at two motor racing tracks.

Motor racing track	Lap length
Brands Hatch	2.3 miles
Silverstone	3.67 miles

At Brands Hatch, there are 85 laps in a race.
At Silverstone, there are 60 laps in a race.
The organisers want to make both races the same length.

a Find the percentage change required in the length of the race at Brands Hatch to make this happen.

b Find the percentage change that would be required at Silverstone.

Extension

Reviewing skills

1) Work out these percentages.

a 9 as a percentage of 90
b 54 as a percentage of 72
c 75 as a percentage of 125
d 20.8 as a percentage of 26

2) Write the first quantity as a percentage of the second quantity.

a 8 mm, 3.2 cm **b** 240 m, 1 km **c** 8.4, 20 **d** 84 ml, 0.75 litres

3) Work out the profit and percentage profit for each item.

	Item	Cost price	Selling price	Profit	Percentage profit
a	Dress	£80	£84		
b	Pencil	80p	£1.08		
c	Dressing gown	£120	£84		
d	Notebook	£2	96p		

4) Sean bought 3 cars.
Car A cost £5000, car B cost £6100 and car C cost £6300.
He sold the 3 cars for £5100, £5800 and £6000, respectively.

a Work out his total percentage loss to the nearest whole number.

b On which car did he make the greatest percentage loss?

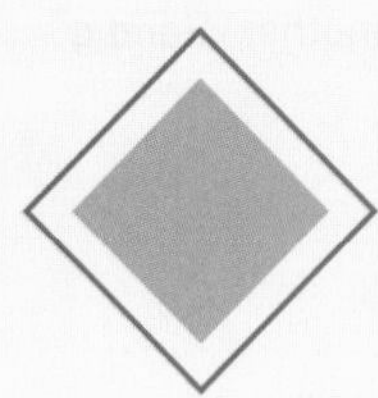

Number Strand 6 Ratio and proportion

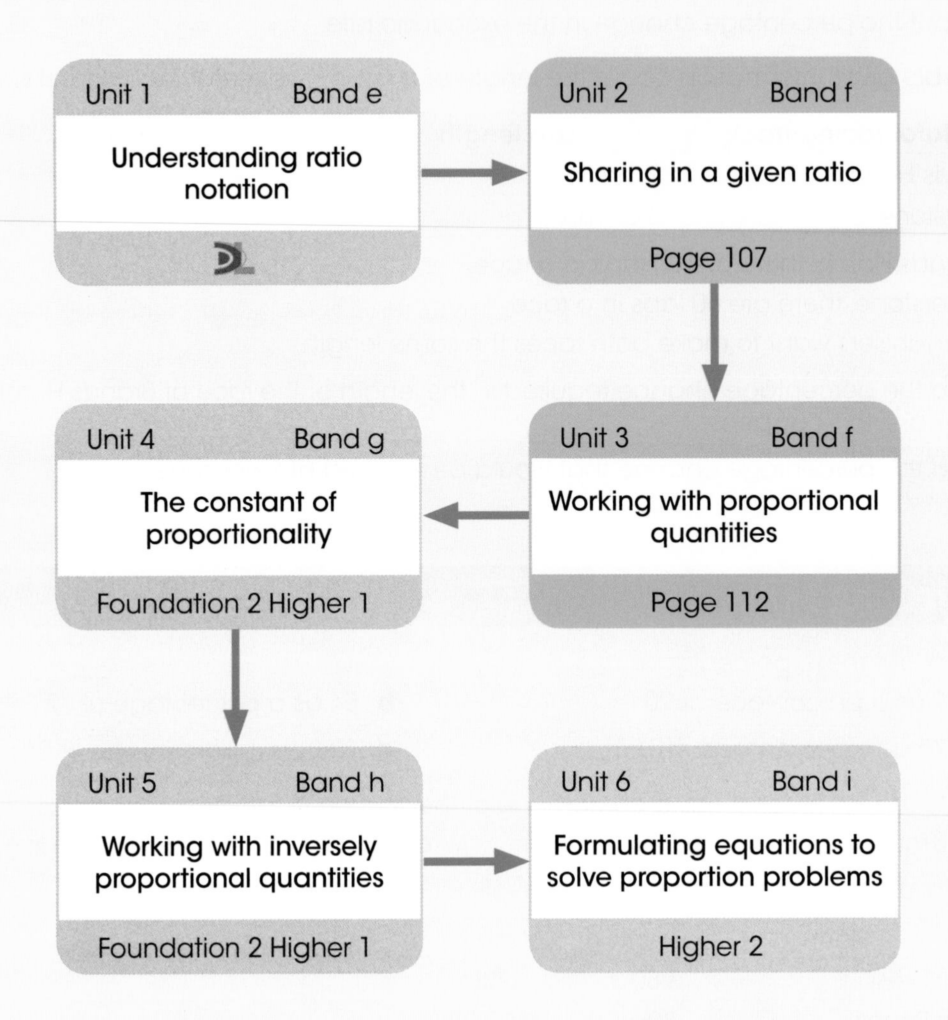

Unit 1 is assumed knowledge for this book. Its content is explained further in Unit 2.

Unit 2 • Sharing in a given ratio • Band f

Outside the Maths classroom

Mixing concrete

How do builders make sure they end up with the right mix?

Toolbox

In sharing problems, start by finding the total number of 'parts'. Then decide how the parts are allocated.

You can use a rectangular bar as in this example.

Kate and Pam share the cost of their £8.40 pizza in the ratio 3 : 4. How much does each pay?

The ratio is 3 : 4

3 + 4 = 7 so there are 7 parts.

Draw a bar with 7 equal parts.

K	K	K	P	P	P	P
£1.20	£1.20	£1.20	£1.20	£1.20	£1.20	£1.20

Kate has 3 shares and Pam has 4.

Divide £8.40 into 7 shares £8.40 ÷ 7 = £1.20

So Kate's share is 3 × £1.20 = £3.60

Pam's share is 4 × £1.20 = £4.80

Alternatively, you can use proportions.

These fractions are called proportional

Kate pays $\frac{3}{7}$ of £8.40 = £3.60

Pam pays $\frac{4}{7}$ of £8.40 = £4.80

$\frac{1}{7}$ of £8.40 is £1.20

Example – Sharing in a given ratio

Matthew and Harry go to a car boot sale.

They have a lucky find and buy a box of old Star Wars figures for £10.

Matthew pays £4 and Harry pays £6.

There are 40 figures in the box.

a Do you agree with Harry? Why?

b The ratio of their money is 2 : 3.
They share the figures in this ratio.
Find how many each of them gets using
i proportions **ii** a rectangular bar.

Solution

a Yes. Harry paid more than Matthew so it seems right that Harry should get more of the figures than Matthew.

b The ratio is 2 : 3

2 + 3 = 5 so there are 5 parts

i Matthew gets $\frac{2}{5}$ of 40 = 16 figures

$\frac{1}{5}$ of 40 is 8

Harry gets $\frac{3}{5}$ of 40 = 24 figures

ii

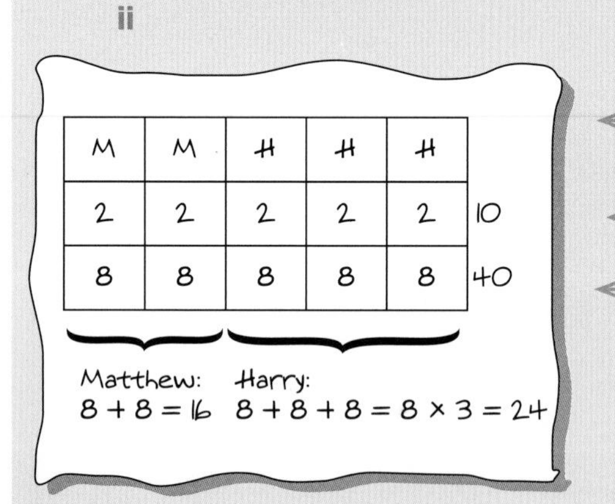

There are two parts for Matthew and three parts for Harry so there are five parts altogether.

This row shows how the £10 they paid in total was shared between them. Matthew paid £4 and Harry paid £6.

There are 40 figures to share. 10 × 4 = 40 so multiply each number in the row above by 4.

Matthew gets 16 figures.
Harry gets 24 figures.

Practising skills

1 £36 is shared between Joe and Jane in the ratio 1 : 2.

a How many equal parts are there?

b How much is each equal part?

c How much does Joe get?

d How much does Jane get?

e What is the total of Joe and Jane's amounts?

2. Jared and Raheem share a packet of 45 stickers in the ratio 2:7.
   - **a** How many equal parts are there?
   - **b** How many stickers are in each equal part?
   - **c** How many does Jared get?
   - **d** How many does Raheem get?
   - **e** What is the total of Jared and Raheem's amounts?

3. Jalel mixes blue and yellow paint in the ratio 1 : 5.
   He ends up with 480 ml of green paint.
   - **a** How many equal parts are there?
   - **b** How much is each equal part?
   - **c** How much blue paint did he use?
   - **d** How much yellow paint did he use?
   - **e** How much more blue than yellow did he use?

4. Work these out.
   - **a** Share £60 in the ratio 2 : 3.
   - **b** Share £96 in the ratio 5 : 1.
   - **c** Share 160 in the ratio 3 : 5.
   - **d** Share 126 ml in the ratio 7 : 2.

5. The ratio of red apples to green apples on a market stall is 3 : 7.
   - **a** What fraction of the apples are red?
   - **b** What fraction of the apples are green?
   - **c** How much bigger is the fraction of green apples than red apples?

6. Work these out.
   - **a** Share £90 in the ratio 3 : 2.
   - **b** Share £120 in the ratio 1 : 7.
   - **c** Share 225 in the ratio 5 : 4.
   - **d** Share 480 m in the ratio 7 : 3 : 2.

Developing fluency

Exam-style

1. £336 is shared between Alex and Abbie in the ratio 5 : 7.
   How much more does Abbie get than Alex?

2. A café sold 168 meals one day.
   The ratio of beef meals to chicken to fish was 4 : 2 : 1.
   How many more beef than fish meals were sold on that day?

Exam-style

3 Jeff is 12 years old and Joseph is 20 years old.
Plan A is to share £5280 in the ratio of their ages now.
Plan B is to wait until Jeff is 18 years old and to share £5280 in the ratio of their ages then.
How much will Jeff gain by waiting until he is 18 years old?

4 In a rectangle the ratio of the length to width is 9 : 5.
The perimeter of the rectangle is 196 cm.
What are the length and width of this rectangle?

5 A canteen serves glasses of milk or water at lunchtime.
On Monday they serve 128 glasses.
The ratio of glasses of milk to glasses of water is 5 : 3.
On Tuesday the ratio is 4 : 3.
The number of glasses of milk is 12 more than on Monday.
How many more glasses of water are served on Tuesday than on Monday?

Problem solving

1 Martin says he could share this bar of chocolate in the ratio 1 : 3.
Sam says he could share it in the ratio 5 : 7.
How many different ratios could the bar be shared into?

Exam-style

2 Ninety girls are asked what their favourite sport is from hockey, tennis and netball.
The ratio of the results of hockey to tennis to netball is 1 : 3 : 5.

a How many girls chose tennis as their favourite sport?

The pie chart shows these results.

b Work out the angle of each sector of the pie chart.

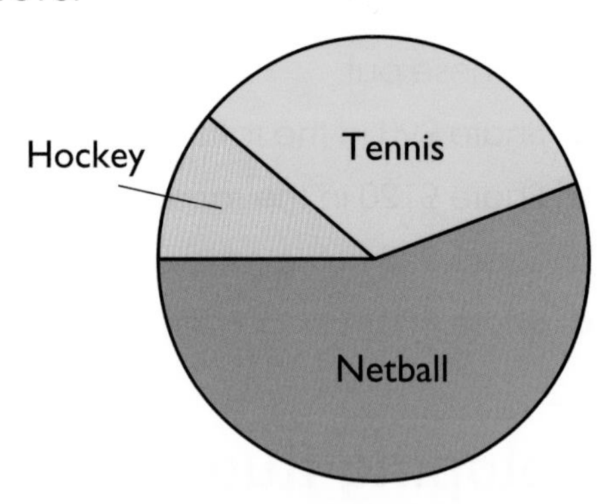

Exam-style

3 Christine and Jenny are sisters.
At Christmas 2004, Christine was 3 years old and Jenny was 12 years old.
Their father gave them a present of £300 shared in the ratio of their ages.

a How much money did Christine receive?

Each Christmas, their father carried on giving them a sum of money shared in the ratio of their ages.

b In which year did Jenny receive twice as much money as Christine?

Exam-style

Extension

(4) Lesley takes the penalties for her football team.
During her career, she has taken 56 penalties.
There are three possible outcomes from a penalty:

a goal is scored
it is saved by the goalkeeper
it misses the target

For Lesley's penalties, the ratio of these outcomes is 5 : 2 : 1 respectively.
Lesley's brother Lionel also takes penalties for his football team.
Lionel has scored 30 goals from penalties.

a Who has scored more goals from penalties?

Of the penalties that Lionel has taken, 6 were saved by the goalkeeper and 4 missed the target.

b Compare Lionel's record of taking penalties with Lesley's record of taking penalties.

Exam-style

Extension

(5) Arfan and Benny each have a box of chocolates.
In each box, there are milk chocolates, dark chocolates and white chocolates.
Arfan's box contains 27 chocolates.
The ratio of milk to dark to white chocolates in Arfan's box is 4 : 3 : 2.
Benny's box contains 28 chocolates.
The ratio of milk to dark to white chocolates in Benny's box is 3 : 3 : 1.
Arfan says, 'I have more milk chocolates than Benny has.'
Is Arfan right?

Reviewing skills

(1) The ratio of oranges to lemons in a recipe for marmalade is 1 : 4.

a How many equal parts of fruit are there?

b What fraction of the fruit are oranges?

c What fraction of the fruit are lemons?

(2) **a** Share 105 g in the ratio 4 : 3.

b Share 330 in the ratio 7 : 8.

c Share 286 g in the ratio 1 : 2 : 10.

d Share 245 in the ratio 1 : 2 : 4.

(3) Andrew goes to the gym on Mondays and Fridays.
He always divides his time between using the treadmill and doing weights in the ratio 3 : 2.
On Monday he was there for $1\frac{1}{2}$ hours.
On Friday he spends the same time on weights as he had spent on the treadmill on Monday.
How long is he at the gym on Friday?

Unit 3 • Working with proportional quantities • Band f

Outside the Maths classroom

Supermarket prices

Why do supermarkets often give the price per 100 g?

Toolbox

It is important to choose the most suitable method to solve a problem involving ratio and proportion.

- A ratio table can be very helpful for understanding a situation.
- The unitary method is good when the numbers are not quite straightforward and can be helpful for comparing quantities using unit costs.

Example – Comparing unit costs

Orange juice can be bought in different-sized cartons.
Holly needs 6 litres of orange juice for a party.
Which size carton should Holly buy?

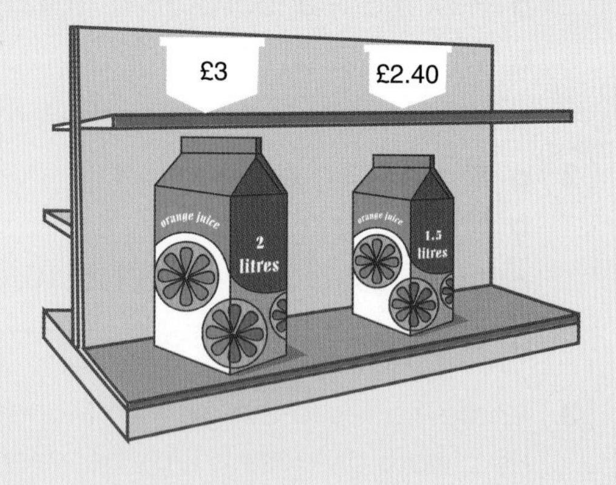

Solution

Smaller carton

1.5 litres cost £ 2.40

1 litre costs $\frac{1}{1.5} \times$ £ 2.40 = £ 1.60

Unit cost = £ 1.60

Larger carton

2 litres cost £ 3.00

1 litre costs $\frac{1}{2} \times$ £ 3.00 = £ 1.50

Unit cost = £ 1.50

The larger carton has a lower unit cost.
Holly should buy the larger carton.

Example – Using the unitary method

These ingredients for apple crumble make enough to serve two people.

Apple Crumble (serves 2)
1 large cooking apple
25 g white sugar
$\frac{1}{4}$ teaspoon cinnamon
90 g wholemeal flour
40 g butter
75 g brown sugar

Inga needs to make an apple crumble for five people.
How much of each ingredient should she use?

In the unitary method you find for 1.

Then you scale up by multiplying.

Solution

Ingredients	For 2	For 1 (÷2)	For 5 (×5)
Cooking apple	1	$\frac{1}{2}$	$2\frac{1}{2}$
White sugar	25 g	12.5 g	62.5 g
Cinnamon	$\frac{1}{4}$ tsp	$\frac{1}{8}$ tsp	$\frac{5}{8}$ tsp
Wholemeal flour	90 g	45 g	225 g
Butter	40 g	20 g	100 g
Brown sugar	75 g	37.5 g	187.5 g

This column gives all the answers.

Example – Solving problems with a ratio table

40 blank CDs cost £20.

a Find the cost of 50 blank CDs.
b Find the cost of 24 blank CDs.

Solution

a

Number of CDs	40	20	10	50
Cost (£)	20	10	5	25

40 + 10 = 50

20 + 5 = 25

50 CDs cost £25.

b

Number of CDs	40	20	4	24
Cost (£)	20	10	2	12

20 + 4 = 24

10 + 2 = 12

24 CDs cost £12.

Practising skills

1. 4 apples cost 96p.

   a How much does 1 apple cost? **b** How much do 3 apples cost?

2. 6 carrots cost 90p.

   a How much does 1 carrot cost? **b** How much do 8 carrots cost?

3. 100 litres of heating oil cost £60.

   a How much do 10 litres of oil cost? **b** How much do 230 litres cost?

4. Look at the prices on these two market stalls.

Bev's Bargains
6 oranges £1.86
8 cans of cola £2.80
8 bananas £2.60

Dan's Discounts
4 oranges £1.28
6 cans of cola £1.80
12 bananas £2.88

a **i** How much does 1 can of cola cost in Bev's Bargains?
ii How much does 1 can of cola cost in Dan's Discounts?
iii Which market stall is better value for cans of cola?

b **i** How much do 4 bananas cost in Bev's Bargains?
ii How much do 4 bananas cost in Dan's Discounts?
iii Which is better value for bananas?

c **i** How much do 12 oranges cost in Bev's Bargains?
ii How much do 12 oranges cost in Dan's Discounts?
iii Which is better value for oranges?

5. Vincent has a recipe for shepherd's pie which serves 5 people.
   Copy and complete this table to help him find the quantities needed for 8 people.

Ingredient	Quantity for 5 people	Quantity for 1 person	Quantity for 8 people
Minced beef	900 g		
Stock	480 ml		
Onion	2		
Tin of tomatoes	1		
Potatoes	700 g		
Worcestershire sauce	40 ml		

Developing fluency

1. Three concert tickets cost £42.
   How much do eight of these tickets cost?

2 Nine cups of coffee cost £15.30.
How much do four cups of coffee cost?

Exam-style

3 Last Saturday Mrs Boston paid £4.30 for five lemonades and £14.40 for six burgers at a café.
She plans to go back this Saturday and order four lemonades and four burgers.
How much less will it cost this Saturday?

4 For each of these work out which is better value. Explain your answer.

a 6 kg for £14.70 or 7 kg for £17.50

b 100 ml for £18 or 150 ml for £24

c 80 g for £16.40 or 60 g for £12.06

Reasoning

5 A window cleaner charges a fixed rate of £5 plus an amount for each window he cleans.
He cleaned a house with 6 windows and charged £9.80.
Tom's house has 12 windows.
He thinks the window cleaner will charge him £19.60.
Is Tom right?
Explain your answer.

Reasoning

6 Paula's favourite perfume comes in 3 sizes.

She sees these two offers online.
Which is the best offer? Explain fully.

OFFER 1
Buy two 40 ml bottles and
get one 40 ml bottle free
USE CODE TWO40 AT CHECKOUT

OFFER 2
Buy one 50 ml bottle and
get one 50 ml bottle half price
USE CODE ONE50 AT CHECKOUT

Exam-style

7 David is paid the same rate from Monday to Friday.
On Saturday he is paid time and a half.
Last week he worked 7 hours on Tuesday and 8 hours on Wednesday.
His pay was £123.
This week he is due to work 6 hours on Monday, 5 hours on Thursday and 8 hours on Saturday.
How much more will his pay be this week?

Problem solving

Exam-style

1. Farah works in the retail industry.

   The table shows some information about the hours she worked and her pay for the last three weeks.

Day	Hours worked		
	Week 1	Week 2	Week 3
Monday	8	9	8
Tuesday	6	4	8
Wednesday	10	10	
Thursday	4	8	10
Friday	10	12	4
Total pay	£475		£450

   a Work out Farah's pay for Week 2.

   b How many hours did Farah work on Wednesday of Week 3?

Exam-style / Extension

2. The ingredients for apple crumble make enough to serve two people.

   Inga is going to make some apple crumble using this recipe.

   In her kitchen, Inga has the following quantities of each ingredient:

   24 large cooking apples

   200 g of white sugar

   6 teaspoons of cinnamon

   1.2 kg of wholemeal flour

   300 g of butter

   500 g of brown sugar.

   Apple Crumble (serves 2)
   1 large cooking apple
   25 g white sugar
   $\frac{1}{4}$ teaspoon cinnamon
   90 g wholemeal flour
   40 g butter
   75 g brown sugar

   Work out the maximum number of servings of apple crumble that Inga can make.

Exam-style / Extension

3. Here are three boxes of the same type of chocolates.

   Melanie and her friend go shopping for chocolates.

   Melanie says, 'You always get the best value for money by buying the largest box'.

   Is Melanie right?

Exam-style

4 Albert and Harvey are salesmen. They work for different companies.
They are both allowed to claim for the use of their car for work.
During one month, Albert claimed £108 for driving 240 miles.
During the same month, Harvey claimed £135 for driving 270 miles.
In the next month, Albert's company increased the rate of car allowance by 10%.
Whose company now has the higher rate of car allowance?

Extension

Exam-style

5 There are 28 children in Mrs Davies' class.
She is going to buy a pen, a pencil and an eraser for each of the children in her class.
In a stationery catalogue, she sees the following information.

Pens: 30p each or a pack of 5 for £1.20

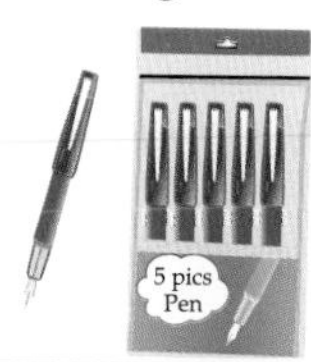

Pencils: 20p each or a pack of 3 for 50p

Eraser: 12p each or a pack of 10 for £1

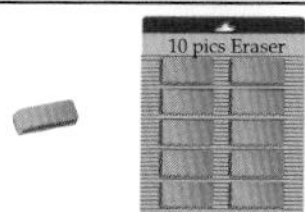

Mrs Davies has just £15 to spend.
Has she got enough money to buy a pen, a pencil and an eraser for each of the children in her class?

Extension

Exam-style

6 Malcolm is working overseas.
He posts 120 letters.
Some of the letters are 1st class.
Some of the letters are 2nd class.
The number of 1st class letters is 5 times the number of 2nd class letters.
Malcolm knows that it costs 32 cents to post a 2nd class letter.
The total cost of posting all 120 letters is \$47.40. There are 100 cents in 1 dollar.
Work out the cost of posting one 1st class letter.

Extension

Exam-style

7 Here is a list of ingredients for making a cheese soufflé.
Frances is going to make 12 cheese soufflés.
She has 150 grams of flour and 2 litres of milk.
Does she have enough flour and milk?

1 oz = 28 grams and 1 pint = 568 millilitres.

Extension

Exam-style

Extension

8 Ten 5p coins weigh 33.5 grams.
Five 2p coins weigh 35.6 grams.
Tom works in a bank.
A customer gives him a bag of 2p coins weighing 3.026 kg.
The customer also gives him 7 bags of 5p coins.
Each bag of 5p coins weighs 402 g.
Tom gives the customer a receipt for the total amount of money.
The receipt is for £50.50.
Show that the amount shown on this receipt is correct.

Reviewing skills

1 5 adult bus tickets cost £11.70.

a How much does 1 adult bus ticket cost?

b How much do 9 adult bus tickets cost?

2 For each of these work out which is better value. Explain your answer.

a 2 m for £8.10 or 60 cm for £2.49

b 600 g for £5.40 or 750 g for £7.20

c 2 litres for £22.12 or 800 ml for £8.92

3 A particular brand of shampoo comes in 3 sizes.

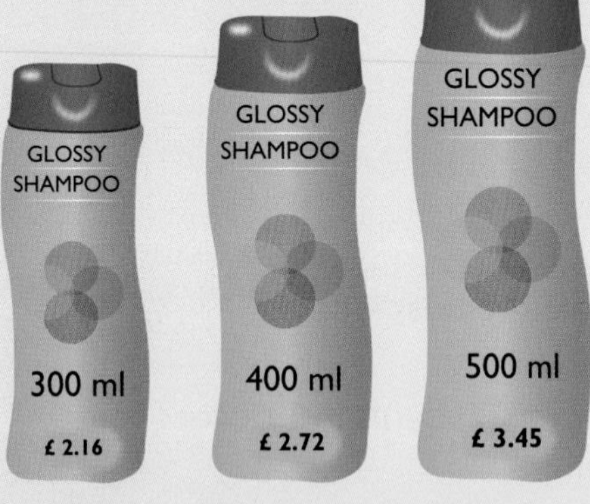

a Work out the cost of 100 ml for each size of bottle.

b Which size is the best value?

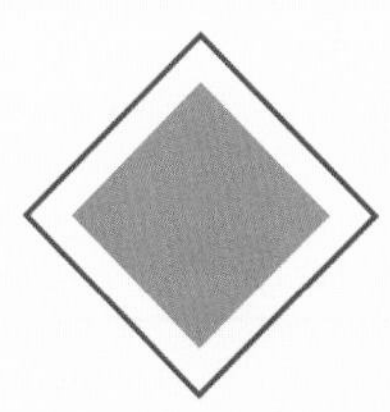

Number Strand 7 Number properties

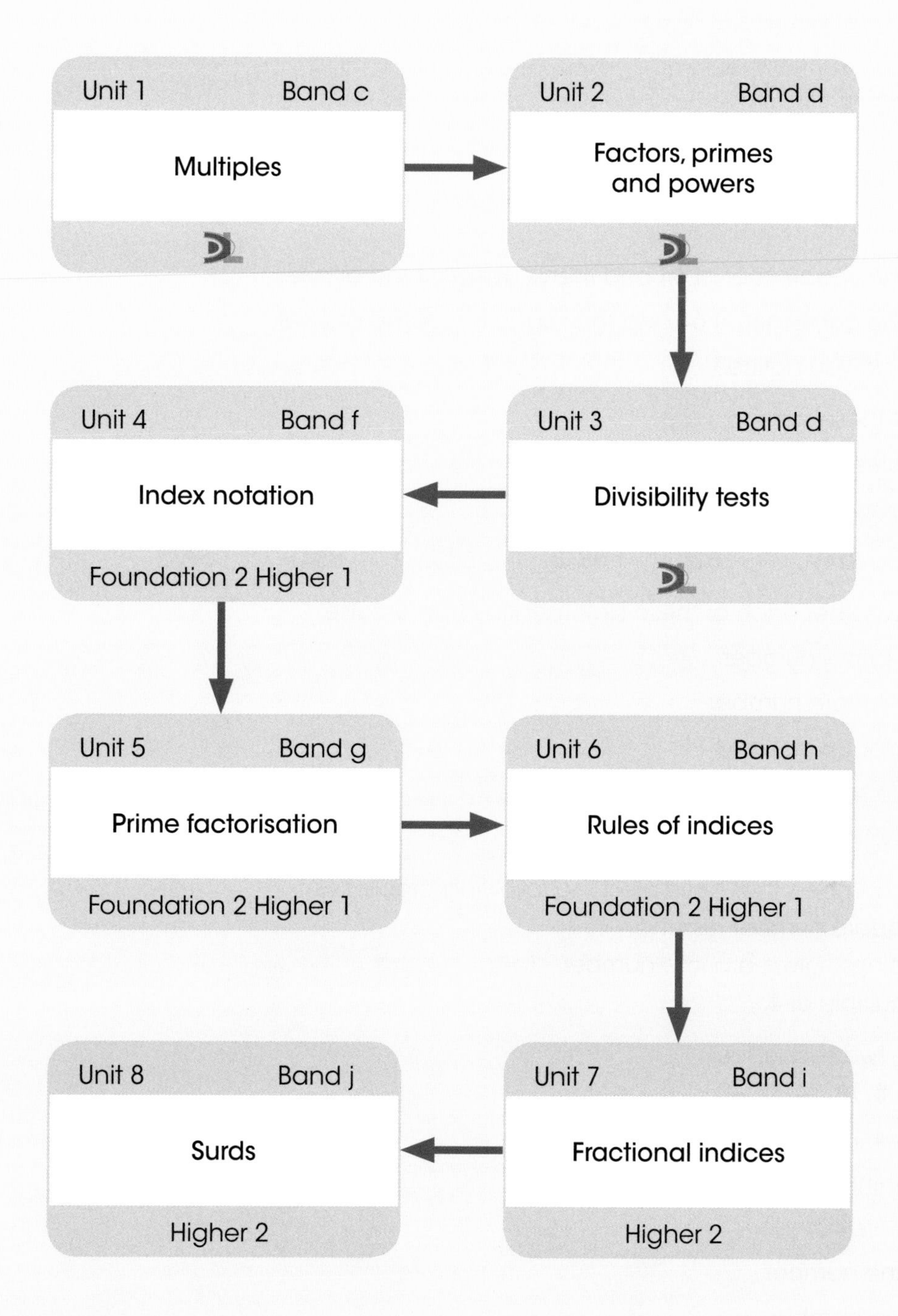

Units 1–3 are assumed knowledge for this book. They are reviewed and extended in the Moving on section on page 120.

7 Strand 7 • Moving on

1. Find all the factors of
   a 36
   b 72
   c 144

2. **a** Find the square root of 64 and then cube root the answer.
   b Find the cube root of 64 and then square root the answer.
   c What do you notice?

3. Which of these numbers are
   a multiples of 4
   b multiples of 9?

   4221 3249 1628 3843 6336 8341 5432

4. Decide whether each of these statements is true or false.
   a 1 is a prime number.
   b 1 is a square number.
   c 1 is a factor of 9 and 11.
   d 1 is the square of –1.
   e 1 is divisible by 2.

5. Find the numbers that have all of the following properties
   a square number less than 40
   adding 1 gives a prime number
   a multiple of 4.

Exam-style

6. Here are two sequences.
   i …, 8, 16, 24, 32, 40, …
   ii …, 4, 7, 10, 13, 16, …

   Do you think each sequence will contain
   a a square number
   b a prime number
   c a cube number
   d a factor of 100
   e a multiple of 21?

   In each case, explain how you know. If you cannot be sure, say why.

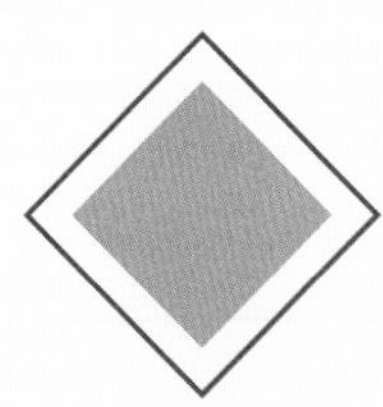

Algebra Strand 1 Starting algebra

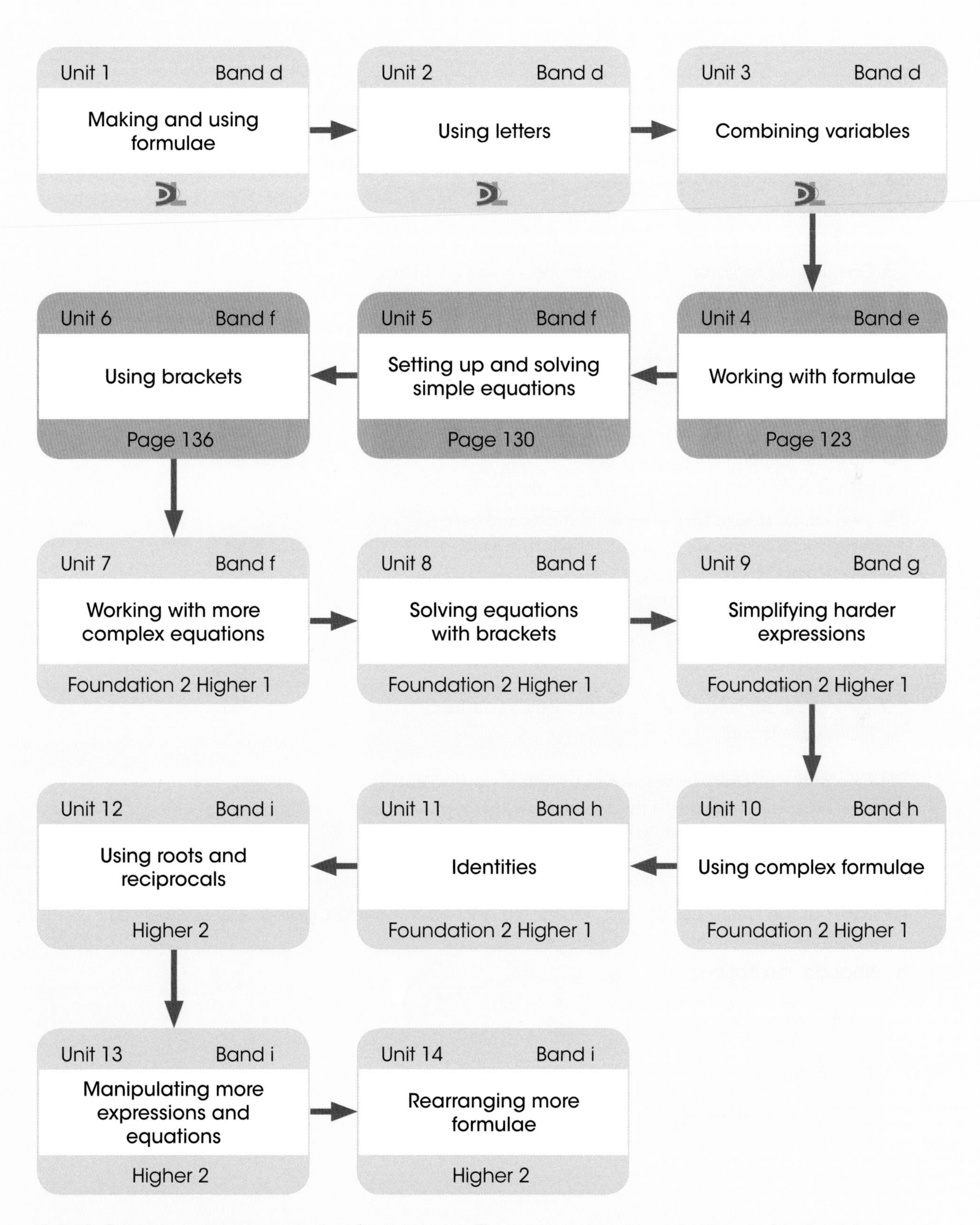

Units 1–3 are assumed knowledge for this book. They are reviewed and extended in the Moving on section on page 122.

Units 1–3 • Moving on

1. The sentences below can be written as algebraic expressions.
Match up the sentences to the expressions that have been given.
In a PE lesson all the students play hockey. There are n students in the PE lesson.

A Every student wears a pair of boots. How many boots are there?

B One boy has forgotten his shin-pads. How many shin-pads are there?

C Every student has a hockey stick and so do the two teachers. How many hockey sticks are there?

D How many sports socks are there?

E Two students have forgotten their shorts and have to wear tracksuit trousers instead. How many pairs of shorts are there?

F The school PE shirts have two stripes on them. How many stripes are there?

G Two of the students play in goal. How many outfield players are there?

H One of the teachers brings two extra water bottles, but every student has brought their own. How many water bottles are there?

$2n$

$n + 2$

$n - 2$

$2n - 2$

2. In this diagram there are four rectangles and three L-shapes.
   - **a** Work out the perimeter of each rectangle and each L-shape, giving your answers as simplified as possible.
   - **b** What do you notice?

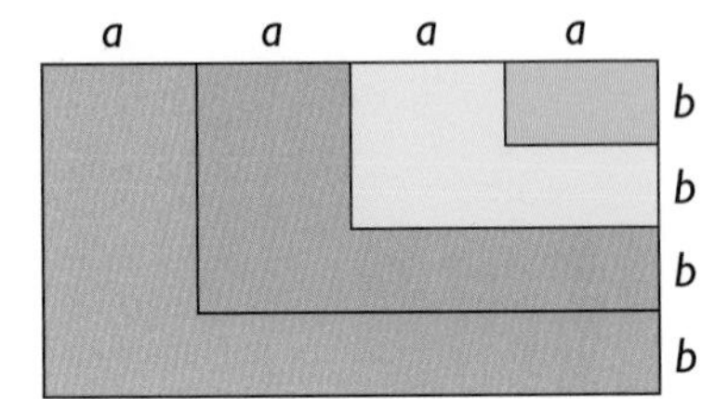

Unit 4 • Working with formulae • Band e

Outside the Maths classroom

Building regulations

What are the regulations about the size of steps?

Toolbox

Number machines can help you work with formulae.

The cost, C pence, of a bus ticket for a journey of m miles is given by the formula

$$C = 20m + 50$$

You can write this using a number machine.

m ⟶ $\times 20$ $\xrightarrow{20m}$ $+50$ ⟶ $20m + 50 = C$

For a journey of 7 miles, $m = 7$.

7 ⟶ $\times 20$ $\xrightarrow{140}$ $+50$ ⟶ 190

The cost is 190 pence or £1.90.

You can use a number machine in reverse.

This will tell you the number of miles you can travel for a certain amount of money.

Work from right to left with the **inverse operations**.

So, for a fare of £1.30, the number machine looks like this.

4 ⟵ $\div 20$ $\xleftarrow{80}$ -50 ⟵ 130

You can travel 4 miles.

Example – Changing a formula

The Williams family are going on holiday to Florida.

a The exchange rate is 1.6 dollars to the pound.
Write a formula to represent this information.

b They wish to take \$1200 with them.
How many pounds should they change into dollars?

Solution

a Number of dollars = number of pounds × 1.6
If D is the number of dollars and P is the number of pounds then $D = 1.6P$

b The formula is used in reverse to find the number of pounds.

Number of pounds = number of dollars ÷ 1.6
= 1200 ÷ 1.6
= 750

To find the number of dollars you multiplied by 1.6, so to find the number of pounds you must divide by 1.6.

They need to change £750 into dollars.

Example – Substituting values into a formula

This formula gives p in terms of e and g.

$$p = \frac{3e - g}{4}$$

Find the values of p when

a $e = 4$ and $g = 0$ **b** $e = 5$ and $g = 7$ **c** $e = 8$ and $g = 20$
d $e = 4$ and $g = 12$ **e** $e = 100$ and $g = 296$

Solution

a $\frac{3e-g}{4} = \frac{3 \times 4 - 0}{4}$ ← Substitute the values given for the variables.

$= \frac{12-0}{4} = \frac{12}{4} = 4$ ← Calculate the numerator first.

b $\frac{3e-g}{4} = \frac{3 \times 5 - 7}{4}$

$= \frac{15-7}{4} = \frac{8}{4} = 2$

c $\frac{3e-g}{4} = \frac{3 \times 8 - 20}{4}$

$= \frac{24-20}{4} = \frac{4}{4} = 1$

d $\frac{3e-g}{4} = \frac{3 \times 4 - 12}{4}$

$= \frac{12-12}{4} = \frac{0}{4} = 0$

e $\frac{3e-g}{4} = \frac{3 \times 100 - 296}{4}$

$= \frac{300-296}{4} = \frac{4}{4} = 1$

Example – Using number machines

A train company uses a formula to work out the time, t minutes, for a rail journey from London to Plymouth.
The number of stops along the way is s.

$$t = 5s + 165$$

a Use a number machine to find the time for a journey with seven stops.

b A train takes 3 hours and 10 minutes to travel from Plymouth to London.
How many stops did it make?

Solution

a Here is the number machine for this formula.

s → $\times 5$ →($5s$)→ $+ 165$ → $5s + 165 = t$

When there are seven stops,

7 → $\times 5$ →(35)→ $+ 165$ → 200

It takes 200 minutes or 3 hours and 20 minutes.

b For a journey of 190 minutes, reverse the number machine. ← Convert 3 hours and 10 minutes into minutes.

5 ← $\div 5$ ←(25)← $- 165$ ← 190

The train made five stops.

Practising skills

1 Find the output for each of these number machines.

a 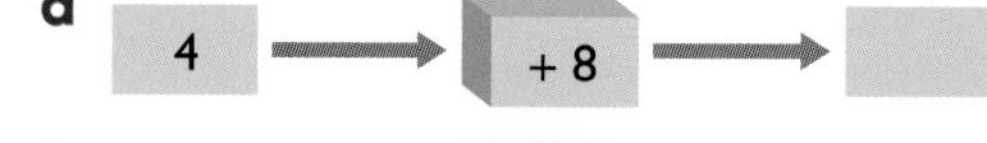

b 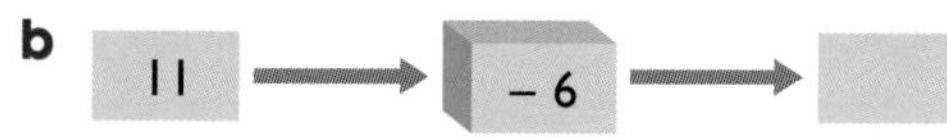

c 

d 21 → $\div 3$ → ☐

2 Work out the addition or subtraction rules for these number machines.

a 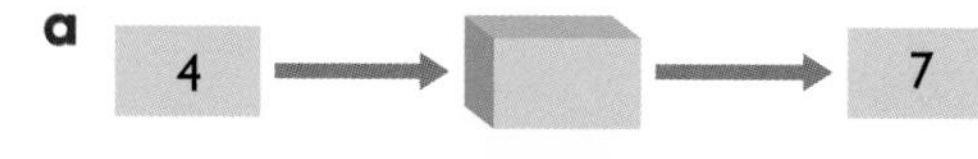

b 10 → ☐ → 1

c 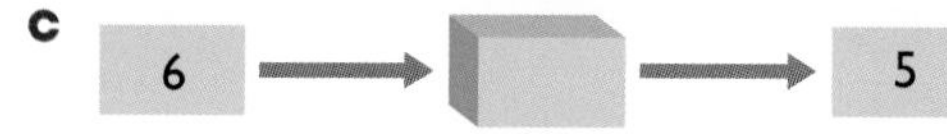

d 17 → ☐ → 31

3 Write down the opposite of these.

a turn left

b sit down

c turn 68° anti-clockwise

d walk 6 steps forward

4 Write down the inverse of each of these operations.

a $+ 7$ **b** $- 4$ **c** $+ 129$ **d** $\times 5$

e $\times 3.9$ **f** $\div 8$ **g** $\div 0.5$

5 The number machines in question 2 can be written the other way.
Work out the new addition or subtraction rules for these number machines.

a 4 ← [ ] ← 7

b 10 ← [ ] ← 1

c 6 ← [ ] ← 5

d 17 ← [ ] ← 31

6 Jeff pays £1.40 for each litre of petrol.

a How much does he pay for 37 litres?

b Copy and complete this number machine to work out how much he pays for his petrol.

c Jeff pays £78.40 for petrol. Reverse your number machine to find out how many litres he buys.

d **i** Andrew pays £62.16 for diesel costing £1.48 per litre. Draw a number machine to help you work out how many litres he buys.

ii Mai pays £42.18 for diesel. How much does she buy?

Developing fluency

1 **a** Work out the value of y in each of the four cases below when $x = 9$.

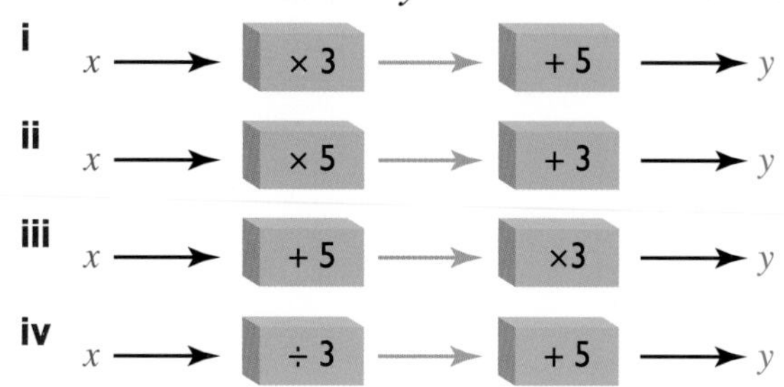

b Match the four number machines in part **a** with the equivalent formulae below.

A	B	C	D
$y = (x + 5) \times 3$	$y = 3x + 5$	$y = \frac{1}{3}x + 5$	$y = 5x + 3$

2 **a** Draw a number machine for $y = 4x - 3$.

b Draw the reverse number machine for your answer to part **a**.

c Use your answer to part **b** to find the value of x which gives $y = 65$.

Exam-style

3 **a** Copy and complete the number machine below.

b Use the number machine to find the output when the input is 7.

c Draw the reverse number machine. Use it to find the input when the output is 55.

(4) Here is a number machine. Use it to complete the tables in parts **a** and **b**.

input → × 4 → −7 → output

a

Input	Output
3	
10	
7	
1	

b

Input	Output
	41
	17
	73
	−7

Exam-style

(5) In a city the cost of a taxi journey in euros is worked out using these steps.

Multiply the number of kilometres by 2.

Add 3.

a Find the cost of these journeys.

i 10 km **ii** 20 km

b Find the length of a journey costing €27.

c Write down a formula connecting the cost €c and the length of the journey, d km.

(6) The exchange rate for pounds and euros is £1 = €1.15.

a Work out how many euros you will get for these amounts in pounds.

i £100 **ii** £25 **iii** £40 **iv** £2000

b Work out how many pounds you will get for these amounts in euros.

i €23 **ii** €460 **iii** €1840 **iv** €9200

(7) This formula gives the cost, C(£), of buying n tickets for a football match.

$C = 18n$

a Work out the cost of buying these numbers of tickets.

i 30 **ii** 45 **iii** 124 **iv** 356

b Work out how many tickets can be bought for these amounts.

i £1548 **ii** £414 **iii** £2556 **iv** £634

Exam-style

(8) Here is a formula to work out the time, T minutes, to cook a piece of meat of weight, w kg.

$T = 15w + 12$

a For how long should a 4 kg piece of meat be cooked?

b What weight of meat takes 102 minutes to cook?

Problem solving

Exam-style

(1) Euler developed a relationship that connects the number of faces, vertices and edges of a solid shape. He worked out that

faces + vertices = edges + 2

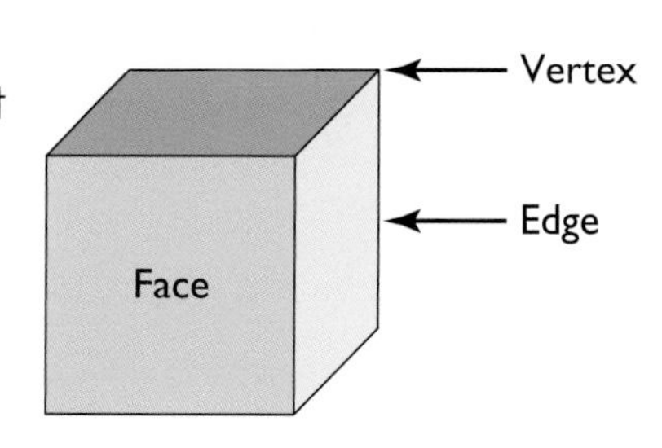

a Show that this works for a cube.

b A shape has five faces and six vertices.

How many edges does it have?

Draw this shape.

2 Jeff uses this formula to find the perimeter of a rectangle.

$P = 2l + 2w$

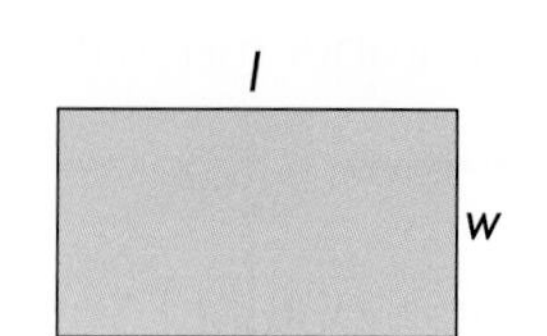

a Find the value of P when $l = 10$ and $w = 6$

b Find the value of l when P is 20 and $w = 4$

c A square has a perimeter of 36 cm
Find the length of one of the sides.

3 **a** Ben hires a cement mixer from Mixers R Us.
He hires the cement mixer for 10 days.
Work out the cost.

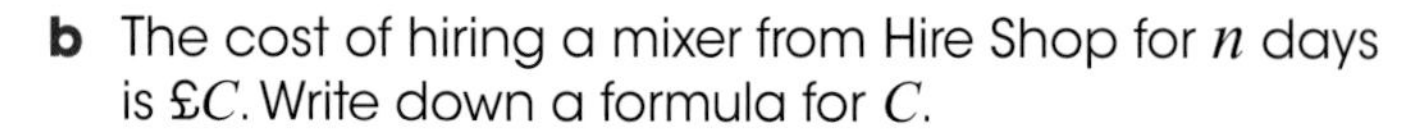

b The cost of hiring a mixer from Hire Shop for n days is £C. Write down a formula for C.

c Show that the costs for 4 days are the same.

4 Here are three patterns made from sticks.

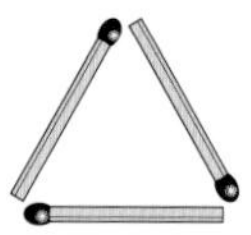

Pattern 1

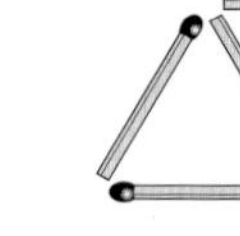

Pattern 2

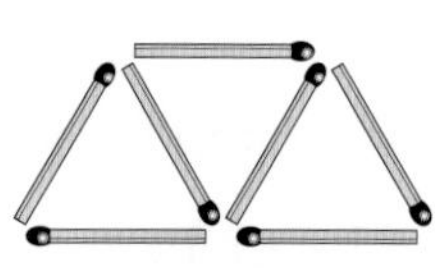

Pattern 3

a Write down the number of sticks in pattern 3.

b Write down the number of sticks in pattern 4.

c Write down the number of sticks in pattern n.

d Write down the number of sticks in pattern 20.

5 Here is a table that gives information about how long it takes to cook a turkey.

Weight of turkey (w), in kg	3	4	5	6
Time (t), in minutes	140	180	220	260

Jenny is cooking a turkey that weighs 7.6 kg.
She wants the turkey to have finished cooking at 7 p.m.
At what time should she put the turkey into the oven?

6 In the grid on the next page you are told which quantity to work out, having been given the values of the other letters.
Start in the bottom left corner.
If an answer is a prime number then move → to the next square.
If an answer is a multiple of 10 then move ← to the next square.
If an answer is a square number then move ↑ to the next square.
If an answer is a factor of 31 then move ↓ to the next square.

	work out $W = FD$ when $F = 14$ and $D = 17$	work out $t = \frac{d}{s}$ when $d = 14$ and $s = 4$	work out $s = \frac{d}{t}$ when $d = 15$ and $t = 6$	**FINISH**
	work out $V = IR$ when $I = 0.3$ and $R = 31$	work out $A = \frac{1}{2}bh$ when $b = 11$ and $h = 2$	work out $F = \frac{W}{D}$ when $W = 68$ and $D = 4$	work out $v = at + u$ when $a = 8$ and $u = 9$ and $t = 5$
	work out $b = \frac{A}{h}$ when $h = 2$ and $A = 17$	work out $d = st$ when $s = 27$ and $t = 3$	work out $P = VI$ when $I = 2.5$ and $V = 8$	work out $A = b \times h \div 2$ when $b = 5$ and $h = 7$
START here →	work out $y = 2x + 3$ when $x = 2$	work out $P = \frac{F}{A}$ when $F = 39$ and $A = 3$	work out $A = b \times h$ when $b = 4$ and $h = 9$	work out $s = \frac{d}{t}$ when $t = 15$ and $d = 45$

a Which of the formulae are equivalent to each other?

b Describe your path from the start to the finish.

Reviewing skills

1 **a** Draw a number machine for $C = 60d + 30$

b Find the value of C when

i $d = 4$ **ii** $d = 10$

c Draw the reverse number machine.

d Find the value of d when

i $C = 390$ **ii** $C = 570$

2 To convert a temperature in Fahrenheit into Celsius (centigrade), you take these steps.

Subtract 32.
Multiply by 5.
Divide by 9.

a Convert these Fahrenheit temperatures into Celsius.

i 212° **ii** 32° **iii** 95°

b Convert these Celsius temperatures into Fahrenheit.

i 10° **ii** 30° **iii** 205°

3 The cost, £C, of a meal for n people at a fixed price restaurant is given by

$C = 15n + 5$

a Work out the costs of meals for

i 4 people **ii** 10 people

b A meal costs £125. How many people are there?

c Explain the numbers 5 and 15 in the formula.

Unit 5 • Setting up and solving simple equations • Band f

Outside the Maths classroom

Fencing enclosures

What factors affect the shape of a play park?

Toolbox

An **equation** says that one expression is equal to another.
For example:

$$4x - 3 = 17$$

Solving an equation means finding the value of x that makes the equation true.

You can solve an equation using the balance method.
You must keep the equation balanced, like a pair of weighing scales, by doing the *same* operation to *both* sides.

$$4x - 3 = 17 \quad (+3 \text{ both sides})$$
$$4x = 20 \quad (\div 4 \text{ both sides})$$
$$x = 5$$

The inverse of subtract 3 is add 3.
Make sure you add 3 to *both* sides.

The inverse of multiply by 4 is divide by 4.

$x = 5$ is the solution.

Example – Solving 'think of a number' problems

Amy thinks of a number. What is her number?

Amy

Solution

You can write an equation and solve it to find Amy's number.

Let n represent her number.

$$n \times 9 - 11 = 52$$

$$9n - 11 = 52$$

Using the balance method rewrite this using correct algebraic notation.

$$9n - 11 = 52$$ ← The inverse of subtract 11 is add 11.

$+11$ / $+11$ ← Add 11 to both sides.

$$9n = 63$$

$\div 9$ / $\div 9$ ← The inverse of multiply by 9 is divide by 9.

$$n = 7$$

Check: $9 \times 7 - 11 = 63 - 11 = 52$ ✓ ← Always check your work.

Example – Solving word problems

Three friends run a relay race.
Altogether their time is 65 seconds.
Harry takes 4 seconds longer than Dan.
Millie takes 5 seconds less than Dan.
How many seconds does each person take?

Solution

First write the expressions for the time each person takes.
Let s stand for the number of seconds that Dan takes.

Dan: s seconds
Harry: $s + 4$ seconds
Millie: $s - 5$ seconds
Total: 65 seconds

So,

$s + s + 4 + s - 5 = 65$ ← The total of the three expressions is 65.

$3s - 1 = 65$ ← Collect like terms.

$3s = 66$ ← Add 1 to both sides.

$s = 22$ ← Divide both sides by 3.

So Dan takes 22 seconds.

Harry takes 26 seconds. ← Harry's time is $s + 4 = 22 + 4 = 26$ seconds.

Millie takes 17 seconds. ← Millie's time is $s - 5 = 22 - 5 = 17$ seconds.

Check: $22 + 26 + 17 = 65$ ✓

Practising skills

1 Work out the weights of the boxes.

a

b

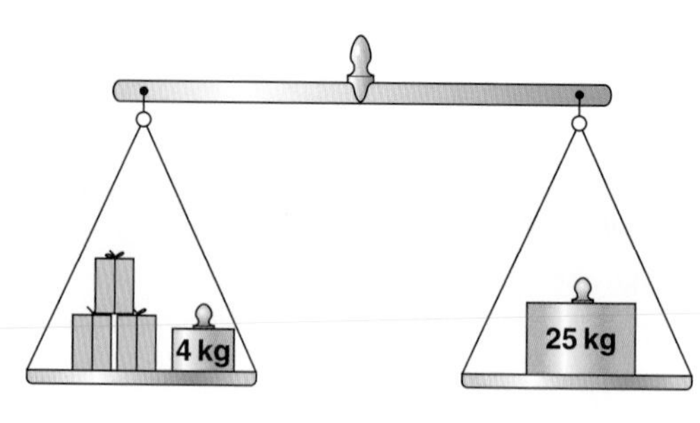

c

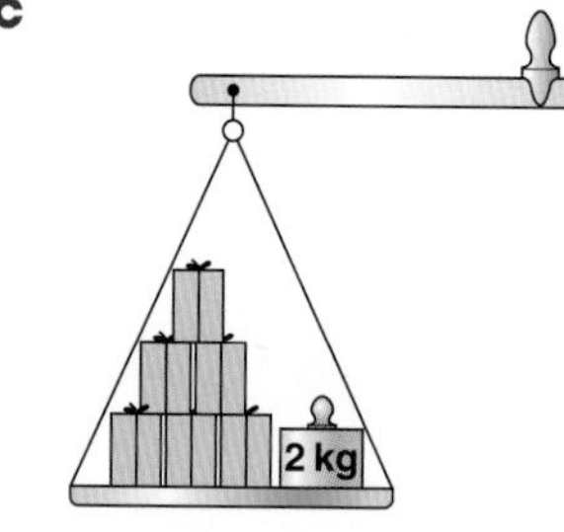

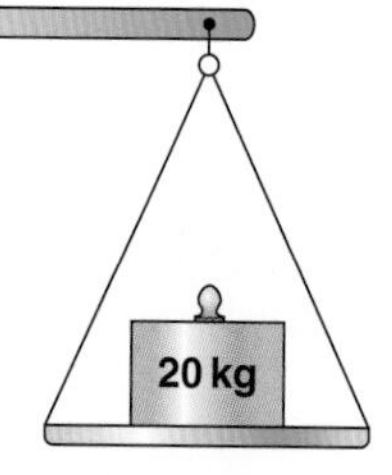

d

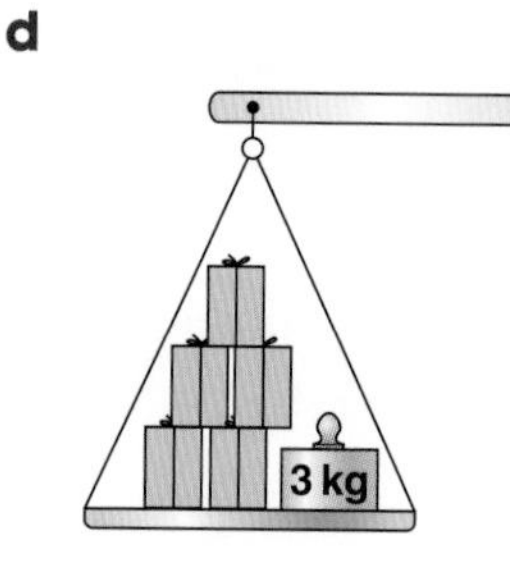

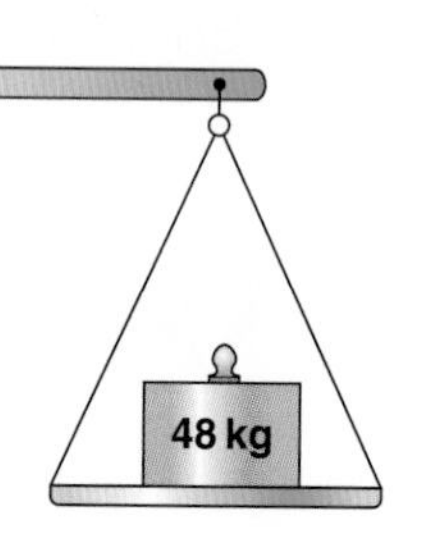

2 Solve these equations by subtracting the same number from both sides.

a $d + 4 = 9$ **b** $f + 6 = 17$ **c** $3 + x = 7$ **d** $m + 2 = 5$

3 Solve these equations by adding the same number to both sides.

a $x - 3 = 8$ **b** $n - 6 = 2$ **c** $p - 7 = 7$ **d** $e - 10 = 9$

4 Solve these equations.

a $f - 6 = 1$ **b** $g + 3 = 11$ **c** $d + 1 = 6$ **d** $b - 7 = 4$

5 Solve these equations by dividing both sides by the same number.

a $2b = 18$ **b** $5f = 30$ **c** $3x = 12$ **d** $6g = 72$

6 Solve these equations by multiplying both sides by the same number.

a $\frac{t}{2} = 5$ **b** $\frac{y}{5} = 3$ **c** $\frac{h}{4} = 6$ **d** $\frac{d}{9} = 2$

7 Solve these equations.

a $6s = 42$ **b** $d - 7 = 3$ **c** $a + 2 = 8$ **d** $5h = 40$

e $\frac{x}{7} = 8$ **f** $u - 3 = 14$

Reasoning

8 Cynthia and Hardip both try to solve the equation $4x - 2 = 32$.
This is what they write.

Cynthia

$4x - 2 = 32$

$4x - 2 + 2 = 32 + 2$

$4x = 34$

$\frac{4x}{4} = \frac{34}{4}$

$x = 8.5$

Hardip

$4x - 2 = 32$

$x - 2 = 8$

$x - 2 + 2 = 8 + 2$

$x = 10$

Whose answer is correct? Explain what is wrong with the incorrect solution.

9 Solve these equations.

a $3g + 4 = 25$ **b** $5h - 3 = 17$ **c** $10y - 2 = 58$ **d** $4d - 4 = 4$

e $6s + 7 = 25$ **f** $2k + 5 = 23$ **g** $8r - 5 = 11$ **h** $7u + 9 = 37$

Developing fluency

Reasoning

1 Andrea is a years old and Benny is b years old.

Explain what these mean in words.

a $a + b = 21$

b $a = 2b$

c $b - a = 7$

Exam-style

2 Write an equation for this situation and solve it to find the number.

I think of a number, multiply it by 5 and add 12 and the answer is 47.

3 Solve these equations.

a $-5x = -10$ **b** $-5x = 10$ **c** $10 = 5x$

d $10 + 5x = 0$ **e** $-5x - 10 = 0$ **f** $0 = 10 - 5x$

4 Match each equation with its solution.

i $2x - 9 = 15$ **ii** $8x + 3 = 19$ **iii** $5x - 1 = 54$

iv $4x + 16 = 4$ **v** $3x - 1 = -16$ **vi** $6x - 24 = 0$

$x = 2$	$x = 4$	$x = 11$	$x = -3$	$x = 12$	$x = -5$

5 Solve these equations.

a $\frac{s}{5} = 12$ **b** $\frac{f}{2} = 8$ **c** $\frac{g}{3} = -3$ **d** $\frac{x}{6} + 3 = 6$

e $\frac{t}{4} - 1 = -2$ **f** $\frac{u}{10} + 6 = 18$ **g** $\frac{v}{12} + 4 = 4$ **h** $\frac{y}{12} + 4 = -4$

6 Grace buys eight text books and a teachers' guide costing £24. The total cost is £138. Write an equation for this situation and solve it to find the price of a text book.

Exam-style

7 Alice bought a £28 railcard and two adult train tickets and the total came to £204. Write an equation for this situation and solve it to find the price of an adult ticket.

8 Bruce buys two oranges at 32p each and four apples. The total comes to £1.72. Write an equation for this situation and solve it to find the price of an apple.

9 Solve these equations.

a $s + 6.8 = 14.2$ **b** $5f = 7.8$ **c** $g - 0.48 = 7.3$

d $3d = 56.4$ **e** $32 - 3x = 17$ **f** $10 - 3b = 16$

10 Amy is a years old. Her sister Fran is 8 years older.

a Write down an expression for Fran's age.

Their brother Damien is 10 years old. The total of their ages is 30 years.

b Use their information to write an equation for part **a**.
Solve your equation.
How old are Amy and Fran?

Problem solving

Exam-style

1

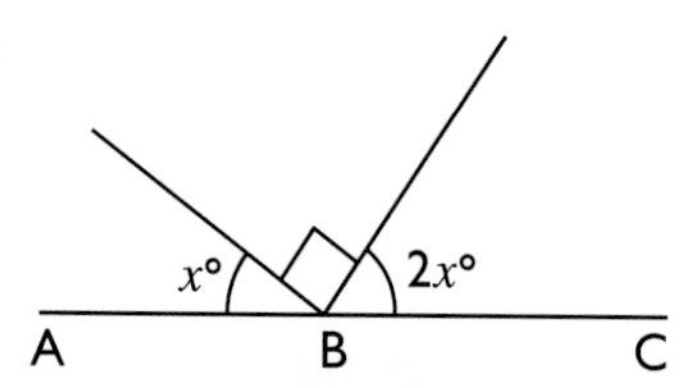

ABC is a straight line.
Find the value of x.

Exam-style

2 ABCD is a quadrilateral.

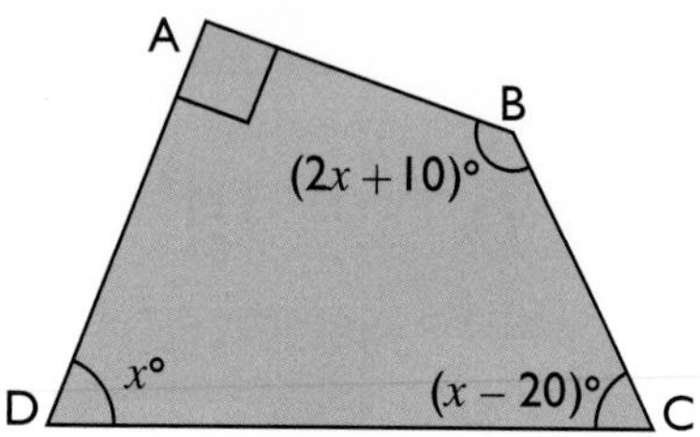

Find the value of the largest interior angle of the quadrilateral.

Exam-style

3 Ami is x years old.
Ben is twice as old as Ami.
Ceri is four years younger than Ami.
The total of their three ages is 36.
Write this information as an equation.
Solve the equation to find x.
How old are Ami, Ben and Ceri?

Exam-style

4 Sam has a field. It is rectangular, with a length l m.
The perimeter of the field is 400 m.
The width of the field is 50 m less than the length.
Write this information as an equation.
Solve the equation to find l.
Find the area of the field.

5 Explain why it is not possible for the angles of an equilateral triangle to have values $2x + 40°$, $6x°$ and $7x - 24°$.

Exam-style

6. A small bottle of water contains l litres.
A large bottle of water contains five times as much water as the small bottle.
The difference in the amount of water they contain is 2 litres.
Write this information as an equation.
Solve the equation.
How much water does each bottle contain?

7. Lily hires a car from Auto Hire.
a Write a formula for the cost, £C, of hiring a car for n days.
b Lily pays £280. How many days does she hire the car for?

8. The equal angles of an isosceles triangle are 30° greater than the third angle, a°.
a Write an equation for a.
b Solve the equation.
c Find all three angles of the triangle.

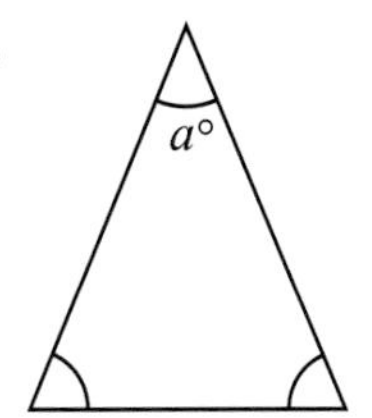

Reviewing skills

1. Solve these equations.
a $3c - 6 = 15$
b $4f + 7 = 19$
c $2x + 3 = -11$
d $6 + 4y = 50$
e $6t + 9 = 3$
f $9u + 14 = -40$
g $5y + 2 = 17$
h $3p + 18 = 12$

2. Solve these equations.
a $4u + 94 = 15.6$
b $8.1 - 2k = 10.5$
c $\frac{h}{6} - 12.5 = 2.9$
d $24 - \frac{m}{4} = 8$

3. So-shan needs 1kg of chickpeas for a soup recipe.
She buys 3 packets and 2 large tins to give her the correct amount.
Each packet contains 230g.
Write an equation for the total amount of chickpeas and solve it to find the size of a large tin.

Unit 6 • Using brackets • Band f

Outside the Maths classroom

Sales forecasting

Why do companies need to predict sales?

Toolbox

When you **expand** an expression you multiply out the brackets.
When you rewrite an expression using brackets you are **factorising.**

$$5a + 10 = 5(a + 2)$$

- $5a + 10$ is the expanded expression.
- $5(a + 2)$ is the factorised expression.

When you expand (or multiply out) brackets you must multiply every term inside the bracket by the term outside the bracket.

$$b \times (a + 3b - 2) = b \times a + b \times 3b + b \times -2$$
$$= ab + 3b^2 - 2b$$

There are three terms inside the bracket so there will be three terms in your answer.

When you factorise an expression, look at the numbers first and then the letters.

$$2fg + 6f^2 = 2 \times fg + 2 \times 3f^2$$ ← 2 goes into 2 and 6.

$$= 2f \times g + 2f \times 3f$$ ← f goes into fg and f^2

$$= 2f(g + 3f)$$ ← so $2f$ goes outside the bracket.

$2f$ is the highest common factor (HCF) of both terms.

Example – Expanding brackets

Expand the brackets in these expressions.

a $3(4a - 7)$

b $t(s - u)$

Solution

a $3(4a - 7) = 3 \times 4a - 3 \times 7$ ← Multiply every term inside the brackets by the term outside the brackets.

$= 12a - 21$

b $t(s - u) = t \times s - t \times u$

$= st - tu$ ← Remember to use correct algebraic notation. You don't write the × sign and you arrange the letters alphabetically.

Example – Factorising expressions

Factorise these expressions.

a $12x - 18$

b $6y^2 + 2y$

Solution

a $12x - 18 = 6 \times 2x - 6 \times 3$ ← 6 goes into $12x$ and 18.

$= 6 \times (2x - 3)$ ← Write the 6 outside the brackets. Be careful to keep the minus sign.

$= 6(2x - 3)$ ← Remember to use correct algebraic notation. You don't write the × sign.

b $6y^2 + 2y = 2y \times 3y + 2y \times 1$ ← 2 and y go into $6y^2$ and $2y$.

$= 2y \times (3y + 1)$ ← Don't forget to write the 1 in the brackets.

$= 2y(3y + 1)$

Check: $2y(3y + 1) = 2y \times 3y + 2y \times 1$

$= 6y^2 + 2y$ ✓

It is always a good idea to check your factorising by expanding the expression again.

Practising skills

1 Lexmi and Kabil need to work out $8 \times (30 + 4)$. Here is some of what they wrote.

Lexmi
$8 \times (30 + 4)$
$= 8 \times 30 + 8 \times 4$
$=$ ____ $+$ ____
$=$ ____

Kabil
$8 \times (30 + 4)$
$= 8 \times$ ____
$=$ ____

a Copy and complete their answers.

b Are they both right?

c Whose method is better?

2 Work out each of these

i by expanding the brackets first

ii by working out inside the brackets first.

a $6 \times (20 - 1)$
b $3 \times (40 + 2)$
c $4 \times (30 - 3)$
d $8 \times (6 + 4 - 3)$
e $2 \times (9 - 5 + 2)$
f $5 \times (10 + 3 - 6)$

3 Work these out by first expanding the brackets.

a $6 \times 29 = 6 \times (30 - 1)$
b $4 \times 52 = 4 \times (50 + 2)$
c $5 \times 37 = 5 \times (40 - 3)$
d $8 \times 93 = 8 \times (90 + 3)$

4 Factorise these expressions and work out their values.

a $6 \times 3 + 6 \times 7$
b $4 \times 8 - 4 \times 7$
c $3 \times 3 + 3 \times 8$
d $45 \times 3 + 45 \times 9 + 45 \times 8$

5 **a** **i** Find the area of each of these rectangles.

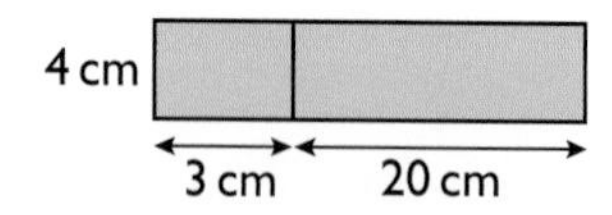

ii Use your answers to write down the volume of $4 \times (20 + 3)$.

b **i** Find the area of each of these rectangles.

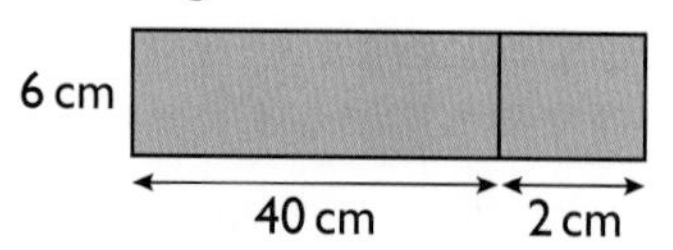

ii Use your answers to write down the volume of $6 \times (40 + 2)$.

c **i** Find the area of each of these rectangles.

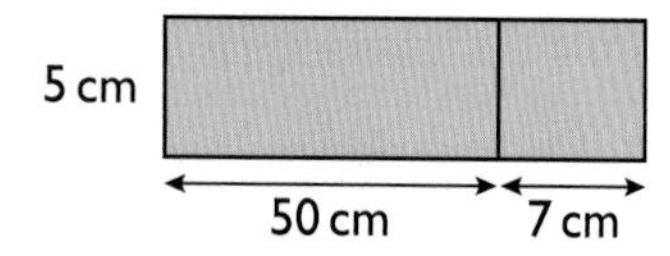

ii Use your answers to write down the volume of $5 \times (50 + 7)$.

d **i** Find the area of each of these rectangles.

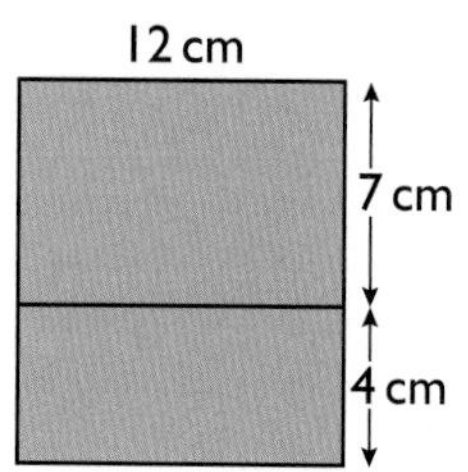

ii Use your answers to write down the volume of $12 \times (7 + 4)$.

6) For each of these rectangles
   - **i** write down an expression for the area using brackets
   - **ii** expand the bracket and find the area when $x = 7$.

a

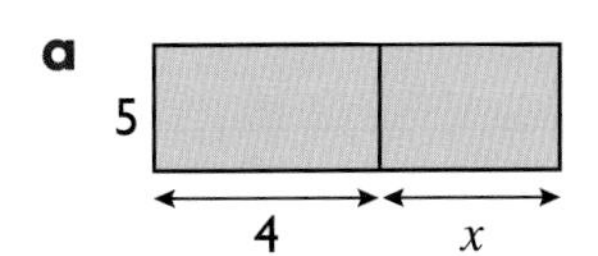

b

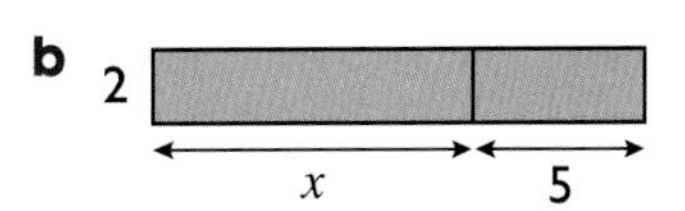

c

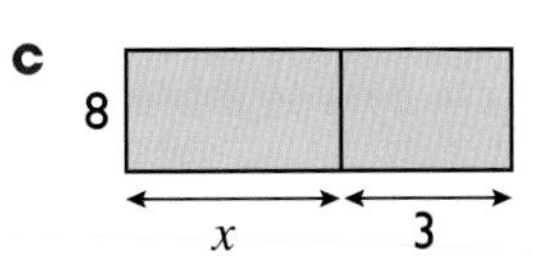

d

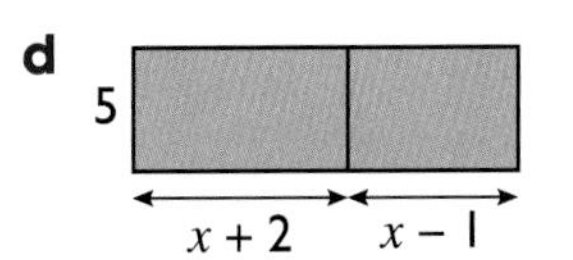

7) Expand these brackets.

a $3(2a + 7)$ **b** $6(7 - 4b)$ **c** $2(8c - 11)$ **d** $5(1 - 8d)$

e $4(2x + 3y)$ **f** $3(5e + 2f + 6)$ **g** $7(2p + 4q)$ **h** $5(8g - 3h + 2)$

Developing fluency

1) Match each expression with its equivalent.

$5(x-3)$	$15x+5$	$3(5x+2)$	$8(x+2)$	$12x+18$
$8-2x$	$2(4x-1)$	$12-2x$	$4(2x-1)$	$3(2x+5)$
$6(2x+3)$	$2(4-x)$	$5x-15$	$6x+15$	$8x-2$
$8x-4$	$8x+16$	$5(3x+1)$	$15x+6$	$2(6-x)$

2) Factorise these expressions fully.

a $4x + 8$ **b** $3y - 12$ **c** $16 - 8f$

d $12g + 18$ **e** $15m - 10$ **f** $7a + 21b$

Exam-style

3) Simplify these expressions and factorise your answers.

a $5x + 4y - 3x + 8y$

b $6a + 2b + 4a + 3b$

c $4x + 9y - 4 + 7 - x - y + 2x - 3y + 2$

4) Find the value of $5(3x - 2y)$ when

a $x = 4$ and $y = -3$ **b** $x = 2$ and $y = 3$ **c** $x = y = 1$ **d** $x = 0$ and $y = 4$

5) In each of these pairs choose which expression is larger and give reasons why.

a $5n - 1$ and $5(n - 1)$

b $3(2n + 3)$ and $2(3n + 5)$

c $5(3n + 7)$ and $7(2n + 5)$

6 Find the height of each of these rectangles.

a
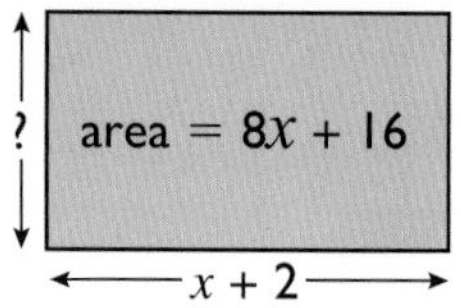

b
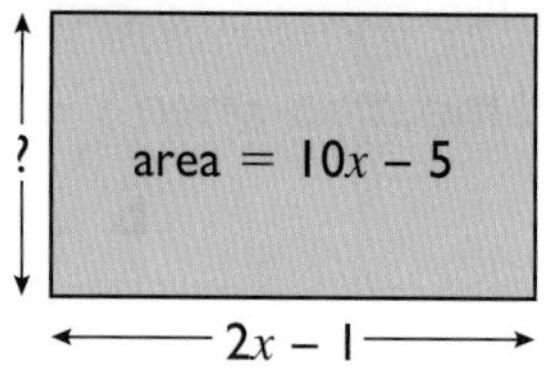

c
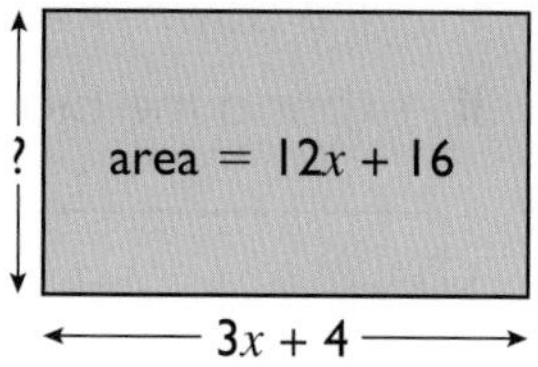

7 Expand the brackets and simplify the expression.

a $3(2x + 4) + 5(3x + 1)$ **b** $4(2x + 6) - 2(3x + 1)$ **c** $6(x + 3) - 2(2x - 3)$

8 Expand these brackets.

a $2(4x + 10)$ **b** $x(4x + 10)$ **c** $3x(x - 3)$

d $4x(2x + 5)$ **e** $3x^2(x - 2)$ **f** $6x(3x + 2y)$

9 Factorise these expressions fully.

a $4x^2 - 4x$ **b** $10x^2 + 5$ **c** $10x^2 + 5x$ **d** $6x^2 - 12x$

e $12cd - 8c$ **f** $6x^2 + 4x$ **g** $4x^2 - 8xy$ **h** $6c^2d + 18cd^2$

10 Chocolate flakes cost 99p.
During one week a family buy these numbers of flakes:
Monday 6 Tuesday 3 Wednesday 0 Thursday 4 Friday 8 Saturday 7 Sunday 4
Using brackets to help you, work out how much they spend on flakes.

Problem solving

Exam-style

1 Jeff sells vegetable planters for gardens.
The planters are in the shape of a square, of side s metres, and an equilateral triangle.

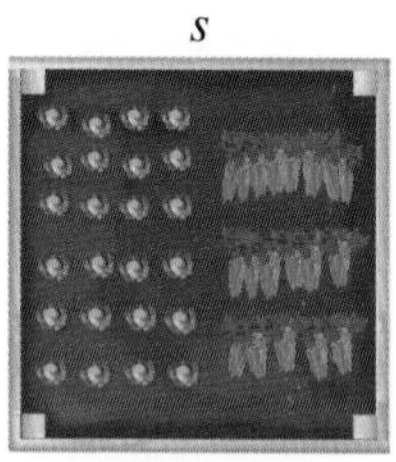

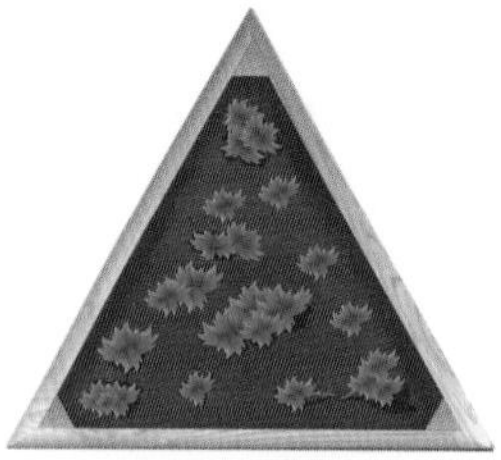

The length of each side of the equilateral triangle planter is 50 cm longer than each side of the square planter.

a Find the perimeter of the triangular planter.
Give your answer in terms of s.

b The perimeters of both shapes are the same.
Find the lengths of their sides.

2 The tile is used in a child's toy.
It is in the shape of parallelogram.

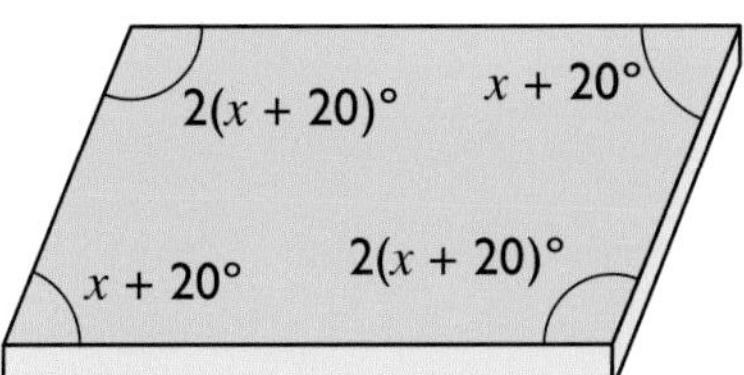

Find the value of the smallest angles in the parallelogram.

3 ABCD is a rectangle.

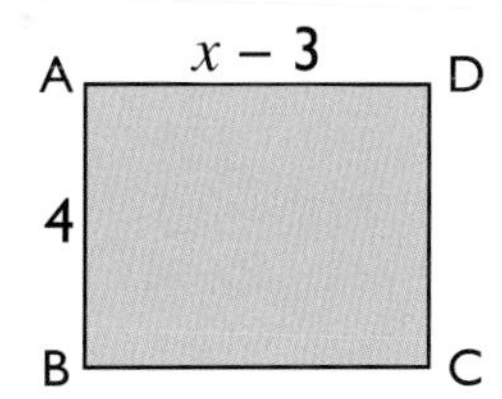

All measurements are in centimetres.
The area of the rectangle is 24 cm^2.
Find the perimeter of the rectangle.

4 Ric is x years old.
Steph is 6 years older than Ric.
Tom is 3 times Steph's age.
The total of their ages is 84.
How old is Tom?

5 Bill hires a car.

a Write an expression for the cost of hiring the car for one week and driving m miles.

b Bill hires a car for 3 weeks and pays a total of £390.
How many miles does he drive each week (on average)?

6 Here are a square and a right-angled triangle.

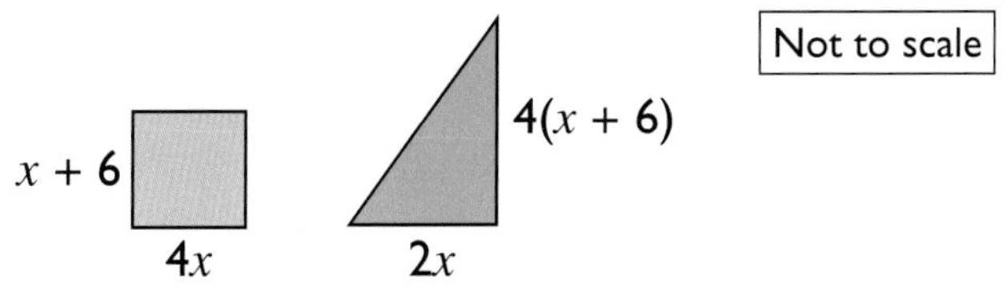

Show that the square and the triangle have the same area.

(7) Here is a rectangle drawn on squared dotted paper. It has a width of 5 and a height of 4.

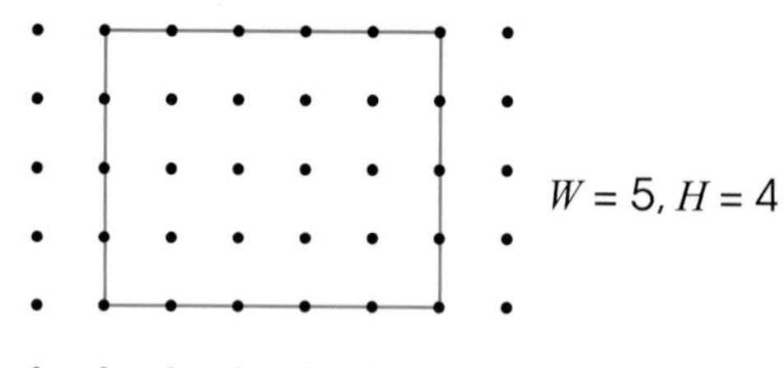

Squares have now been drawn around the inside edge.

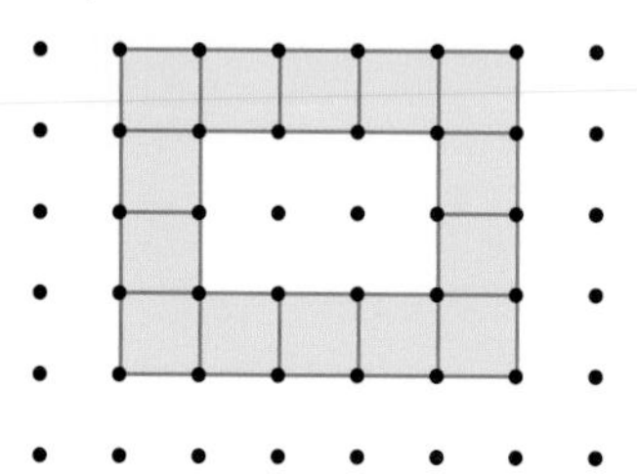

In this picture there are 14 squares.

Here is a general diagram.

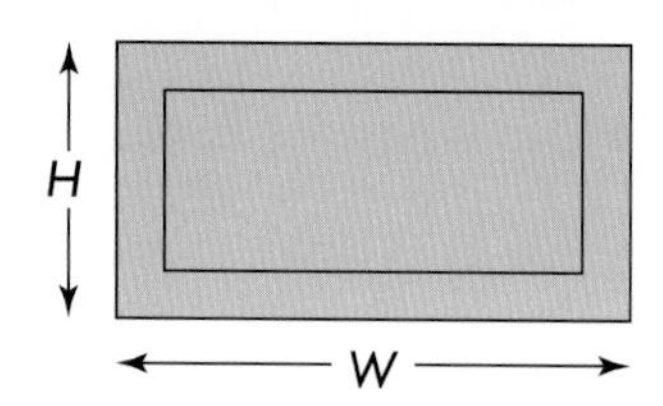

Use it to work out four different expressions for the number of squares around the inside edge.

a Method 1: Add up the number along each side and then subtract the corners (why?).

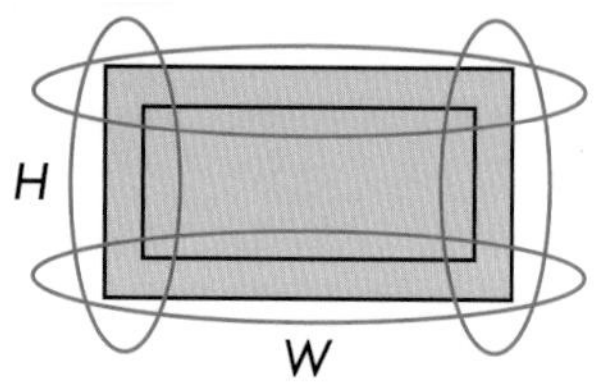

b Method 2: Subtract two from the width and subtract two from the length. Add these together. Add one for the bottom left-hand corner. Double this and add 2 (why?).

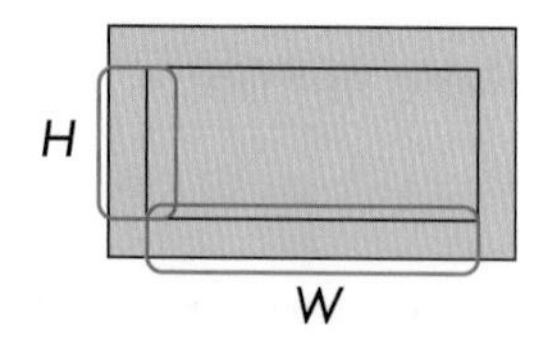

c Method 3: Add the number along the top to the number down the side. Double this and subtract 4 (why?).

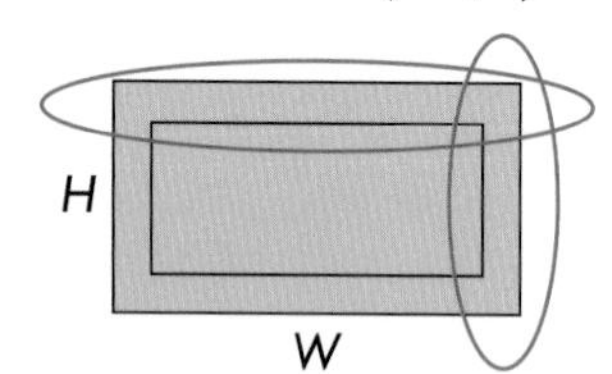

d Method 4: Subtract one from each side and add them all together.

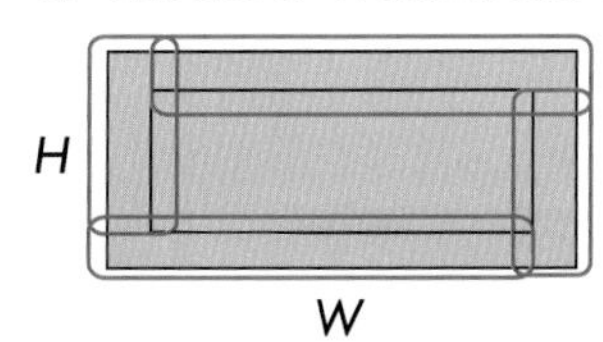

e Show that all of the expressions you get are equivalent.

Reviewing skills

1 Work out each of these

- **i** by expanding the brackets first
- **ii** by working out inside the brackets first.

a $5 \times (100 - 1)$ **b** $6 \times (40 + 5)$ **c** $12 \times (20 + 1)$

d $15 \times (4 + 3 - 1)$ **e** $7 \times (8 - 5 - 3)$ **f** $11 \times (100 + 1 - 2)$

2 Work these out by first expanding the brackets.

a $17 \times 999 = 17 \times (1000 - 1)$ **b** $9 \times 82 = 9 \times (80 + 2)$

c $12 \times 109 = 12 \times (100 + 9)$ **d** $99^2 = 99 \times (100 - 1)$

3 Factorise these expressions and work out their values.

a $5 \times 8 + 5 \times 12$ **b** $7 \times 3 + 7 \times 2 - 7 \times 5$ **c** $11 \times 13 + 11^2 - 11 \times 4$

4 For each of these rectangles

- **i** write down an expression for the area using brackets
- **ii** expand the bracket and find the area when $x = 5$.

a

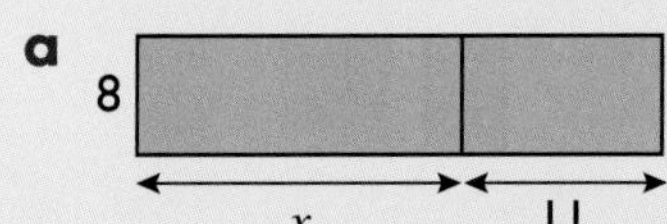

b

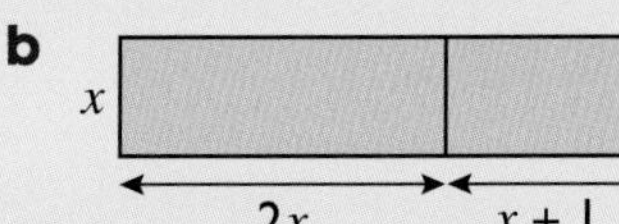

5 Expand these brackets and simplify the expression.

a $5(2a + 3b) + 2(4a + b)$ **b** $3(5a + 6b) - 3(2a + 4b)$

c $8(a - b) + 6(3a + 2b)$ **d** $4(3a + 2b) - 5(a - 3b)$

6 Simplify these expressions and factorise your answers.

a $8f + 3g - 5f + 6g$ **b** $12k - 3m - 4k - 5m$ **c** $7x + 5 + 3x - 4 - 2x + 8$

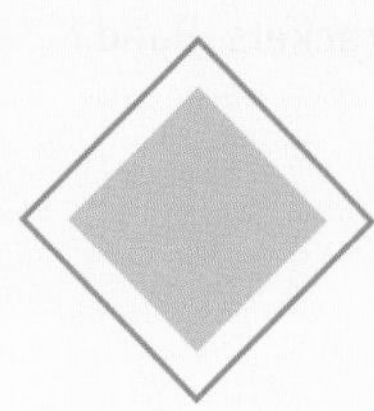

Algebra Strand 2 Sequences

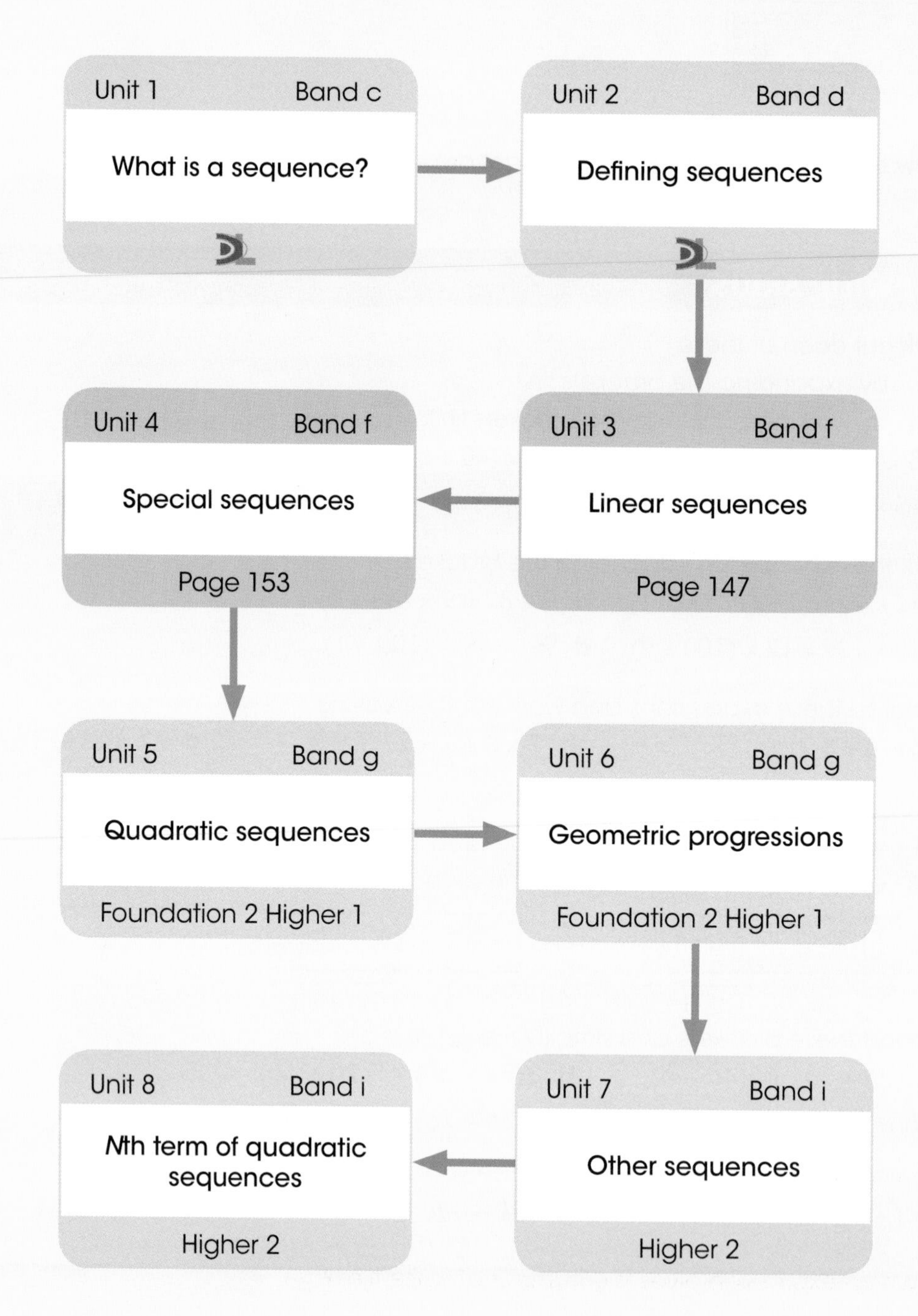

Units 1–2 are assumed knowledge for this book. They are reviewed and extended in the Moving on section on page 145.

Units 1–2 • Moving on

Reasoning

1. Find the next three terms in these sequences.
   - **a** 1160, 580, 290, 145, ☐, ☐, ☐, ...
   - **b** −60, 30, −15, 7.5, ☐, ☐, ☐, ...
   - **c** −128, −192, −288, −432, ☐, ☐, ☐, ...

Exam-style

2. At Samil's bus stop, the 208 bus is due at 10:00 and then every 10 minutes.
   The 119 bus is due at 9:47 and then every 7 minutes.
   When is the first time after 10:00 that both buses are due at the same time?

3. Here are two sequences. Their terms are added to give a third sequence.

1	3	5	7	9
2	4	6	8	10
3	7	11	15	19

   - **a** Find two other sequences that add to give the sequence 3, 7, 11, 15, ...
   - **b** How can you find more sequences that add to give the sequence 3, 7, 11, 15, ... ?

Reasoning

4. Here is a tiling pattern from a cathedral floor.

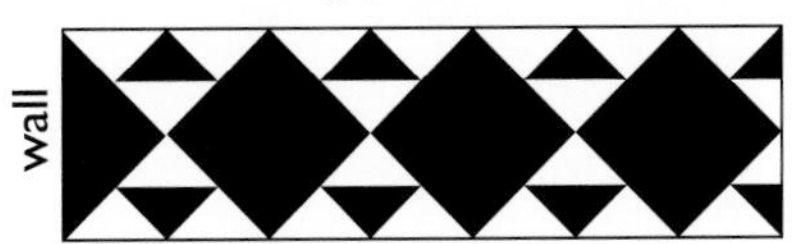

   - **a** Draw the design from the wall up to, but not including, any repetition.
     Call this Pattern 1.
   - **b** Draw Pattern 2, from the wall up to and including the first repetition.
   - **c** Here are three expressions for the nth term.
     - A 9 × pattern number + 1
     - B 3 × pattern number + 1
     - C 6 × pattern number
     - **i** Which expression gives the number of white tiles?
     - **ii** What do the other two expressions represent?
   - **d** **i** What fraction of Pattern 1 is coloured black?
     ii What fraction of Pattern 2 is coloured black?

Reasoning

(5) We are going to use a term-to-term rule of 'add 6'.

a If the first term in the sequence is 1, what is special about all of the terms in the sequence?

b If the first term in the sequence is 2, what is special about all of the terms in the sequence?

c If the first term in the sequence is 3, what is special about all of the terms in the sequence? (Give an answer that is different from the answer you gave to question **a**.)

d **i** If the first term in the sequence is 6, what is special about all of the terms in the sequence? (Give an answer that is different from the answer you gave to question **b**.)

ii Can you explain your answers?

iii If the term-to-term rule is 'add 5', which starting numbers will give you a sequence where the numbers are all special in some way?

Unit 3 • Linear sequences • Band f

Outside the Maths classroom

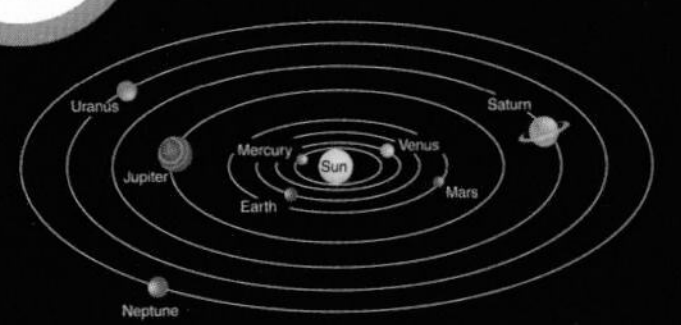

Scheduling

We often schedule events by week or by month.
What is a lunar month?

Toolbox

The sequence 4, 7, 10, 13, ... is a **linear sequence**.
This is because the gap between successive terms, known as the **difference**, is always the same, in this case 3.
All sequences where the **term-to-term rule** is an addition or subtraction of a constant amount are linear sequences.
The sequence

4, 7, 10, 13, ...

has a constant difference of 3. This means the sequence is linked to the 3 times table.
The **position-to-term** rule is therefore of the form

nth term = $3n + c$ where c is a number.

The first term will be

1st term = $3 \times 1 + c$
$= 3 + c$

So in this case, $c = 1$

Example – Finding the position-to-term formula

For each of the sequences below
- **i** find the next three terms
- **ii** find the position-to-term formula

a 8, 16, 24, 32, 40, ... **b** 27, 25, 23, 21, 19, ...

Solution

a i To get the next term, add 8 each time.
So the next three terms are 48, 56, 64.

ii Since the sequence is linear, nth term = $8n$ + a number.
Look at the first term of the sequence.
It is 8 so there is no need to add a number in this case.
So nth term = $8n$.

> The difference is 8 so the sequence is related to the 8 times table.

b i To get the next term, subtract 2 each time.
So the next three terms are 17, 15, 13.

ii Since the sequence is linear, nth term = $-2n$ + a number.
The first term of the sequence is 27.
So nth term = $-2n + 29$.
Or, more neatly, nth term = $29 - 2n$.

> The difference is −2 so the sequence is related to the −2 times table.

> $n = 1$ for the first term.

Example – Generating a sequence using a position-to-term rule

Write down the first five terms of the sequence with this position-to-term rule.

nth term = $14n + 45$

Solution

Method 1 – using nth term formula for all terms:

When $n = 1$, 1st term = $14 \times 1 + 45 = 59$ ← Substitute the term number into the nth term formula.

When $n = 2$, 2nd term = $14 \times 2 + 45 = 73$

When $n = 3$, 3rd term = $14 \times 3 + 45 = 87$

When $n = 4$, 4th term = $14 \times 4 + 45 = 101$

When $n = 5$, 5th term = $14 \times 5 + 45 = 115$

Method 2 – using position-to-term rule for first term only:

When $n = 1$, 1st term = $14 \times 1 + 45 = 59$

From the formula, the common difference is 14.

2nd term = $59 + 14 = 73$ ← Add 14 to previous term.

3rd term = $73 + 14 = 87$

4th term = $87 + 14 = 101$

5th term = $101 + 14 = 115$

Practising skills

1. For each of these describe the sequence by giving the first term and the term-to-term rule.

   a 4, 11, 18, 25, 32, …
   b 6, 11, 16, 21, 26, …
   c 38, 35, 32, 29, 26, …
   d 12, 15, 18, 21, 24, …

2. For each of these position-to-term rules write down the first five terms of the sequence.

   a nth term = 3 × position
   b nth term = 4 × position + 2
   c nth term = 2 × position + 25

3. Look at these linear sequences.

   A 30, ☐, ☐, 54, 62, …
   B 84, ☐, ☐, 75, 72, …

   a Work out the missing terms.
   b The 100th terms are for **A** 822 and for **B** −213.
   Write down the 101st term for each sequence.
   c Complete these position-to-term rules for each sequence.
   i ☐ × position + 22
   ii −3 × position + ☐

4. The first term in a sequence is a and the number added on to each term to get the next term is d.
   In each sequence
   i find the values of a and d
   ii give the next two terms.

   a 23, 27, 31, 35, …
   b 38, 44, 50, 56, …
   c 46, 43, 40, 37, …
   d −4, −2, 0, 2, …
   e 6, 1, −4, −9, …

5 Here are the position-to-term formulae for three sequences.

a $3n + 1$ **b** $6n - 2$ **c** $4n + 3$

In each case

i write down the first 4 terms, and the 100th term

ii write down the differences between each term, and write down the first term

iii what is the connection between your answers to part **ii** and the formulae?

6 Write down the position-to-term formulae for these sequences.

a 5, 8, 11, 14, … **b** 4, 6, 8, 10, … **c** 5, 9, 13, 17, …

7 Use these position-to-term formulae to work out the first five terms and the 20th term for these sequences.

a $3n + 6$ **b** $2n + 5$ **c** $7n - 3$

8 For each of these sequences

i write down the position-to-term formula

ii work out the 100th term.

a 11, 15, 19, 23, … **b** 2, 12, 22, 32, … **c** 11, 18, 25, 32, …

Developing fluency

1 Match the sequences to their rules.

3, 6, 9, 12, …	$5n + 1$
7, 9, 11, 13, …	$3n$
6, 11, 16, 21, …	$2n + 5$

2 For each of these number machines, copy and complete the table and then write down the rule as a formula for the output.

a

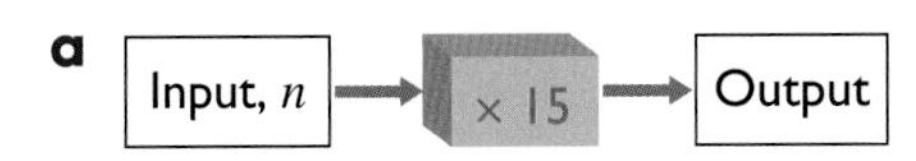

Input, n	2		5		12
Output	30	45		150	

b

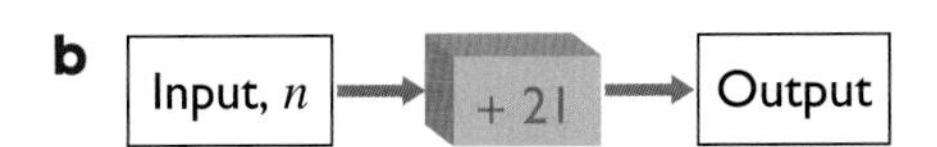

Input, n	6			15	
Output		33	35		53

c

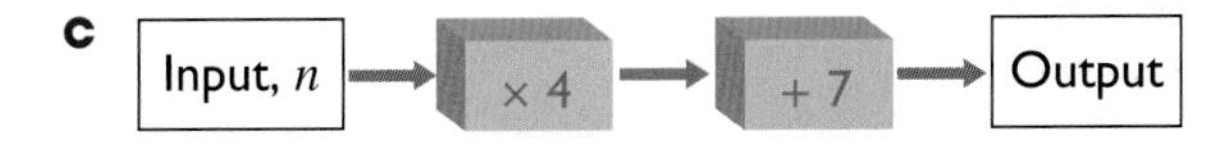

Input, n	1	2	5		
Output	11			47	87

d

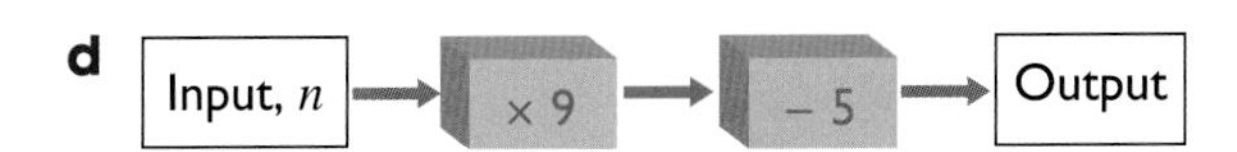

Input, n	1	6		20	
Output			85		202

3 Here is a sequence of pentagonal matchstick patterns.

Pattern 1 Pattern 2 Pattern 3 Pattern 4

a Draw the next two patterns.

b Copy and complete this table for the first six patterns.

Number of pentagons	1	2	3	4	5	6
Number of matchsticks	5					

c Predict the number of matchsticks for seven pentagons.
Explain how you found your answer.

d Write down the position-to-term formula.

e Predict the number of matchsticks for
i 10 pentagons **ii** 20 pentagons.

f How many pentagons will 101 matchsticks make?

4 Write these position-to-term formulae with the correct notation.

a the nth term is found by $n \times 5$ subtract 2

b the nth term is found by $n \times 3$ add 8

c the nth term is found by multiplying n by 6 and subtracting 5

d the nth term is found by dividing n by 2 and adding 3

Reasoning

5 Here are the first four terms of a sequence.

6, 10, 14, 18

a **i** Write down the next term. **ii** Explain how you found your answer.

b Write down the position-to-term formula for this sequence.

c Use your formula to work out the 40th term.

d Which position is the term 350?

Reasoning

6 Here are the first four terms of a sequence.

3, 9, 15, 21

a **i** Write down the next term. **ii** Explain how you found your answer.

b Write down the position-to-term formula for this sequence.

c Use your formula to work out the 40th term.

d Which position is the term 267?

Exam-style

(7) Here is a sequence of square matchstick patterns.

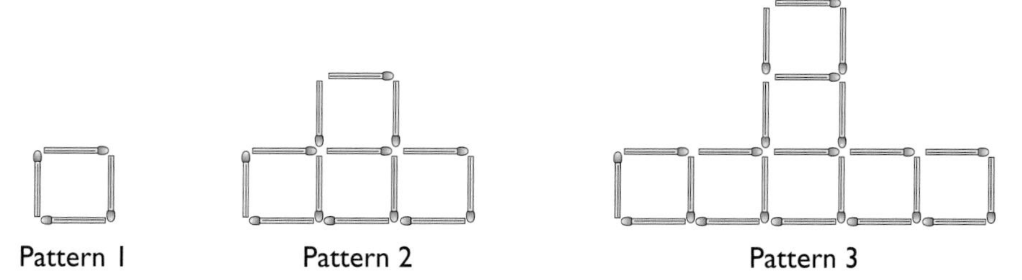

a Draw the next two patterns.

b Copy and complete the table for the first six patterns.

Pattern number	1	2	3	4	5	6
Number of matchsticks	4					

c **i** Predict the number of matchsticks for pattern number seven.
ii Explain how you found your answer.

d Write down the position-to-term formula.

e Predict the number of matchsticks for
i pattern 10 **ii** pattern 20.

f Why is the number of matchsticks for pattern 20 **not** double that for pattern 10?

g How many **squares** will 85 matchsticks make?

Problem solving

Exam-style

Extension

(1) A school uses hexagonal tables in its dining hall.
The tables are always laid out according to this pattern.
One chair is placed at each open edge of each table.

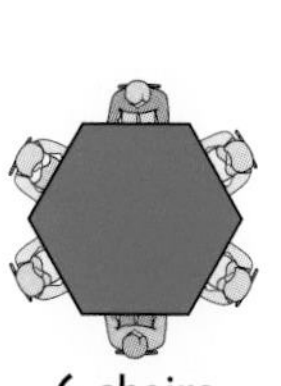

6 chairs

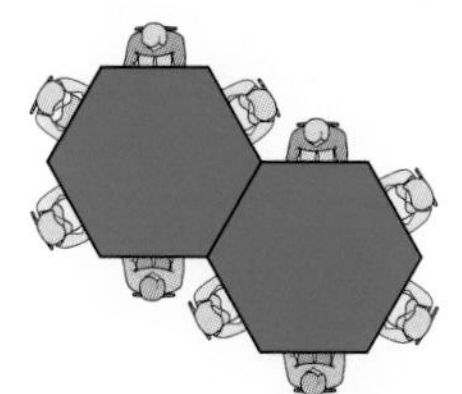

10 chairs

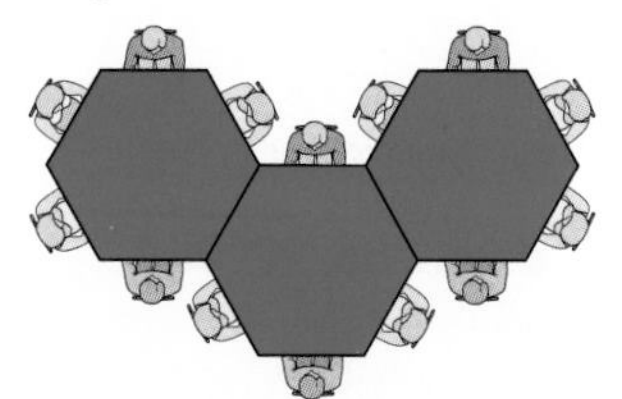

14 chairs

a How many chairs are needed for a pattern with 4 tables?

b How many chairs are needed to fit around a pattern with n tables?

c 77 students need to be seated on tables and chairs laid out in this pattern.
Show that there will be one empty chair if the least number of tables are used.

Exam-style

Extension

(2) Andy makes a sequence by counting back from 40 in threes.
The sequence starts 40, 37, 34, ….

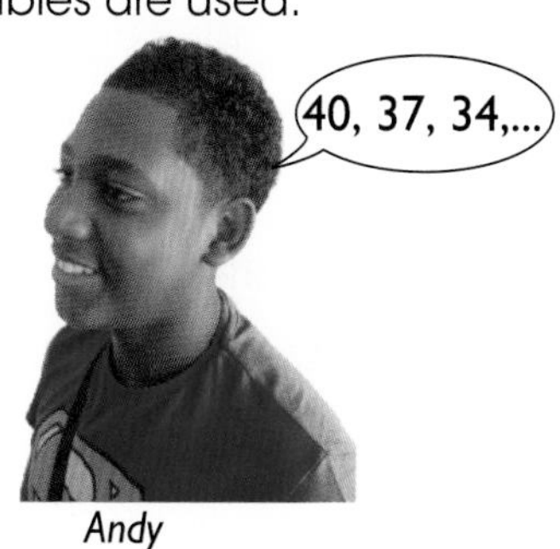

Andy

a State which one of the following is the formula for the nth term of Andy's sequence.

i $40 - 3n$ **ii** $40 + 3n$

iii $43 - 3n$ **iv** $43 + 3n$

b Explain whether 2 is a member of this sequence.

Extension

(3) The mth term of sequence A is $4m - 3$.
The nth term of sequence B is $63 - 7n$.
How many terms do the sequences have in common?

(4) A group of primary school pupils stand in a ring and play catch, always throwing the ball in the same direction around the ring.
When they throw the ball they say a number, with the first person (person A) saying '1', the second (person B) saying '2', and so on.
The game stops with the person saying '100' being the winner.
When four people play, Billy works out that he can win if he stands in position D.

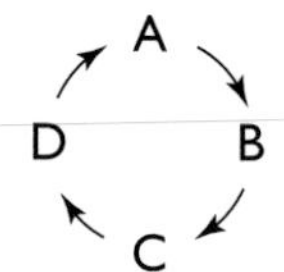

a Explain how he knows this.

b Where should Billy stand if there are five people?

c What about

i six people **ii** seven people?

Reviewing skills

(1) Here are the first few terms of a sequence.
9, □, □, □, 41, □, 57, ...

a Work out the term-to-term rule.

b Write down the position-to-term formula for this sequence.

c Find the 50th term.

d Which position is the term 449?

(2) For each of these sequences

i write down the position-to-term formula **ii** work out the 100th term.

a 80, 76, 72, 68, ... **b** 100, 95, 90, 85, ...

c 60, 58, 56, 54, ... **d** 80, 86, 92, 98, ...

(3) Here are the first three of a sequence of cross-shaped patterns.

a How many dots are there in patterns 4 and 5?

b Give a formula for the number of dots, d, in pattern number n.

c How many dots are there in

i pattern 90 **ii** pattern 120?

d What pattern has 645 dots?

Unit 4 • Special sequences • Band f

Outside the Maths classroom

Patterns in nature

What do the spirals on these pine cones have in common with flower petals?

Toolbox

There are some important number sequences that are not linear (i.e. that do not have a constant difference between terms).

- **The triangular numbers:** 1, 3, 6, 10, ... (that is, 1, 1 + 2, 1 + 2 + 3, 1 + 2 + 3 + 4, etc.).

  The difference between successive terms increases by one each time.

  The position-to-term formula for the triangular numbers is nth term = $\frac{n(n+1)}{2}$.
- The **square numbers**: 1, 4, 9, 16, 25, ... (that is 1 × 1, 2 × 2, 3 × 3, 4 × 4, 5 × 5, etc.).

  The difference between successive terms is 1, 3, 5, 7, 9, etc. and the position-to-term formula is nth term = n^2.
- The **Fibonacci numbers**: 1, 1, 2, 3, 5, 8, 13, 21, ... where each term (after the initial two) is the sum of the two previous terms.

  The difference between successive terms is the Fibonacci sequence itself!

Many sequences are variations of these.
For example,

101, 104, 109, 116, ... is the sequence of square numbers plus 100.
3, 12, 27, 48, ... is the sequence of square numbers multiplied by 3.

Example – Investigating triangular numbers

Look at this sequence of patterns.

Triangle 1 Triangle 2 Triangle 3

a Draw the next three triangles.

b How many dots are there in each triangle?

c What pattern do you notice in the number sequence?

d The formula for the nth term of the sequence is $\frac{n(n+1)}{2}$.

Use this formula to check your answer for triangle 5.

e Find the 10th triangular number.

Solution

a

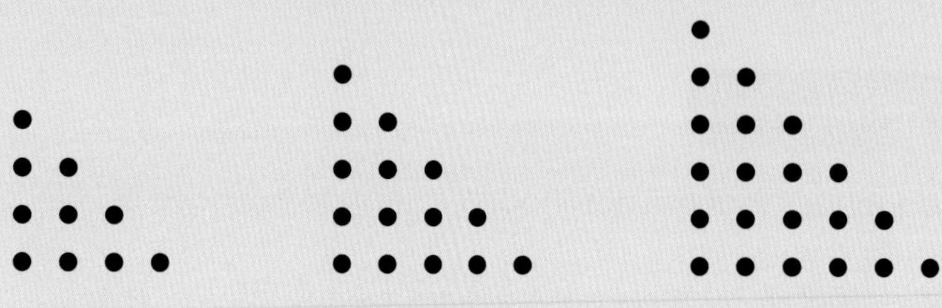

Triangle 4 Triangle 5 Triangle 6

b Triangle 1 has 1 dot.

Triangle 2 has 3 dots – 2 dots are added.

Triangle 3 has 6 dots – 3 dots are added.

Triangle 4 has 10 dots – 4 dots are added.

Triangle 5 has 15 dots – 5 dots are added.

Triangle 6 has 21 dots – 6 dots are added.

c The number of dots added increases by one each time.

d For $n = 5$,

$$\frac{n(n+1)}{2} = \frac{5 \times 6}{2} = 15$$

Substitute $n = 5$ into the formula.

So the answer for triangle 5 is right.

e When $n = 10$,

$$\frac{n(n+1)}{2} = \frac{10 \times 11}{2} = 55$$

So the 10th triangular number is 55.

Practising skills

1. For each of these sequences
   - **i** describe the sequence
   - **ii** draw the first 4 patterns in a sequence that gives these numbers
   - **iii** find its position-to-term formula.

   a 2, 4, 6, 8, 10, 12, … **b** 1, 4, 9, 16, 25, 36, … **c** 1, 8, 27, 64 , 125, …

2. Find the missing numbers in each of these sequences.

   a

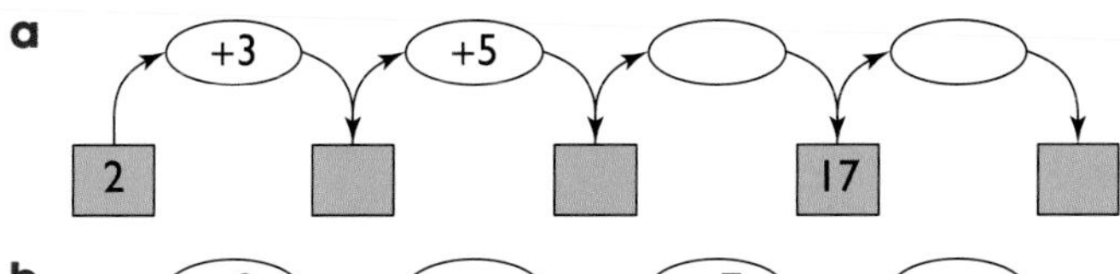

   b

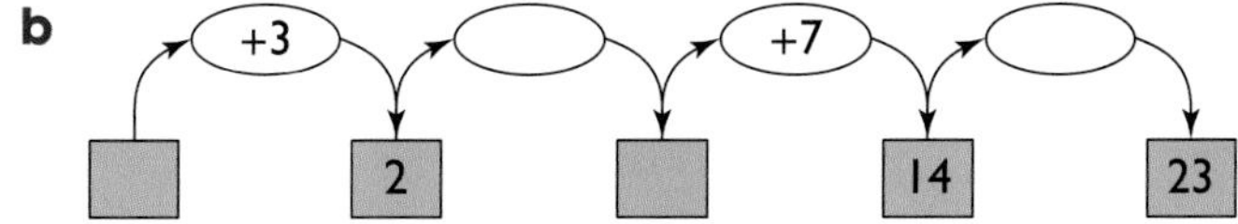

   c

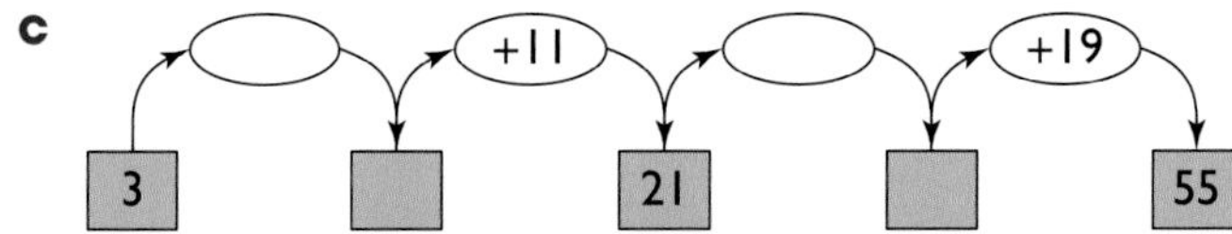

   d

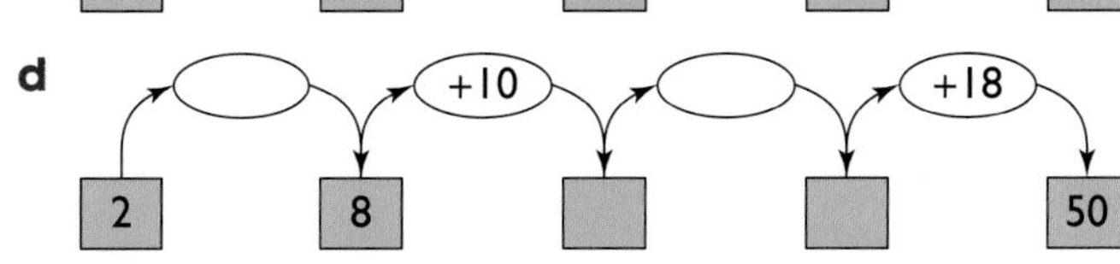

3. Write down the first five terms of these sequences.

   a $n^2 + 10$ **b** $n^3 - 1$ **c** $\frac{(n^2 + n)}{2}$

   What name is given to the numbers in the last sequence?

4. Find the position-to-term formulae for these sequences.

   a 1, 4, 9, 16, 25, … **b** 2, 5, 10, 17, 26, … **c** 0, 3, 8, 15, 24, … **d** 2, 8, 18, 32, 50, …

5. Find the position-to-term formulae for these sequences.

   a 11, 14, 19, 26, 35, … **b** 1, 8, 27, 64, 125, … **c** 6, 13, 32, 69, 130, … **d** 2, 16, 54, 128, 250, …

Developing fluency

1 Here are the first three patterns in a sequence made from triangles.

Pattern 1 Pattern 2 Pattern 3

a Draw pattern number 4.

b Copy and complete the table.

Pattern number	1	2	3	4	5
Number of red triangles	1	3			
Number of green triangles	0	1			
Total number of triangles, T	1				

c What are the names of the sequences in the table?

d Work out the total number of triangles in the 10th pattern.
In the 10th pattern, how many triangles are

i red **ii** green?

e Find a formula for the total number of triangles, T, in pattern n.

2 Look at these sequences.

i 2, 4, 8, 16, … **ii** 1, 2, 4, 7, 11, …

a What is the rule for finding the next term in each sequence?

b Write down the first 16 numbers in each sequence.

c Which numbers are in both sequences?

d Which square numbers are in sequence **i**?

Problem solving

3 Lucy is making tiling patterns.

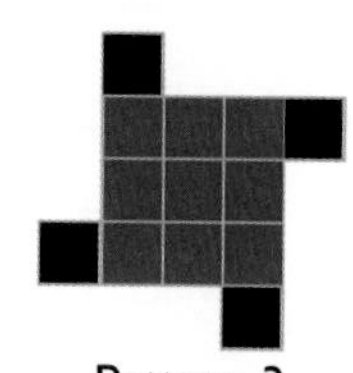

Pattern 1 Pattern 2 Pattern 3

a Draw pattern number 4.

b Copy and complete the table.

Pattern number	1	2	3	4	5
Number of black tiles	4	4			
Number of blue tiles	1	4			
Total number of tiles, T	5				

c Work out the total number of tiles in the 10th pattern.

d Which pattern uses 229 tiles?

e Find a formula for the total number of tiles, T, in pattern n.

f Lucy has 400 tiles.

i Can she use them all to make a single pattern?
Give a reason for your answer.

ii Write down the pattern number she can make.
How many tiles (if any) does she have left over?

4 Here are the first four patterns in a sequence made from coloured counters.

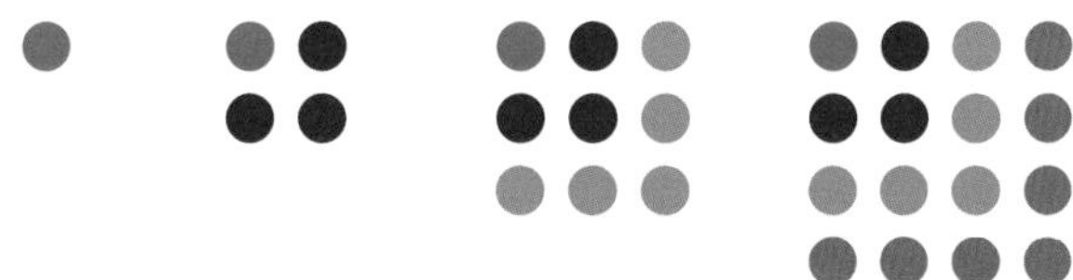

Pattern 1 Pattern 2 Pattern 3 Pattern 4

a Write down how many counters are added to

i pattern 1 to make pattern 2

ii pattern 2 to make pattern 3

iii pattern 3 to make pattern 4.

b How many counters need to be added to make pattern 5?

c How many counters are in pattern 10?

d How many counters are in the nth pattern?

e Which pattern number has (1 + 3 + 5 + 7 + 9 + 11 + 13) counters?

f What is the sum of the first 100 odd numbers?

5 **a** Write down the first five terms of the sequence with this position-to-term formula.

nth term = $n(n + 1)$

b **i** Use your answer to part **a** to write down the position-to-term formula for this sequence.

1, 3 , 6, 10, 15, …

ii Describe the numbers in this sequence.

iii Draw the first five patterns in a sequence that gives these numbers.

iv Find the value of the 40th term.

c **i** Now look at these patterns.

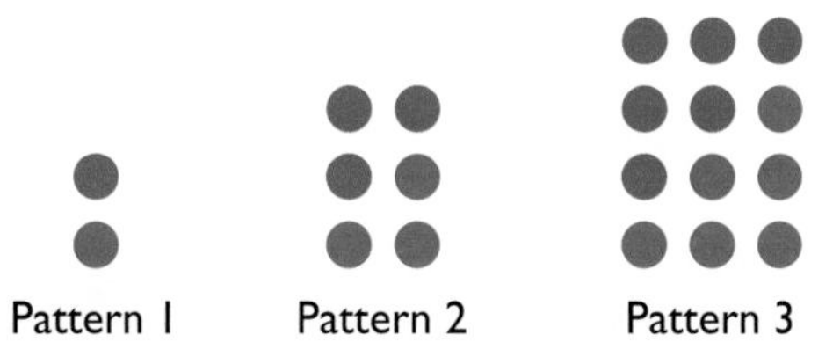

Pattern 1 Pattern 2 Pattern 3

Draw patterns 4 and 5 of this sequence.

ii The number of circles in **pattern 1** is 1 × 2.

In **pattern 2** it is 2 × 3.

How many circles are there in pattern n?

iii In pattern n, how many circles are red?

How many circles are green?

iv How is this connected to part **b** of this question?

Reasoning

6 Look at these matchstick patterns.

Pattern 1 Pattern 2 Pattern 3

a Draw pattern number 4.

b Copy and complete the table.

Pattern number	1	2	3	4	5
Number of matches, M	4	12			

c Work out the total number of matches in the 8th pattern.

d **i** Write down the first five triangular numbers.

ii How is the sequence for the number of matches related to the triangular numbers?

iii The nth triangular number is $\frac{1}{2}n(n + 1)$.
How many matchsticks are in the 20th pattern?

iv Write down a formula for the total number of matches, M.

Problem solving

Exam-style Extension

1 Here is a sequence of Fibonacci numbers.
0, 1, 1, 2, 3, 5, 8, 13, ...

a Continue the pattern for all Fibonacci numbers less than 200.

b List the Fibonacci numbers under 200 that are prime numbers.

c Find the Fibonacci numbers under 200 that can be factorised using other Fibonacci numbers.

Exam-style Extension

2 Here is an old puzzle.
A cow produces its first female calf at age two years and after that she produces another female calf every year.
The table shows the number of cows after six years.

Year	Original cow	First new cow's calf	Second new cow's calf		Total
1	1				1
2	1				1
3	1 + 1				2
4	1 + 1 + 1				
5	1 + 1 + 1 + 1	1			
6	1 + 1 + 1 + 1 + 1	1 + 1	1		

How many female cows are there after 12 years, assuming none die?

Exam-style

3 Jilly is making rectangles out of dominoes. They are always the height of a domino.
When she has two dominoes there are two ways she can make a rectangle.

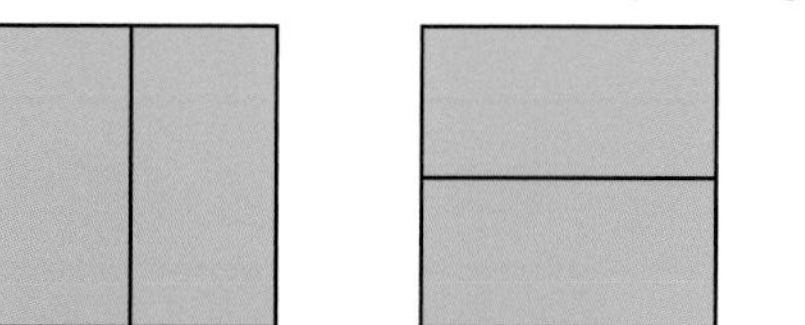

✓

Putting the dominoes end to end doesn't work because the rectangle is not the height of a domino.

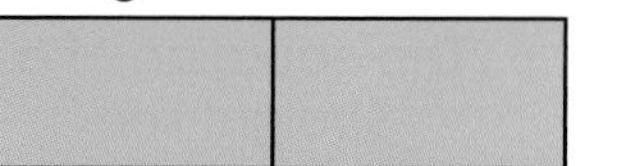

✗

a How many different rectangles that are the height of a domino can Jilly make if she has
 i 3 dominoes **ii** 4 dominoes **iii** 5 dominoes?

b Describe the sequence you have generated.

c How can you explain the pattern?

Reviewing skills

1 Write down the first five terms of these sequences.
 a $n^2 + 5$ **b** $2n^2 - 1$ **c** $n^3 + 3$ **d** $n(n - 1)$

2 Find the position-to-term formulae for these sequences.
 a 3, 6, 11, 18, 27, … **b** −2, 1, 6, 13, 22, … **c** 2, 9, 28, 65, 126, … **d** 3, 12, 27, 48, 75, …

3 Look at these tiling patterns.

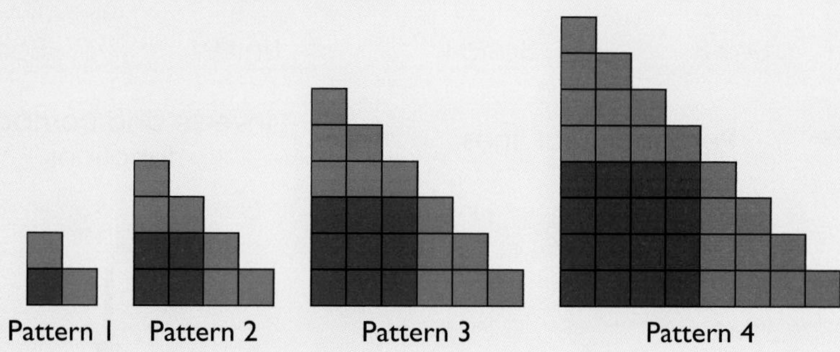

Pattern 1 Pattern 2 Pattern 3 Pattern 4

a Copy and complete the table.

Pattern number	1	2	3	4	5
Number of blue squares					
Number of red squares					
Total number of squares					

b Write down an expression for the number of blue squares in pattern n.
Ben realises that the total number of red squares follows this pattern:

Number of red squares in pattern 1 = 1 × 2 = 2
Number of red squares in pattern 2 = 2 × 3 = 6
Number of red squares in pattern 3 = 3 × 4 = 12

c How many red squares are in pattern 10?

d Write down an expression for the number of red squares in pattern n.

e Ben says that it is impossible for a pattern to have 500 red squares.
Is Ben right? Give a reason for your answer.

f Hence find a rule for the total number of squares. What type of numbers are these?

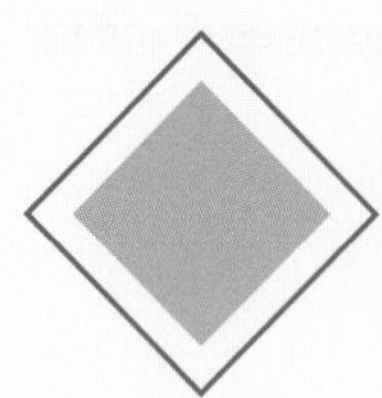

Algebra Strand 3 Functions and graphs

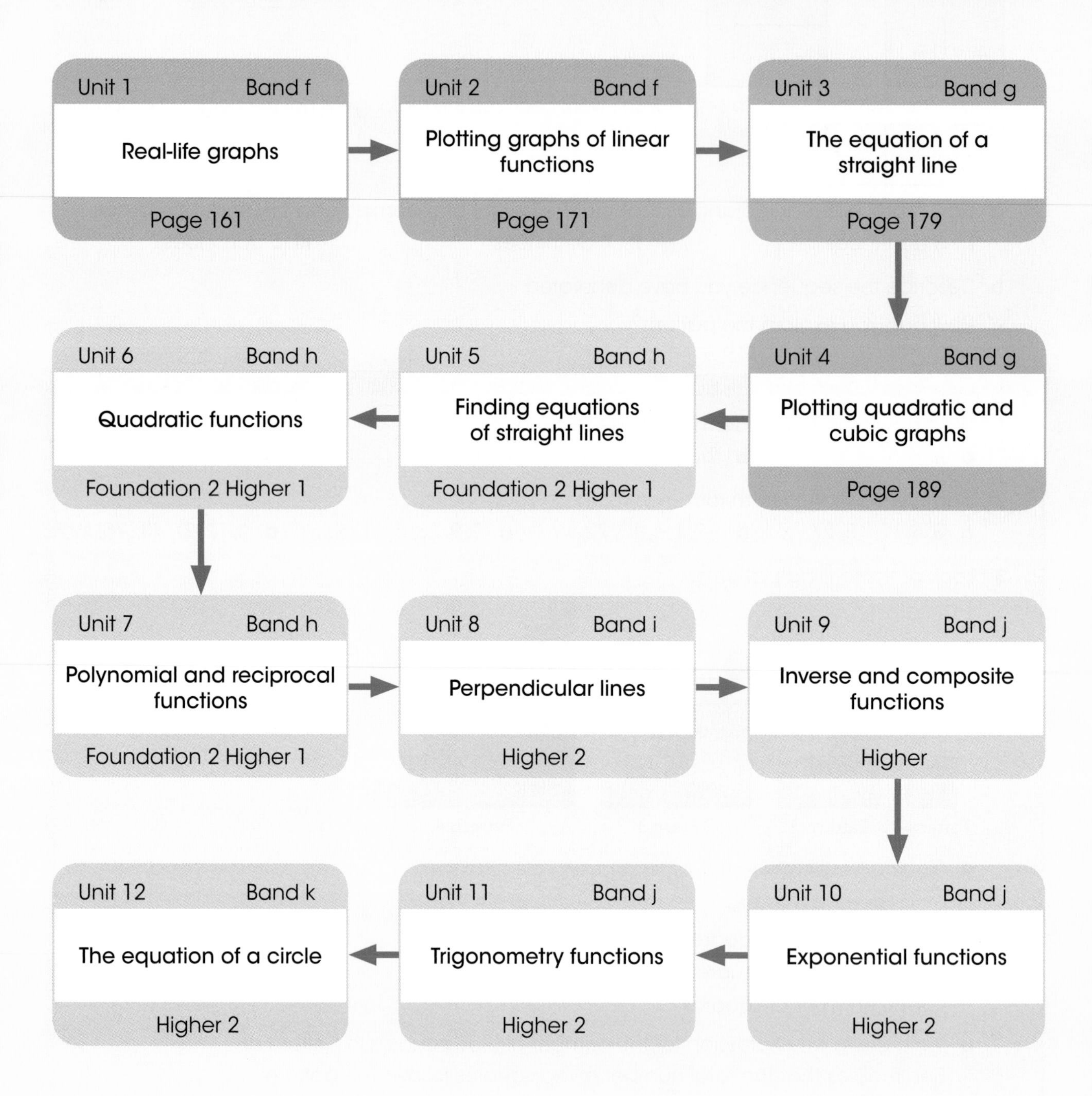

Unit 1 • Real-life graphs • Band f

Outside the Maths classroom

Setting prices

Can companies make more profit by reducing prices?

Toolbox

Graphs are diagrams used to represent data.

They show the relationship between (usually) two things, such as time and distance.

They consist of one or more straight lines or can be curved, depending on the situation.

Data is usually easier to interpret on a graph than in a table.

There are some crucial pieces of information you must use to read the story that a graph tells.

- The title explains what the story is about.
- The labels on the axes tell you what the points represent.
- The shape of the graph tells you how the relationship changes.

Reading graphs sometimes allows you to make predictions about what will happen in any given circumstance.

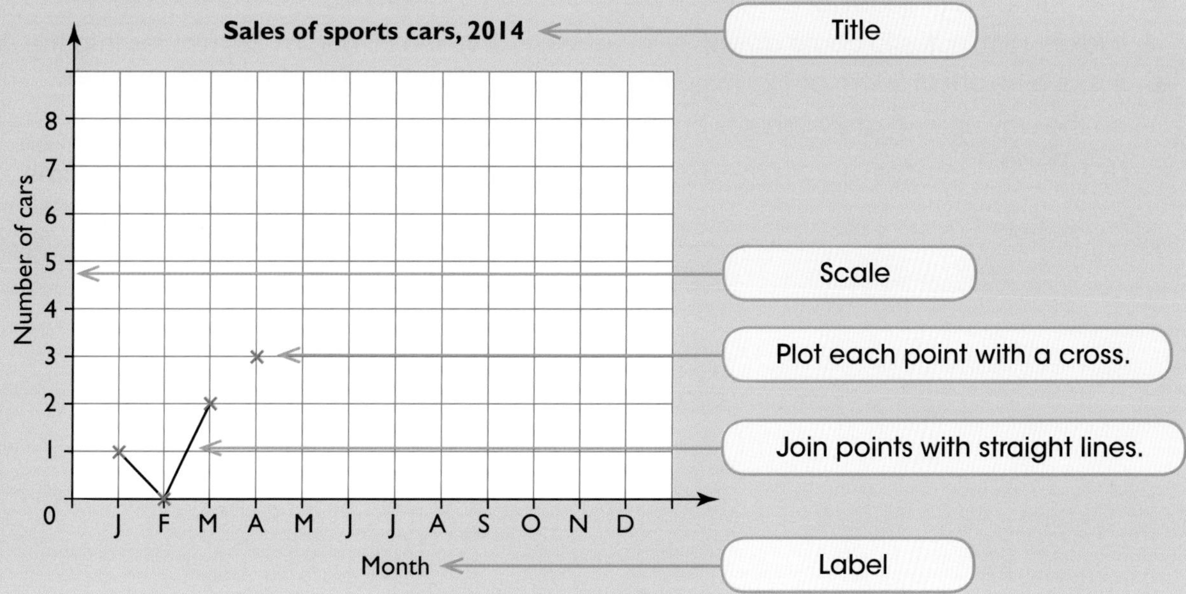

Example – Reading distance–time graphs

Look at this graph showing the height of a hot air balloon.
It is called a travel graph or a distance–time graph.

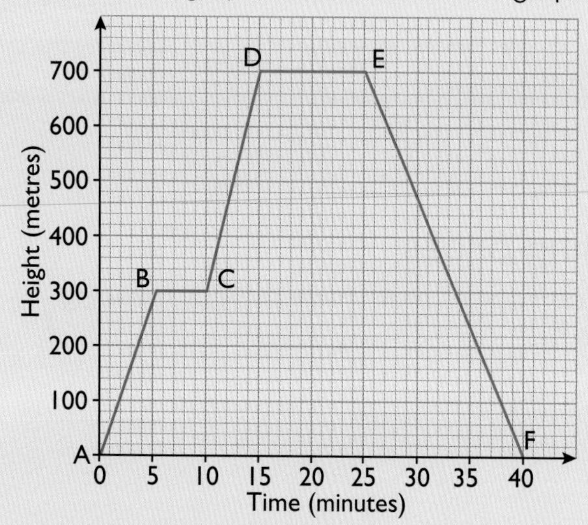

a How high does the balloon go?
b What is shown by the lines BC and DE?
c For how long is the balloon at a height of 300 m?
d How long does the balloon take to land?
e When is the balloon at a height of 500 m?

Solution

a The balloon reaches a maximum height of 700 m.

The maximum height is the largest value on the vertical axis that the graph reaches.

b The lines BC and DE represent times when the balloon remains at the same height.

The lines BC and DE are horizontal lines.

c The balloon is at 300 m for 5 minutes.

given by the length of line BC

d It takes 15 minutes to come down.

The line EF begins 25 minutes into the flight and finishes at 40 minutes.

e It is at a height of 500 m at $12\frac{1}{2}$ minutes on the way up and at 29 minutes on the way down.

Draw a line horizontally from 500 m on the vertical axis and read off the time values of the point where it crosses the graph.

Example – Understanding the shape of graphs

The diagram shows the shape of Jane's bath.

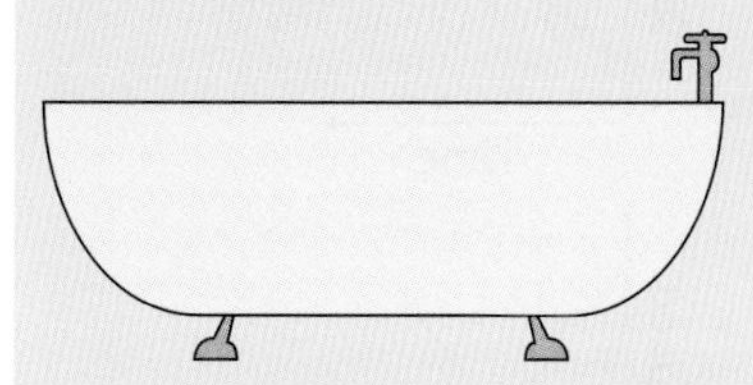

Water runs in at a constant rate.

Which sketch graph best shows the relationship between the depth of the water and time?

Justify your answer.

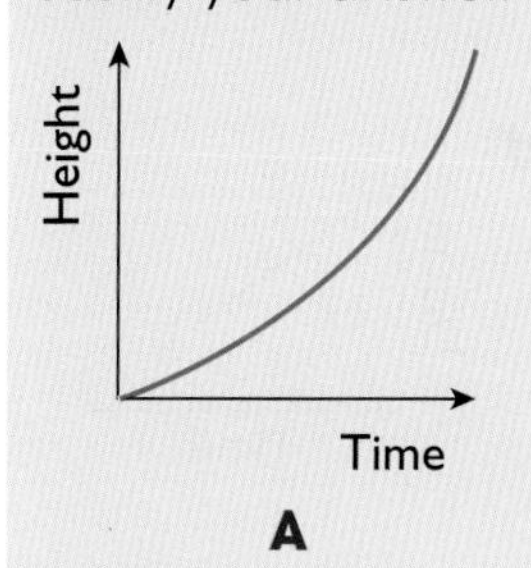

A

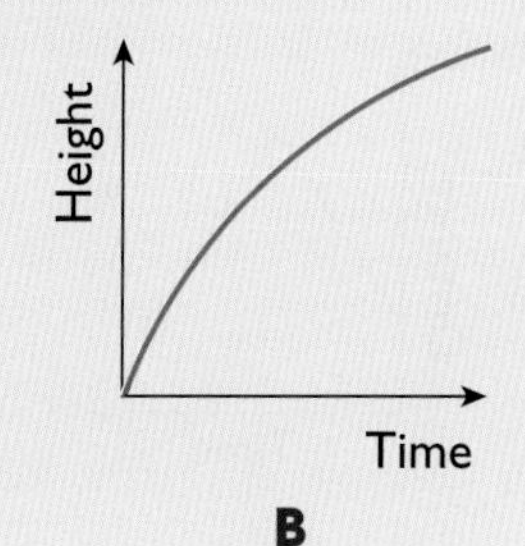

B

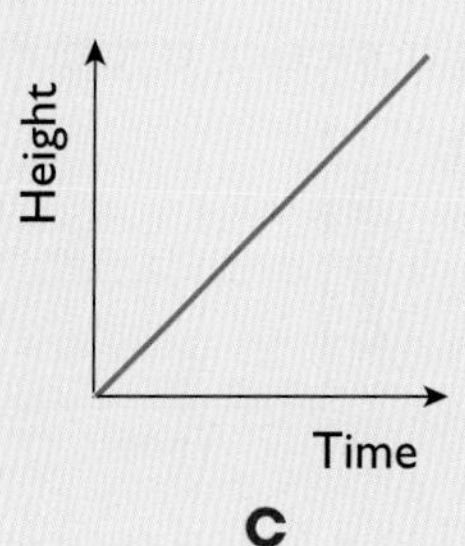

C

Solution

Graph **B** is correct.

Initially the cross-section of the bath is small and the bath fills up more quickly.

As time elapses the cross-sectional area increases thus slowing down the rate of increase in depth.

Practising skills

Exam-style

Extension

1 Each of these situations can be represented by a graph.

1. The temperature of a cup of tea left on a desk for a long period of time.
2. The distance travelled by a car travelling at a constant speed.
3. The population of an ant colony that doubles in size every week.
4. The number of litres of fuel left in a car's fuel tank when it travels at a constant speed.
5. The conversion between two currencies.
6. The height of a ball dropped from a window.
7. The value of a typical house over a long time period.
8. The weight of someone who goes on a diet.

A **B** **C** **D** **E**

Match the statement to the appropriate graph, stating your reasons.

2 The graphs below show the wind speed over one day.

a Speed / Time

b Speed / Time

c Speed / Time

d Speed / Time

e Speed / Time

f Speed / Time

Describe the information about the wind speed given by each graph.

3 The graphs show the distance of a robot from its base.

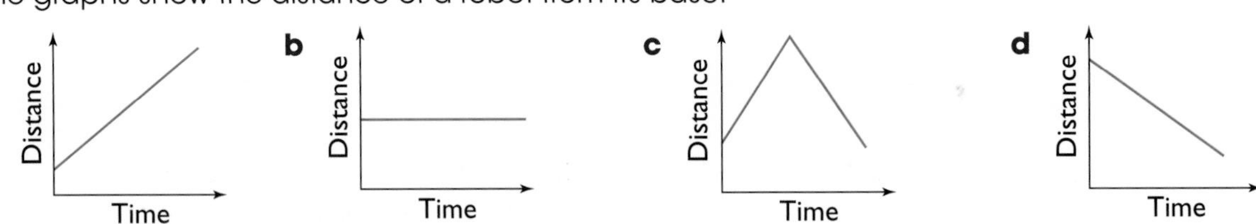

Match a statement **i** to **v** to the graphs **a** to **d**.

One statement is left. Draw the graph that matches this statement.

i The robot is travelling towards its base.

ii The robot is travelling away from its base.

iii The robot is always the same distance from its base.

iv The robot is travelling towards its base and then away from its base.

v The robot is travelling away from its base and then towards its base.

Developing fluency

Exam-style

1 The graph shows the height of a tree, in metres, each year for 10 years after planting.

Use the graph to answer these questions.

a What was the height of the tree when it was planted?

b What was the height of the tree after

i 4 years **ii** 7 years?

c Roughly in what year after planting was the tree

i 3 m **ii** 6 m tall?

d A tree for this species grows 70 cm on average each year.

On a copy of this graph, draw the line to show the average tree growth.

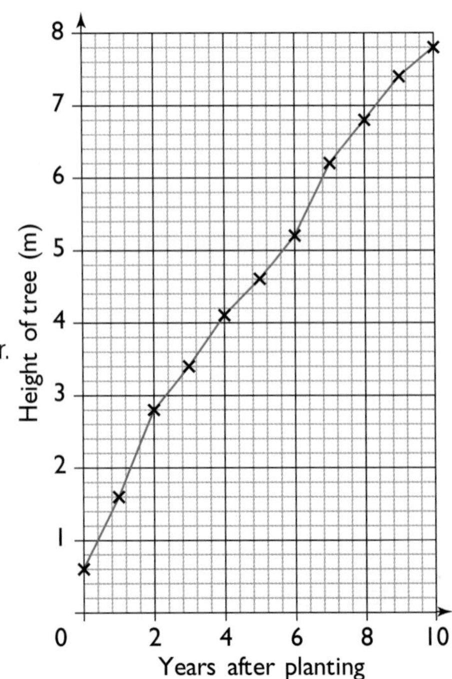

Exam-style

2 The graph can be used to convert between pounds and euros.
Use the graph to answer these questions.

a Convert £80 to euros.

b Convert 50 euros to pounds.

c **i** Explain how to use the graph to convert £150 to euros.

ii Convert £150 to euros.

d Convert 450 euros to pounds.

e Which is worth more, £80 or 90 euros?

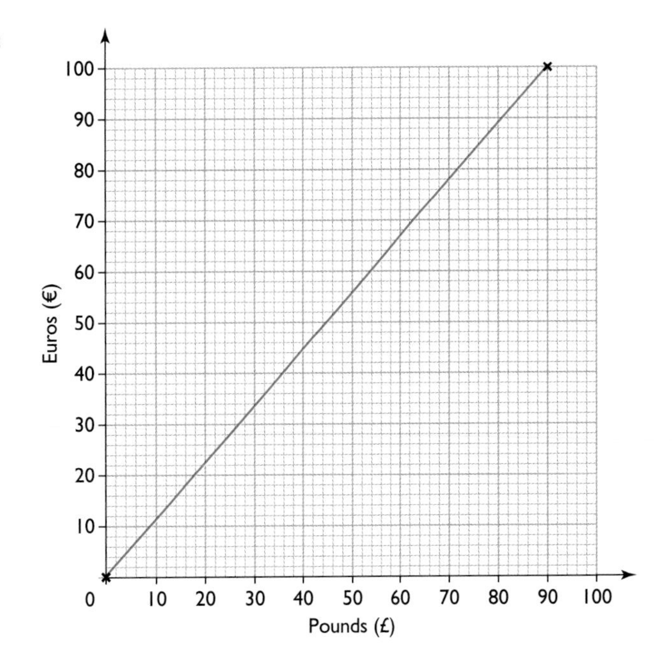

3 Brian's cycle journey is shown in the graph on the right.
Use the graph to answer these questions.

a **i** How many breaks does he have?

ii When are they and how long do his breaks last?

b How far is he from home at

i 10:00

ii 12:30

iii 15:48?

c At what times is he 80 km from home?

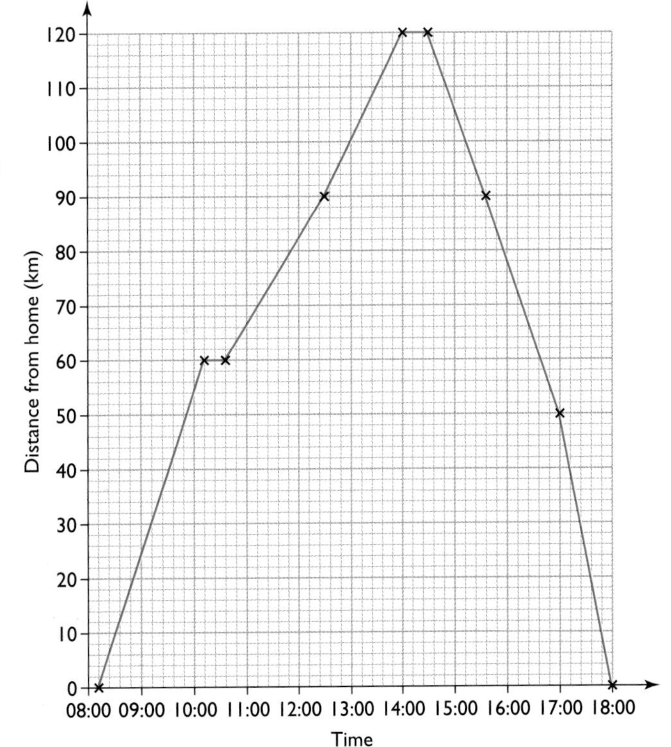

Exam-style

4 The conversion rate for roubles (RUB) and pounds (£) is £1 = 58 RUB.

a Draw a conversion graph with pounds on the horizontal axis from 0 to 100, and roubles on the vertical axis from 0 to 6000.

b Use your graph to convert

i 450 RUB to £

ii £84 to roubles

iii 1500 RUB to £

iv £420 to roubles.

Exam-style

5 The graph shows the journey of a train from Avonford.
Use the graph to answer these questions.

a At what time did the train stop at a station and for how long?

b How far had the train travelled at 10:30?

c There is a level-crossing 70 km from Avonford. When did the train go over it?

d Another train is travelling towards Avonford. At 10:00 it is 90 km from Avonford.

It arrived at Avonford at 11:26, travelling at a constant speed without stopping.

i On a copy of the graph, draw the journey of this train.

ii The two trains travelled along the same route, in opposite directions.

When did the two trains pass each other?

Where did they pass each other?

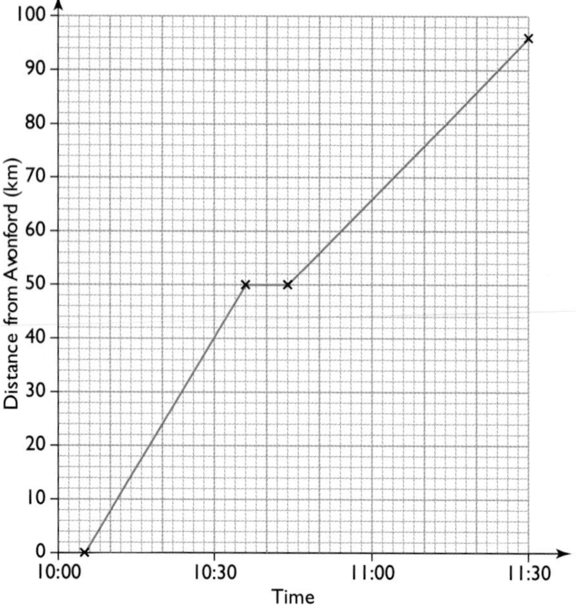

6 Here is a conversion graph for pounds and Swedish kronor (SEK).
Use the graph to answer these questions.

a Convert £50 to SEK.

b Convert 900 SEK to pounds.

c **i** Explain how to convert £250 to SEK.

ii Convert £250 to SEK.

d Convert 6600 SEK to pounds.

e John says, 'The graph shows that £2 has the same value as 20 SEK. So £1 has the same value as 10 SEK.'

Criticise this statement. How many SEK do you think £1 is worth?

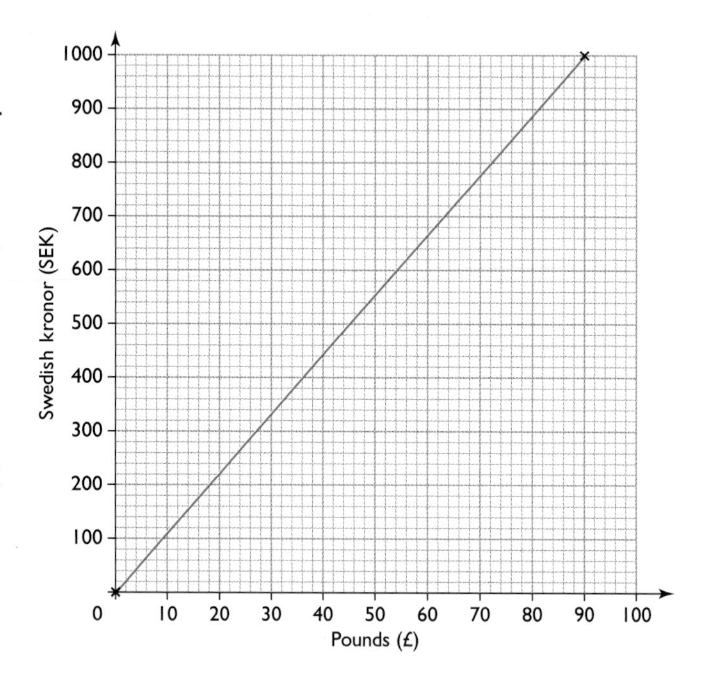

7 Grace goes on a cycling trip.

The graph shows how far she is from the start, with the hours after starting.

Use the graph to answer these questions.

a How far has she gone after 1.5 hours?

b After how many hours is she 50 km from the start?

c When does she stop and for how long?

d She starts at 09:45.

i At what time does she stop?

ii At what time is she 80 km from the start?

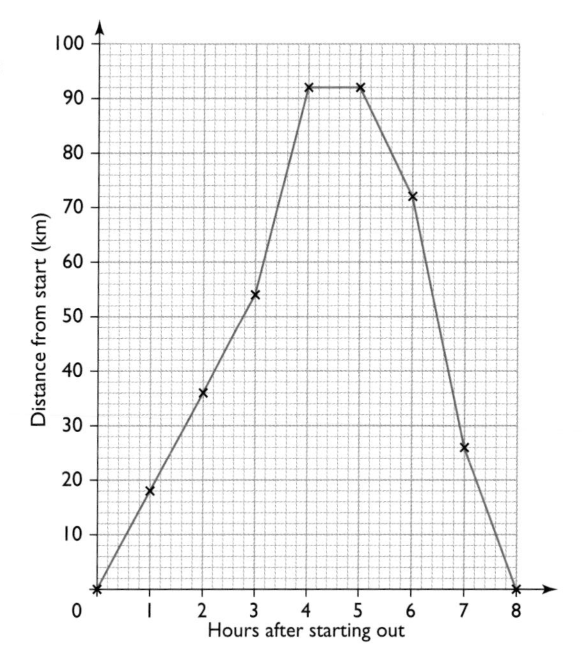

Exam-style

8 Jim and Henry run in a 10 km road race.

The graph shows Jim's progress.

Henry runs for the first 20 minutes at a steady 18 km per hour.

He then runs the rest of the race at a steady speed of 16 km per hour.

He starts at the same time as Jim.

a Copy and complete the graph to show Henry's progress.

b Out of Jim and Henry, who is leading after 5 km? Who finishes first?

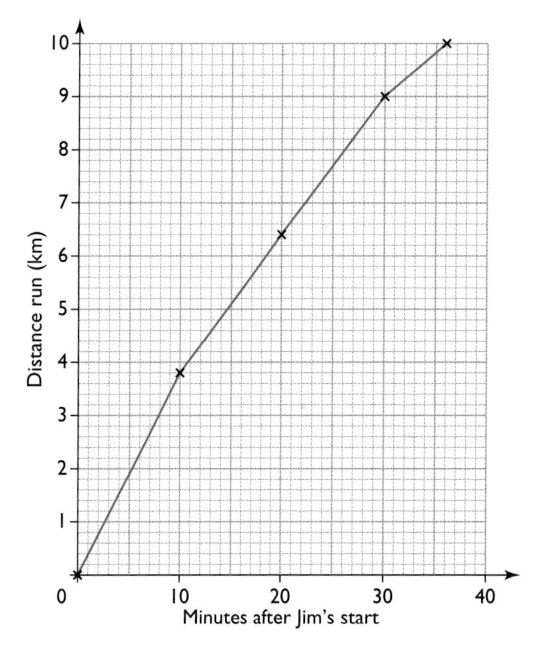

Problem solving

Exam-style

1 Here is a travel graph of June's car journey from her home to her father's house.

June stopped at 11:00 for a rest.

a How long did June stop for?

b Find the distance from June's home to her father's house.

c Work out the average speed of the car on the journey from June's home to her father's house.

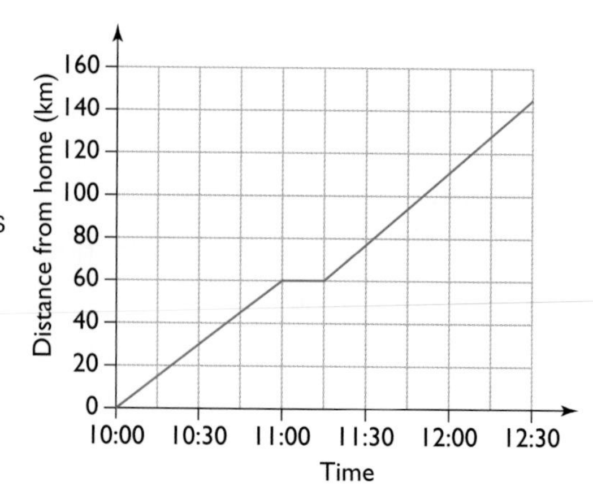

Exam-style

2 The graph shows the cost of hiring a cement mixer.

a How much does it cost to hire a cement mixer for

i 5 days **ii** 3 days?

b Fill in the missing numbers in this statement;

'There is a fixed cost of £☐ and then a charge of £☐ per day.'

c Keith budgets £105 for hiring a cement mixer for building an extension.

For how many days can he hire a cement mixer?

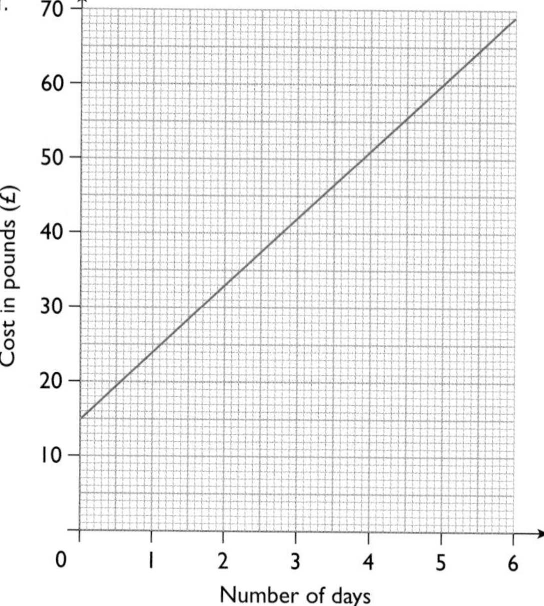

Exam-style

Extension

3 Here are some containers. They can all contain 1 litre of liquid.

Liquid is poured into these containers at a constant rate.

A

B

C

D

E

These graphs show the height of the liquid in the containers as they are filled.

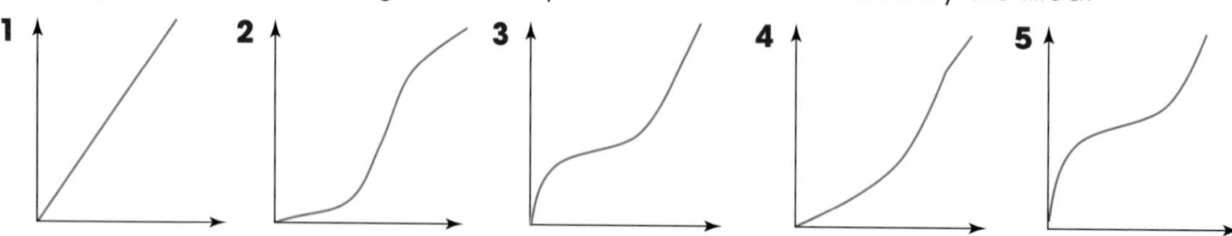

Match the graphs to the containers.

Exam-style

4 Here is a table that can be used to change pounds (£) to euros (€).

Pounds (£)	1	2	3	4	5	6	7	8	9	10	100
Euros (€)	1.21	2.42	3.63	4.84	6.05	7.26	8.47	9.68	10.89	12.10	121.00

a Change £12 to euros (€).

b Change €60.50 to pounds (£).

c Change £345 to euros (€).

d Change €546.92 to pounds (£).

Exam-style

5 Noah hires a water pump from the Pumps are Us company.

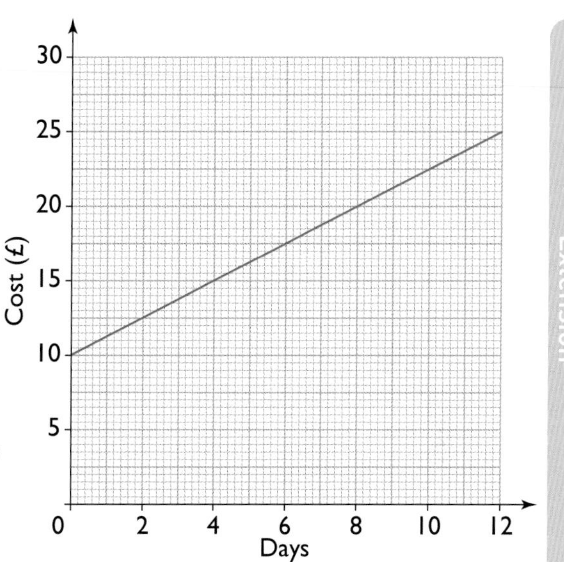

a Find the cost of hiring the water pump for 5 days from Pumps are Us.

b The cost of hiring water pumps from Pumps are Us is £10 plus a daily rate.
Work out the daily rate.

c Noah wants to compare the cost of hiring a water pump from Pumps are Us and from Equipment Hire who charge £5 for each day of hire, but have no fixed charge.
On a copy of the graph, add a line showing the cost of hiring a pump from Equipment Hire.

d Which of the two companies is the cheapest to hire the pump from for 5 days?

Extension

Exam-style

6 Niall uses this conversion graph to change between gallons and litres.

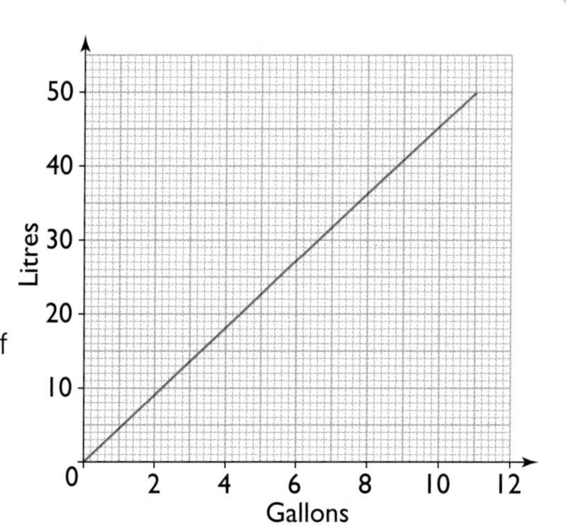

a Change

i 7 gallons to litres

ii 24 litres to gallons.

Niall sells oil for central heating systems.

He stores the oil in two tanks.

There are 3000 **litres** of oil in tank P and 950 **gallons** of oil in tank Q.

Niall delivers 4800 litres of the oil to a customer.

He takes all the 3000 litres of oil from tank P and takes the rest of the oil from tank Q.

b How many gallons of oil are left in tank Q?

Reviewing skills

1. The graph can be used to convert between stones and kilograms.

   Use the graph to answer these questions.

   a Millie weighs 8 stones.
   What is her weight in kilograms?

   b A large dog weighs 30 kilograms.
   What is its weight in stones?

   c A large boulder weighs 25 stones.
   Use the graph to covert 25 stones to kilograms.

   d A trampoline can take a maximum weight of 100 kilograms.
   Mr Smith weighs 15 stones. Can he safely go on the trampoline?
   Show all your working.

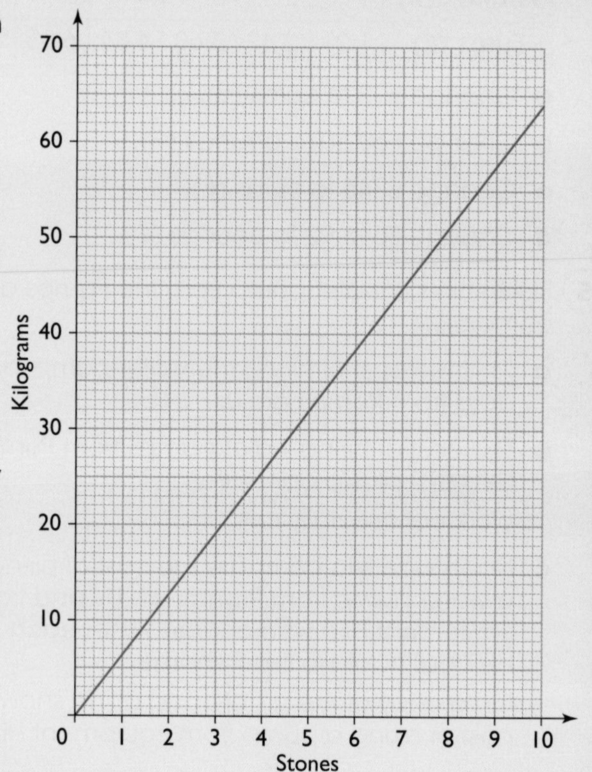

2. Here are some containers. They are all full of liquid.
   Each container has a hole in its base and water pours out at a constant rate.

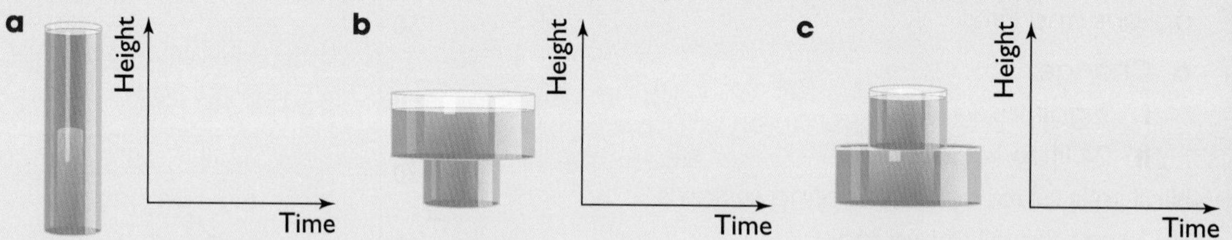

   Sketch a graph for each container to show the height of the liquid as the container empties.

3. Millie walks from her house to the sports centre to meet a friend. She then walks to her friend's house and then returns home on the bus.

   The travel graph shows Millie's journey.

   a How far is the sports centre from Millie's house?

   b How long does Millie stay at the sports centre?

   c What time does Millie arrive at her friend's house?

   d Does Millie walk faster on her way to the sports centre or on her way to her friend's house?
   Explain how you know.

   e What is Millie's average speed for her journey back home?

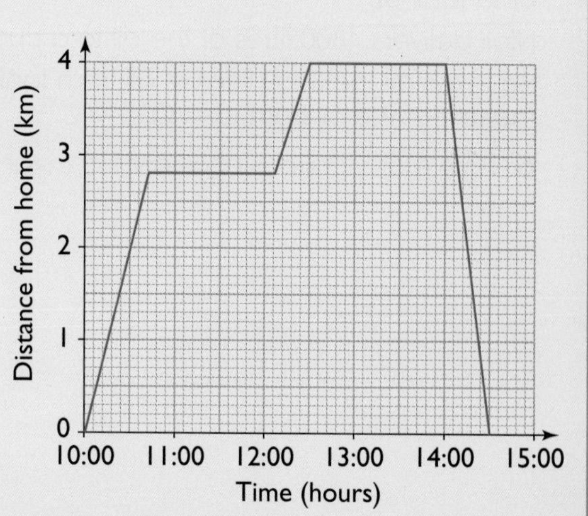

Unit 2 • Plotting graphs of linear functions • Band f

Outside the Maths classroom

Scientific research

Why do scientists try to work with straight line relationships?

Toolbox

Functions can be represented on a mapping diagram.

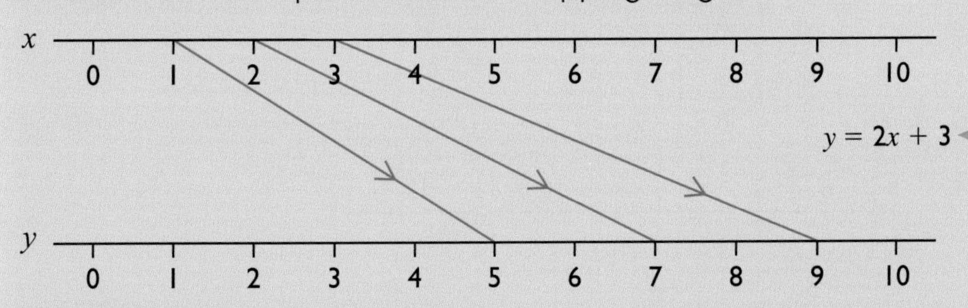

This is sometimes written $x \rightarrow 2x + 3$

Every value on the x line maps to a corresponding value on the y line.
Inverse mappings return you to where you started.
Another way to represent a function is to plot its graph.

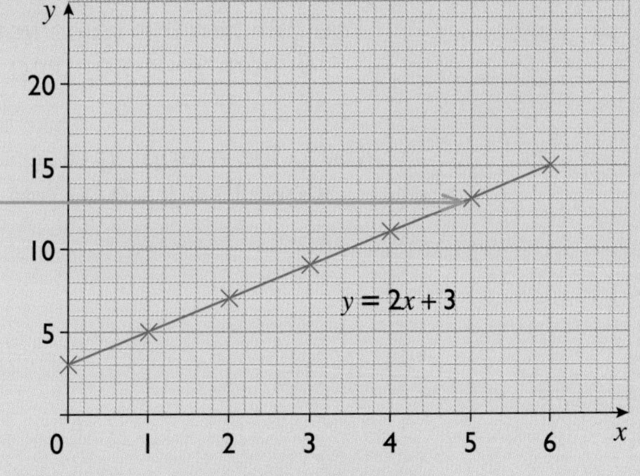

A linear function is represented by a straight line graph.
Conversion graphs are an example of linear functions.

Start by drawing a table of values.
This table is for the function $y = 2x + 3$.

x	1	2	3	4	5	6
$2x$	2	4	6	8	10	12
$+ 3$	3	3	3	3	3	3
$y = 2x + 3$	5	7	9	11	13	15

Example – Drawing and using a mapping diagram

1 For the function $x \to x + 3$:

a make a table of values with x from 0 to 6

b draw a mapping diagram

c write down the value of y when $x = 4$

d find the value of x when $y = 5$.

Solution

a

x	0	1	2	3	4	5	6
$+3$	3	3	3	3	3	3	3
$y = x + 3$	3	4	5	6	7	8	9

b

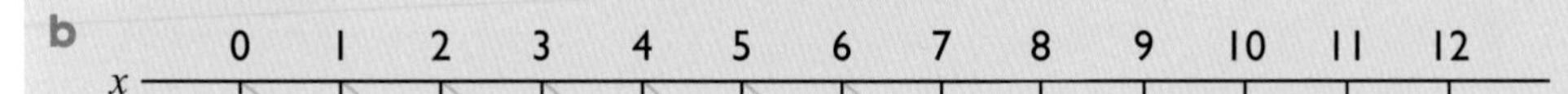

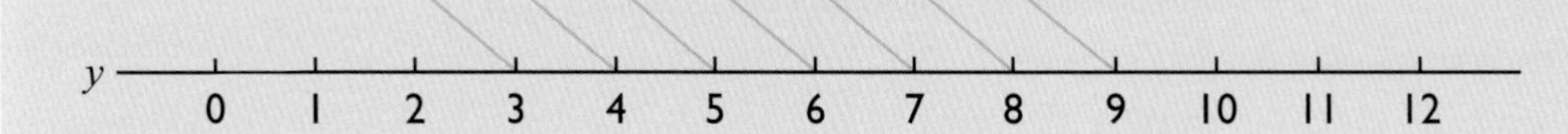

c When $x = 4$, $y = 7$. ← You can find this value from your table or from your mapping diagram.

d When $y = 5$, $x = 2$. ← Using the mapping diagram in reverse shows that $y = 5$ comes from $x = 2$.

Example – Drawing and using a graph of a linear function

In a science experiment, the length of a piece of elastic is recorded when different masses are hung on it.
Here are the results.

Mass (grams), m	0	10	20	30	40
Length of elastic (cm), l	30	40	50	60	70

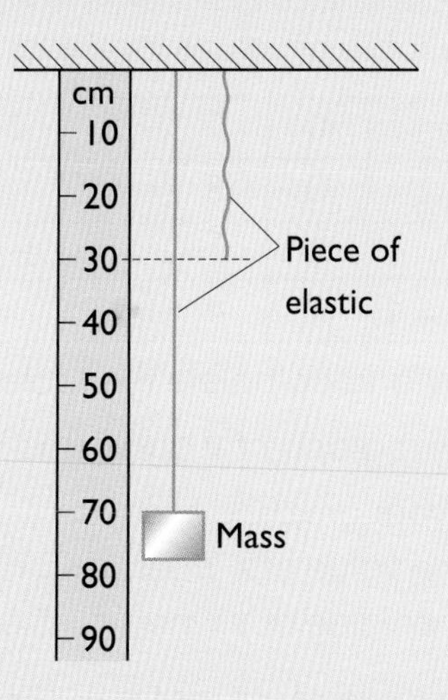

a Plot the graph.
b Find the mass that stretched the elastic to 56 cm.

Solution

a

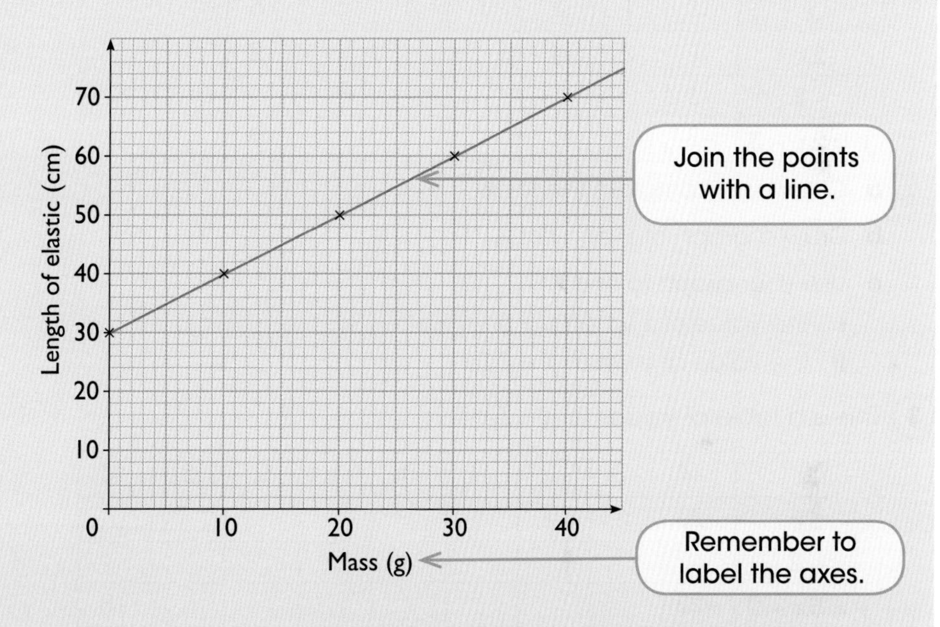

b The mass would have been 26 g. Draw a horizontal line from 56 cm on the vertical axis. Read off the value on the horizontal axis of the point where this line meets the graph line.

Practising skills

1 This is a table of values for $y = 3x + 1$.

x	0	1	2	3	4	5	6
$3x$	0	3	6				18
$+ 1$	1	1					1
$y = 3x + 1$	1	4					19

a Copy and complete the table.

b Draw a graph of y against x.

c Use your graph to find

i the value of x when $y = 11.5$

ii the value of y when $x = 5.5$.

2 This is a table of values for $y = 2x - 3$.

x	0	1	2	3	4	5	6
$2x$	0	2	4				12
$- 3$	−3						
$y = 2x - 3$	−3						9

a Copy and complete the table.

b Draw a graph of y against x.

c Use your graph to find

i the value of x when $y = 5$

ii the value of y when $x = 2.5$.

3 This is a table of values for $y = 5x - 2$.

x	−4	−3	−2	−1	0	1	2	3	4
$5x$	−20								
$- 2$	−2								
$y = 5x - 2$	−22								18

a Copy and complete the table.

b Draw a graph of y against x.

c Use your graph to find

i the value of x when $y = -10$

ii the value of y when $x = 3.5$.

(4) This is a table of values for $y = 4x - 3$.

x	0	1	2	3	4	5	6
$4x$	0	4	8				
-3	−3						
$y = 4x - 3$	−3						

a Copy and complete the table.

b Draw a graph of y against x.

c Use your graph to find

i the value of x when $y = 15$

ii the value of y when $x = 5.5$.

(5) This is a table of values for $y = 2x + 5$.

x	0	1	2	3	4	5	6
$2x$	0						
$+5$	5						
$y = 2x + 5$	5						

a Complete the table.

b Draw a graph of y against x.

c Use your graph to find

i the value of x when $y = 16$

ii the value of y when $x = 3.5$.

Developing fluency

(1) The rule for cooking meat is 40 minutes per kilogram plus 20 minutes.

a Copy and complete the table to show the time to cook some meat.

Weight (kg)	1	2	3	4	5	6	7	8
$40W$	40							
$+ 20$	20							
Time ($T = 40W + 20$)	60							

b Draw a graph to show the time to cook the meat.
Put W on the horizontal axis and T on the vertical axis.

c **i** Where does the line cross the T axis?

ii What does this number represent?

d Use your graph to find the time to cook meat weighing 6.5 kg.

e Use your graph to find the weight of meat that takes 3 hours 20 minutes to cook.

Exam-style

2 Boats are hired for £3 per hour plus £4 fixed charge.

a Complete the table to show the cost £C of hiring a boat for t hours.

Number of hours, t	1	2	3	4	5	6	7	8
$3t$								
+ 4								
$C = 3t + 4$								

b Draw a graph to show the cost of hiring a boat. Put the cost on the vertical axis.

c Use your graph to find the cost of hiring a boat for 4.5 hours.

d Use your graph to find how many hours a boat can be hired for £11.50.

3 Apples are sold for £6 for 5 kg.

a Complete this table.

Weight of apples (kg)	5	10	15	20
Cost (£)	6	12		

b Draw a graph to show the cost of the apples. Put the cost on the vertical axis.

c Use your graph to find the cost of 8 kg.

d Use your graph to find how many packs can be bought for £21.

Exam-style

4 A shop charges £0.80 per kg for potatoes. As a promotion, they give anyone buying potatoes 1 kg free.

a Complete this table to show the cost of some weights of potatoes.

Weights of potatoes (kg)	1	2	3	4	5	6
Cost (£)	0	0.80			3.20	

b Draw a graph to show the cost of potatoes. Put the cost on the vertical axis.

c Use your graph to find the cost of 4.5 kg.

d Use your graph to find the weight of potatoes you get for £2.00.

e **i** Where does the line cross the vertical axis?

ii What does this number represent?

5 **a** Construct a table of values for $y = 4x - 2$ with x from −2 to 5.

b Use your table of values to draw the graph of $y = 4x - 2$.

c Use your graph to find the value of x when $y = 0$.

6 **a** Construct a table of values for $y = 8 - x$ with x from −2 to 10.

b Use your table of values to draw the graph of $y = 8 - x$.

c Use your graph to find the value of y when $x = -1$.

d Use your graph to find the value of x when $y = -1$.

7 **a** Construct a table of values for $y = 12 - 2x$ with x from −2 to 8.

b Use your table of values to draw the graph of $y = 12 - 2x$.

c Use your graph to find the value of x when $y = 15$.

Problem solving

Exam-style

1 The cost, C, in pounds, of n business cards is $20 + 0.05n$.

a Copy and complete this table of values.

n	0	200	400	600
20		20		
0.05n		10		
$C = 20 + 0.05n$		30		

b Draw a graph of C against n. Plot C on the vertical axis.

c Sally has £120 to spend on buying business cards.
How many business cards can she buy?

Extension

Exam-style

2 The thickness, T mm, of a hard-cover book is given by the formula

$$T = 5 + \frac{n}{20}$$

where n is the number of pages.

a Construct a table of values of T for $n = 0, 200, 400$ and 600.

b Draw a line graph from $n = 0$ to $n = 600$ on the horizontal axis to show this information.

c Sonia has a hard-cover book that is 40 mm thick.
How many pages are there in the book?

Extension

Exam-style

3 **a** Construct a table of values for $y = 2x - 1$, taking x from -3 to 3.

b Draw the graph of $y = 2x - 1$.

c Find the value of y when $x = 1.7$.

d Find the value of x when $y = -4.5$.

e Explain why the point $(-1.5, 2)$ does not lie on the line $y = 2x - 1$.

Extension

Reviewing skills

1. **a** Copy and complete this table of values for $y = 2x - 5$.

x	5	4	3	2	1	0	-1	-2
$2x$	10					0	-2	
-5	-5	-5					-5	
y	5						-7	

   b Draw a graph of $y = 2x - 5$.

   c Use your graph to find the value of

   i y when $x = 2.5$

   ii x when $y = -8$.

2. **a** Construct a table of values for $y = 3x + 5$ with x from -3 to 5.

   b Use your table of values to draw the graph of $y = 3x + 5$.

   c Use your graph to find the value of x when $y = 4$.

3. A taxi company makes a fixed charge of £4 and then a certain amount per minute. This gives the formula $C = 4 + 0.5m$, when £C is the cost of a journey in minutes.

   a What is the cost per minute?

   b Construct a table of values of C for journeys of up to half an hour.

   c Draw a graph of C against m. Plot C on the vertical axis.

   d Use your graph to find

   i the cost of a journey of 18 minutes

   ii the time taken when the cost is £11.

Unit 3 • The equation of a straight line • Band g

Outside the Maths classroom

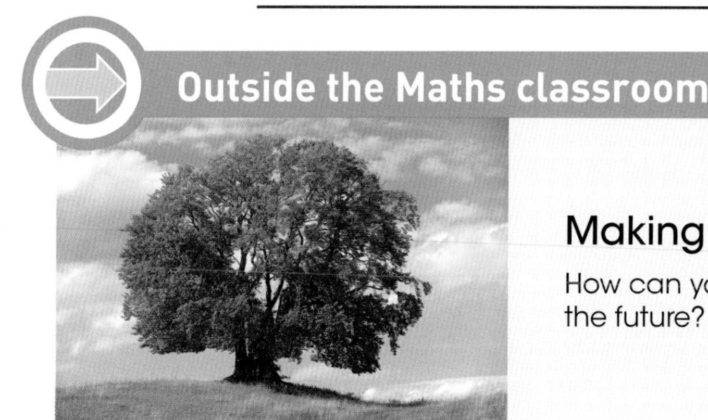

Making predictions

How can you use current data to predict the future?

Toolbox

Straight lines have an equation of the form $y = mx + c$.

- The value of m represents the gradient or 'steepness' of the line.
- The value of c tells you where the line crosses the y axis.

Special cases are:

- Horizontal lines which have an equation of the form $y = a$ where a is a number.
- Vertical lines which have an equation of the form $x = b$ where b is a number.

Parallel lines have the same gradient.

The y intercept is the value where the line cuts the y axis.

To find the gradient of a straight line

1. choose any two points
2. subtract the y co-ordinates
3. subtract the x co-ordinates
4. the gradient is the change in y divided by the change in x.

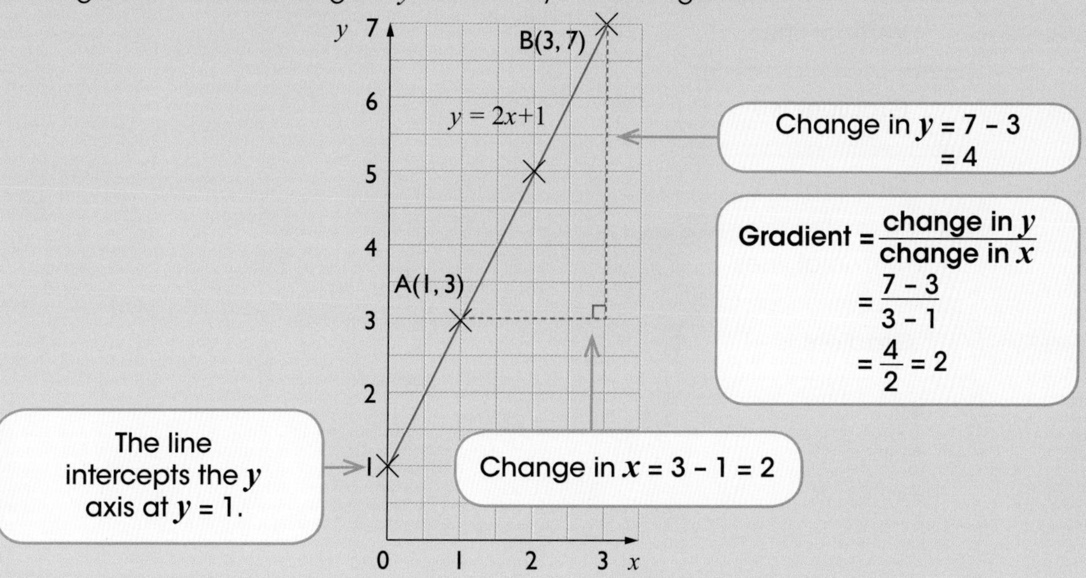

If the line slopes down from left to right the gradient is negative.

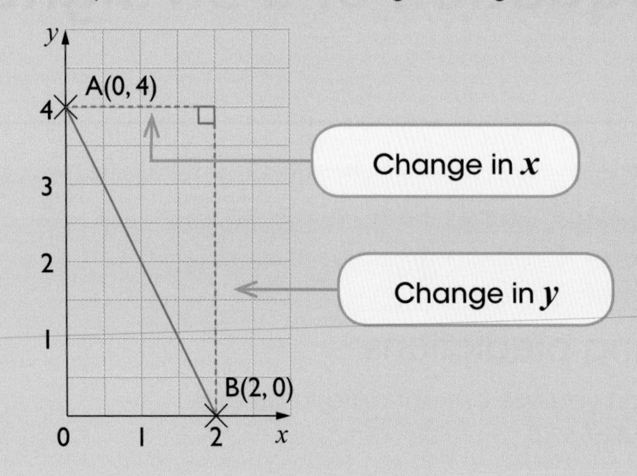

$$\text{Gradient} = \frac{0-4}{2-0} = \frac{-4}{2} = -2$$

Example – Finding the gradient from the equation of a line

Here are the equations of ten lines.
Write down which ones are parallel.

a $x = 7$
b $y = 3 + 7x$
c $y = 2x + 3$
d $y = -5$
e $y = 2x - 4$
f $x = -2$
g $y = 5 - x$
h $y = -x$
i $y = 7x + 5$
j $y = 4$

Solution

a and **f** ← Vertical lines

b and **i** ← Gradient is 7

c and **e** ← Gradient is 2

d and **j** ← Horizontal lines

g and **h** ← Gradient is −1

Example – Finding an equation from a graph

Match the lines in this diagram with the correct equations.

$x = 3$
$x = -3$
$x = 0$
$y = 3$
$y = -3$
$y = 0$
$y = x + 1$
$y = x - 2$
$y = 2x - 1$
$y = 1 - x$

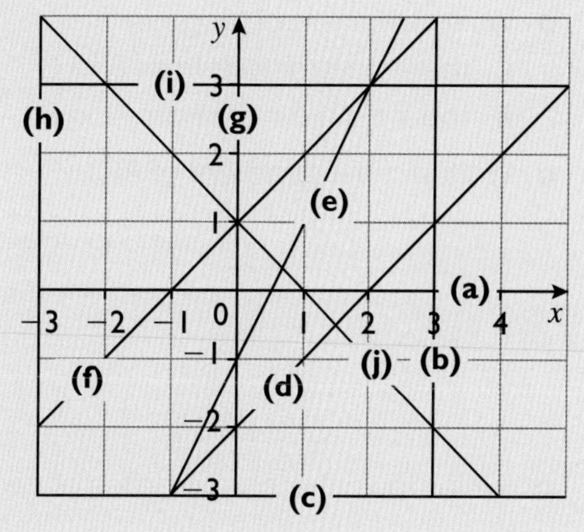

Solution

a is the x axis. Its equation is $y = 0$

b is a vertical line. Every point on the line has an x co-ordinate of 3. The equation of the line is $x = 3$

c is a horizontal line. Every point on the line has a y co-ordinate of -3. The equation of the line is $y = -3$

d has a gradient of 1 and a y intercept of -2. The equation of the line is $y = x - 2$

e has a gradient of 2 and a y intercept of -1. The equation of the line is $y = 2x - 1$

f has a gradient of 1 and a y intercept of 1. The equation of the line is $y = x + 1$

g is the y axis. Its equation is $x = 0$

h is a vertical line. Every point on the line has an x co-ordinate of -3. The equation of the line is $x = -3$

i is a horizontal line. Every point on the line has a y co-ordinate of 3. The equation of the line is $y = 3$

j has a gradient of -1 and a y intercept of 1. The equation of the line is $y = -x + 1$ or $y = 1 - x$

Practising skills

1 Look at this graph.

a Which points lie on these lines?
 i $x = 2$ **ii** $y = 8$

b Write down the equation of these lines.
 i BD **ii** CD

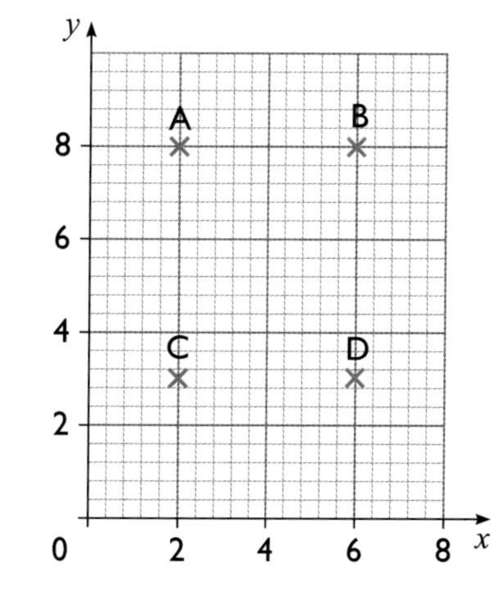

2 Look at this graph.

a Write down the co-ordinates of

i A **ii** B **iii** C.

b Write down the change in

i the y co-ordinate from B to C

ii the x co-ordinate from A to B.

c Work out the gradient of AC.

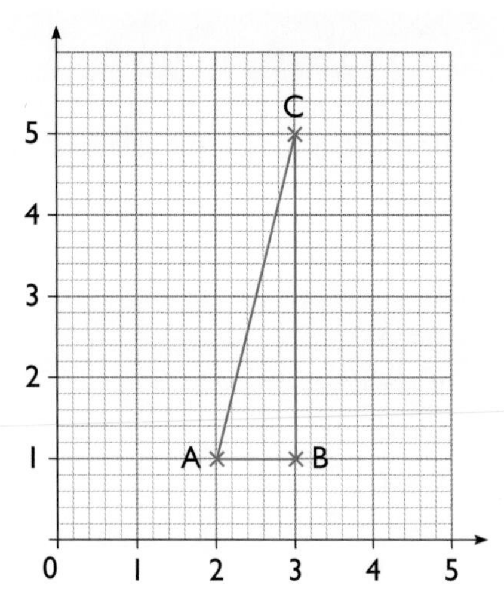

3 **a** Write down the co-ordinates of two points on each of these lines.

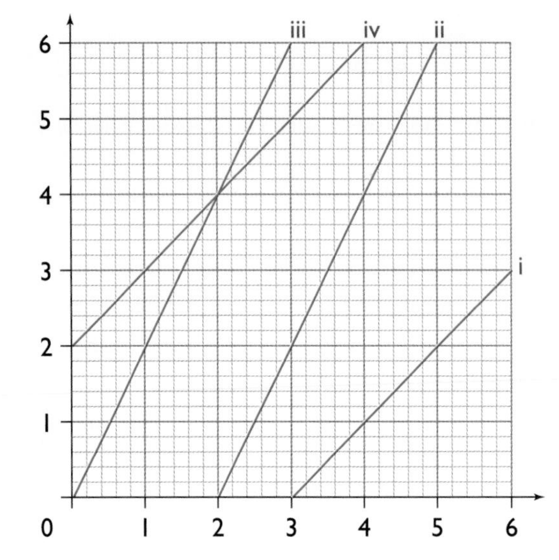

b Work out the gradient of each line.

c Which pairs of lines are parallel? What can you say about the gradients of parallel lines?

4 Work out the gradient and the equation for each of the lines going through these points.

a (1, 2) and (3, 8) **b** (5, 6) and (6, 7) **c** (3, 0) and (5, 0) **d** (–1, –2) and (1, 6)

e (0, 2) and (2, 0) **f** (0, 5) and (3, –1) **g** (0, –3) and (5, –18) **h** (0, 8) and (2, –2)

5 **a** Write down the co-ordinates of two points on each of these lines.

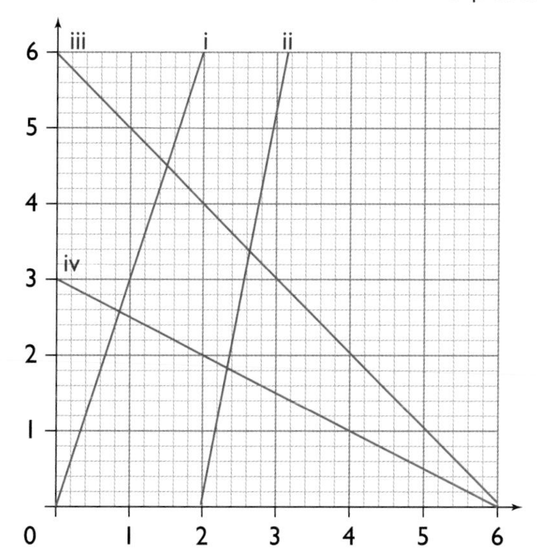

b Work out the gradient of each line.

c What can you say about the gradients of lines that slope down from left to right?

6 **a** Find the co-ordinates of two points on the line $y = 2x - 3$.

b Draw the graph of $y = 2x - 3$ taking values of x from −1 to 4.

c Find

i the gradient of the line

ii the y intercept (the value of y where it crosses the y axis).

d Where can you see the answers to part **c** in the equation of the line?

Reasoning

7 **a** Copy and complete these tables.

x	−2	−1	0	1	2
$4x$	−8				
− 3	−3				
$y = 4x - 3$	−11				

x	−2	−1	0	1	2
$5x$	−10				
+ 2	+2				
$y = 5x + 2$	−8				

b Draw these lines on the same graph.

c Work out the gradient of each line.

d Write down the y intercept of each line.

e Explain why you did not need to draw the graphs of the two lines to answer parts **c** and **d**.

f Write down the gradient and the y intercept of the line $y = 8x - 5$.

8 Look at the five lines on this graph.

a For each line

i write down two points

ii work out the gradient

iii write down the y intercept.

b Match each line with one of these equations.

i $y = 2x - 2$

ii $y = -x + 5$

iii $y = -2x + 8$

iv $y = x + 3$

v $y = 1$ ← This is $y = 0x + 1$

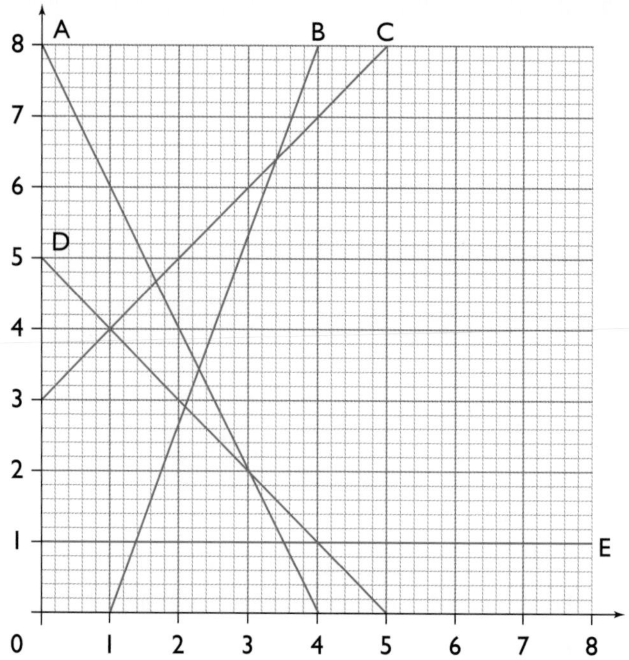

Developing fluency

1 Here are three road signs showing the gradient of three hills.

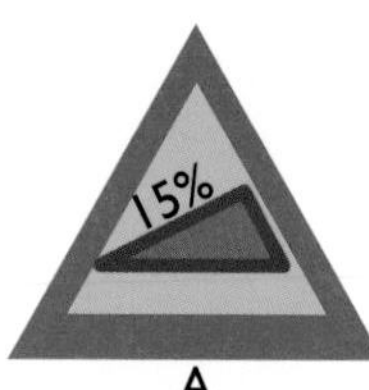

A

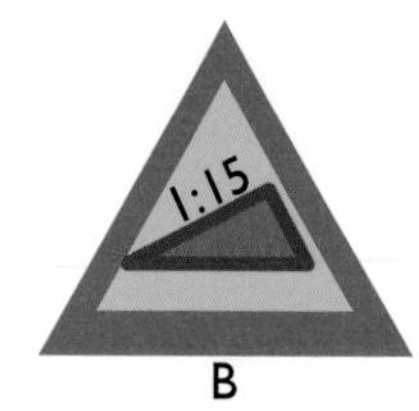

B

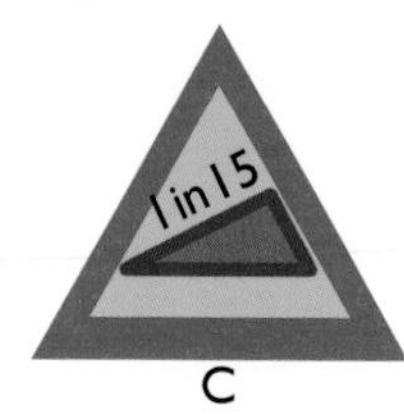

C

Arrange the gradients of these hills in order of size.
Start with the smallest gradient.

2 Here are the equations of some lines.

A $y = 2x + 7$

B $y = -2x + 7$

C $y = 7x$

D $y = 7x + 2$

E $y = -7x - 2$

F $y = 2x - 7$

Which of these line(s) has the property

i goes through the origin

ii parallel to $y = 2x$

iii has y intercept of +7

iv parallel to $y = -7x + 2$

v has y intercept of −2?

3 **a** Draw the lines with these equations on the same axes.

A $y = x + 5$

B $y = 3x - 5$

C $y = 5x + 3$

D $y = 3x$

E $y = x - 5$

F $y = 5x + 5$

b Which lines have the same gradient (are parallel)?

c Which lines have the same y intercept?

4 Find the equations of the lines going through these pairs of points.

a (0, 0) and (5, 15) **b** (0, 3) and (4, 11) **c** (0, 1) and (10, 41) **d** (0, −2) and (6, 4)

5 **a** Draw these lines on a sheet of graph paper.

i $x = 4$ **ii** $y = -3$

iii $x + y = 6$ **iv** $2x + y = 4$

v $y = -2x + 5$ **vi** $y = -x + 5$

vii $y - 3 = 0$ **viii** $x - 2 = 0$

b Which pairs of lines are parallel?

c How can you tell which pairs are parallel from their equations?

Exam-style

6 This graph shows the charges (£C) that ABCars make for hiring a car for m miles.

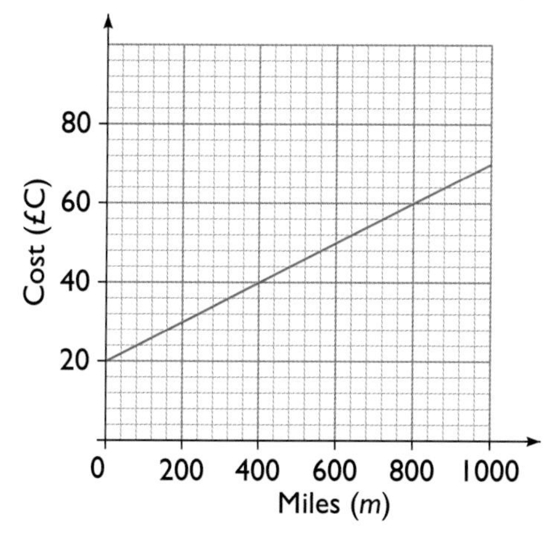

a Find the formula in the form $C =$ _______ .

b Work out the cost for 1200 miles.

c How many miles can you drive a hire car for a cost of £100?

Problem solving

Exam-style

1 Here is the graph that can be used to find the cost C, in pounds, of hiring a hedge-trimmer for d days.

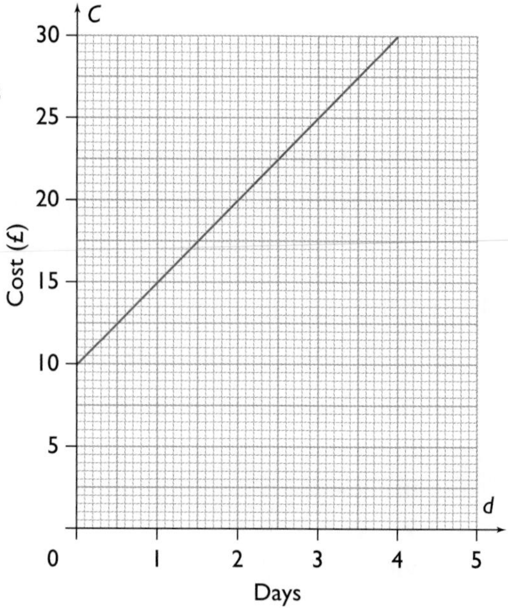

a Write down the equation of the straight line in terms of C and d.

b Write in words the cost C of hiring this hedge-trimmer for d days.

c **i** Explain what the gradient of the straight line represents.

ii Explain what the intercept with the vertical axis represents.

Exam-style

2 Here are the equations of some straight lines.

A $y = -x + 6$
B $y = 2x + 3$
C $y = 3 - 2x$
D $y = -2x + 6$
E $y = 2x - 6$

a Which of these lines are parallel to each other?

b Which of the lines meet at (0, 3)?

Extension

Exam-style

3 Ayesha is comparing the daily cost of van hire. She compares these three firms.

A £50 plus 10p a mile
B £25 plus 30p a mile
C £65 a day with no mileage charge

a Ayesha writes the cost, £C, for m miles for company **A** as the formula $C = 50 + 0.1m$
Write similar formulae for the other two companies.

b Draw a graph showing the three formulae.
Put m on the horizontal axis with values from 0 to 200.
Put C on the vertical axis.

c Ayesha thinks she will drive the van about 160 miles.
Which company should she choose?

Exam-style

4 Look at the straight line l on this graph.

a **i** Find the gradient of l.

ii Find the y intercept of l.

b Write down the equation of l.

c If the line was drawn on a bigger piece of graph paper, would it go through the point (8, 12)?

d Find the equation of the line parallel to l passing through the point (0, 1).

Extension

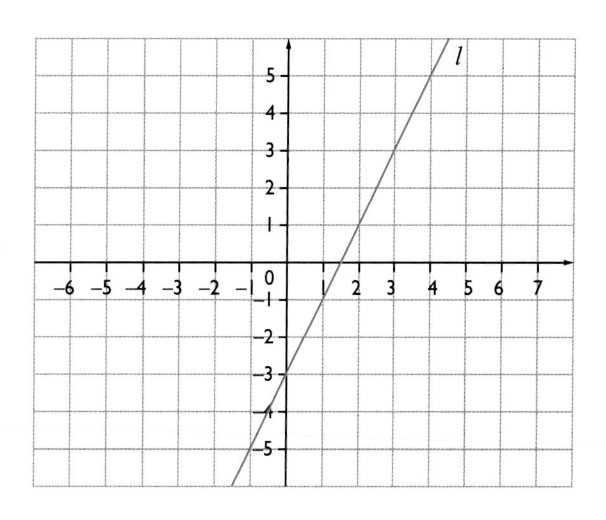

Reviewing skills

1 The table gives information about some lines.

Line	Gradient	y intercept
A	4	3
B	8	2
C	2	–2
D	–5	4
E	–1	–5
F	6	–3

Write down the equation of each line.

2 Write down the equation of each line.

a

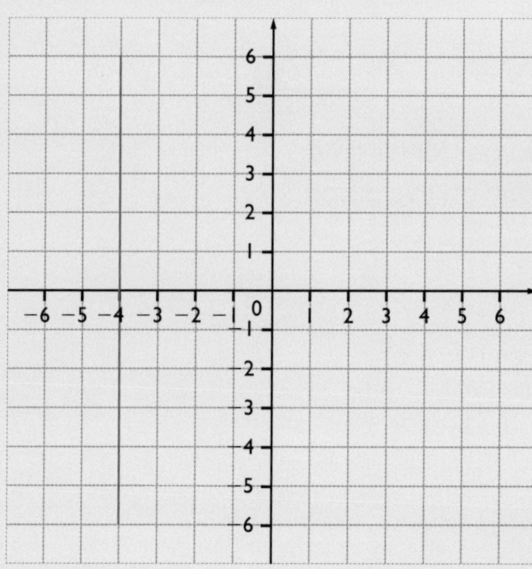

b

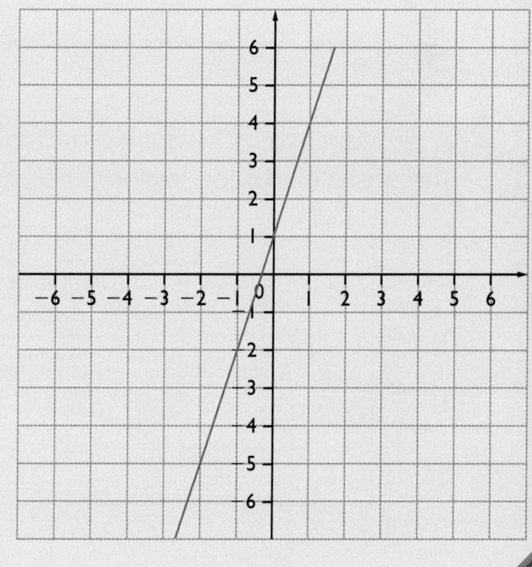

c

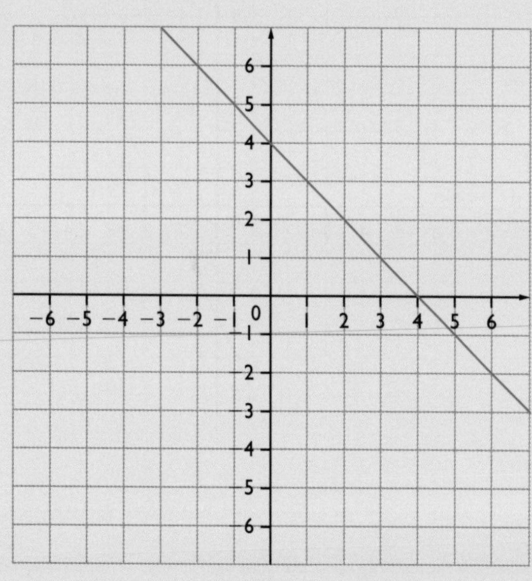

d

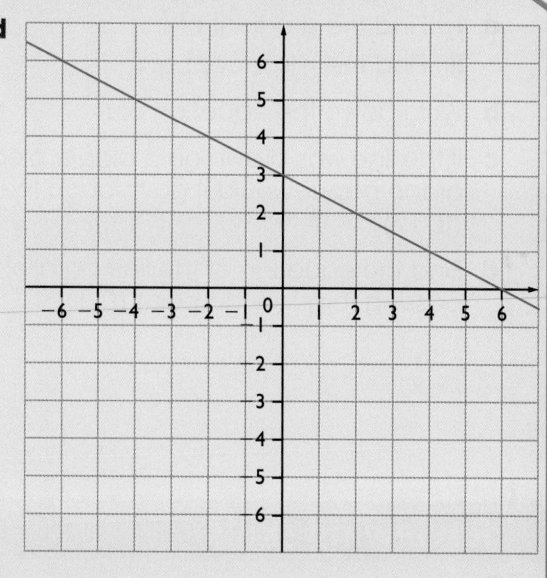

e

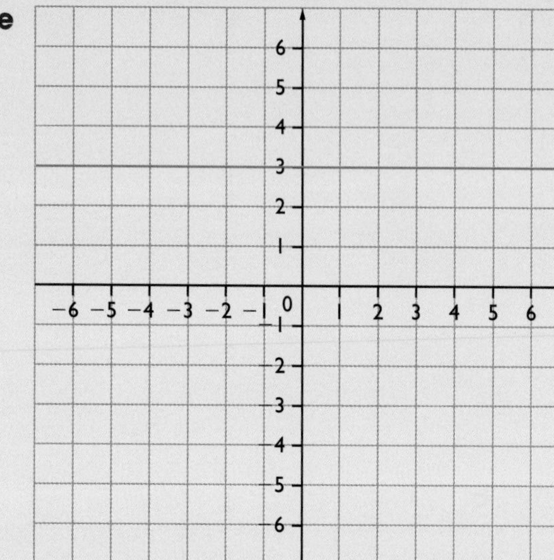

3 A plumber has a call out fee and then charges for his time.
The cost, £C, of a job lasting h hours is shown on this graph.

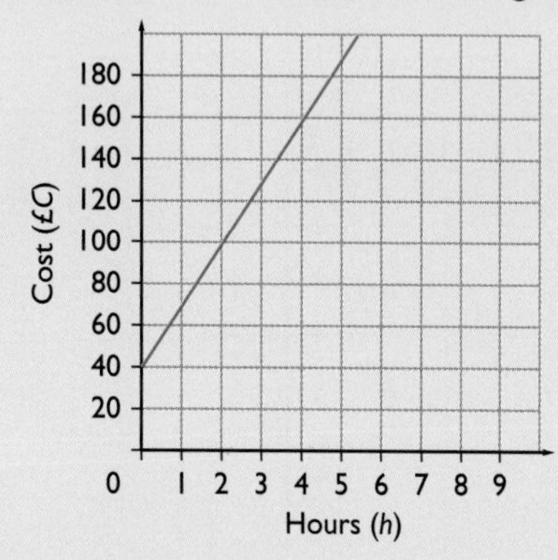

a Find the plumber's formula in the form $C = \square + \square h$

b **i** What is his call out fee?

ii How much does he charge per hour?

c For one job the bill is £250. How many hours did the plumber work?

Unit 4 • Plotting quadratic and cubic graphs • Band g

Outside the Maths classroom

Maximising profit

What would a graph of profit against selling price look like?

Toolbox

Equations which contain powers of x greater than 1 (e.g. x^2, x^3) are not linear functions; instead they produce curved lines.

Equations with x^2 terms but no x^3 terms produce **quadratic** curves.

x	-3	-2	-1	0	1	2	3
$y = x^2$	9	4	1	0	1	4	9

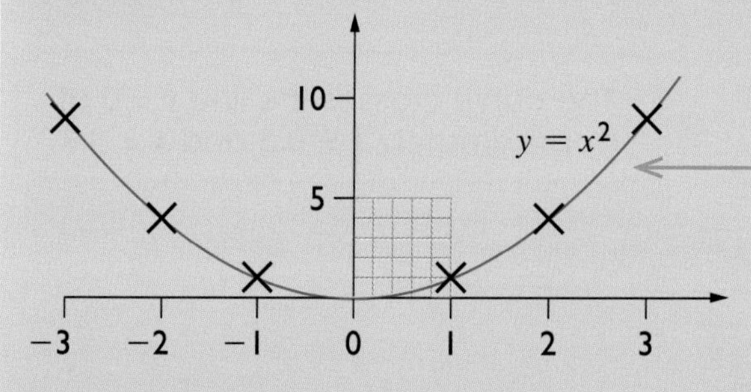

The points are joined by a smooth curve.

Equations with x^3 terms produce **cubic** curves.

x	-3	-2	-1	0	1	2	3
$y = x^3$	-27	-8	-1	0	1	8	27

The points are joined by a smooth curve.

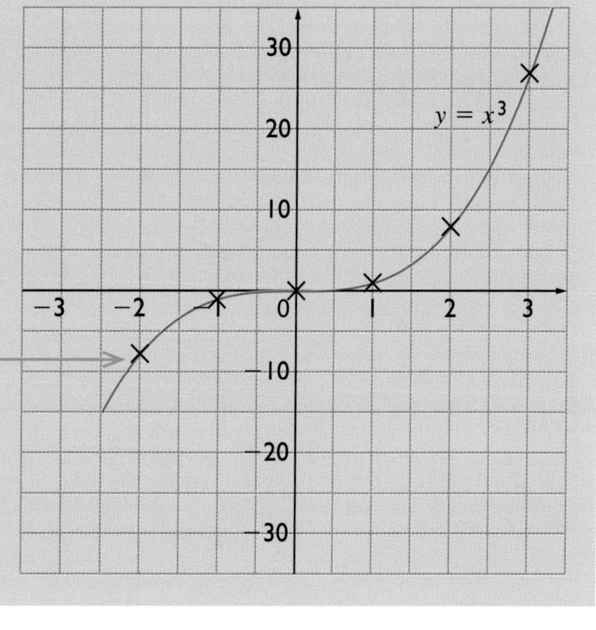

Example – Drawing and using quadratic curves

a Make a table of values for $y = x^2 - 4x + 1$, taking values of x from –1 to 5.

b Draw the graph of $y = x^2 - 4x + 1$.

c Use your graph to solve the equation $x^2 - 4x + 1 = 0$.

Solution

a

x	-1	0	1	2	3	4	5
x^2	1	0	1	4	9	16	25
$-4x$	4	0	-4	-8	-12	-16	-20
$+1$	1	1	1	1	1	1	1
$y = x^2 - 4x + 1$	6	1	-2	-3	-2	1	6

b

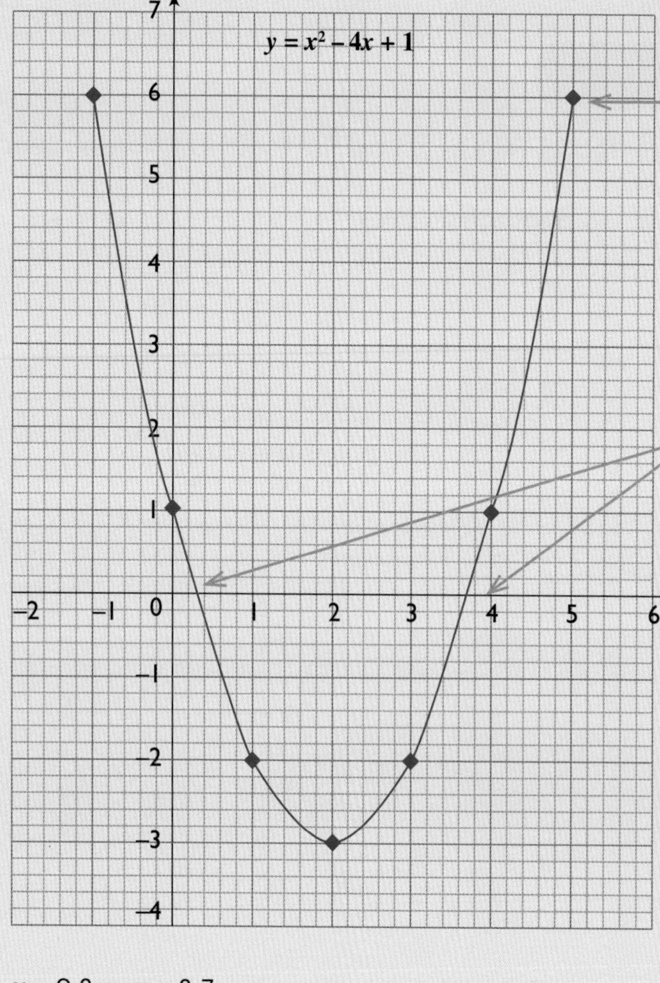

All quadratic equations produce a curve in the shape of a parabola. Draw it as smoothly as you can.

The curve crosses the line $y = 0$ at approximately $x = 0.3$ and $x = 3.7$.

The solution to the equation $x^2 - 4x + 1 = 0$ is given by the x values at the points where the curve crosses the x axis (the line $y = 0$).

Values taken from a graph are only approximate.

c $x = 0.3$ or $x = 3.7$

Example – Drawing and using cubic curves

a Make a table of values for $y = x^3 - 5x$, taking values of x from -3 to 3.

b Draw the graph of $y = x^3 - 5x$.

c Find the values of x where this curve crosses the x axis.

d Write down the solution of the equation $x^3 - 5x = 0$.

Solution

a

x	-3	-2	-1	0	1	2	3
x^3	-27	-8	-1	0	1	8	27
$-5x$	15	10	5	0	-5	-10	-15
$y = x^3 - 5x$	-12	2	4	0	-4	-2	12

b

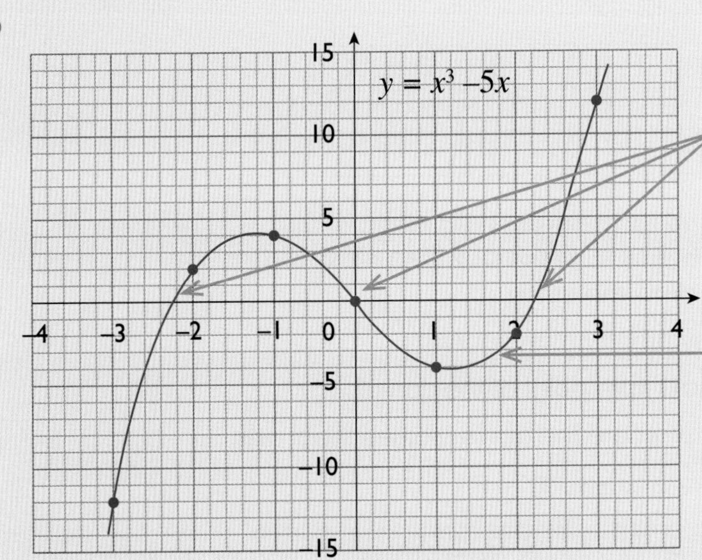

The curve crosses the line $y = 0$ at $x = 0$ and approximately $x = -2.2$ and $x = 2.2$.

All cubic equations produce an S-shaped curve.

Draw it as smoothly as you can.

c $x = 0, x = -2.2$ or $x = 2.2$

The value $x = 0$ is accurate and you could have found it from your table.

The other two values are only approximate.

Practising skills

1. **a** Copy and complete this table for the quadratic curve $y = x^2 + 1$.

x	−3	−2	−1	0	1	2	3
x^2	9						
+1	1						
$y = x^2 + 1$	10						

b Draw axes with values of x from −3 to 3 and values of y from 0 to 10. Plot the points from your table and join them with a smooth curve.

c What is the equation of the line of symmetry of the curve?

d What are the co-ordinates of the minimum point of the curve?

e Use your graph to find the values of x when $y = 2.5$.

2. **a** Copy and complete this table for $y = 12 - x^2$.

x	−4	−3	−2	−1	0	1	2	3	4
12	12								
$-x^2$	−16								
$y = 12 - x^2$	−4								

b Draw axes with values of x from −4 to 4 and values of y from −5 to 15. Plot the points from your table and draw the curve.

c What is the equation of the line of symmetry of the curve?

d What are the co-ordinates of the maximum point of the curve?

e Use your graph to find the values of x where the curve crosses the x axis.

3. **a** Copy and complete this table for $y = x^2 - 4x + 2$.

x	−1	0	1	2	3	4	5
x^2							
$-4x$							
+2							
$y = x^2 - 4x + 2$							

b Draw axes with values of x from −1 to 5 and values of y from −3 to 7. Draw the curve.

c What is the equation of the line of symmetry of the curve?

d What are the co-ordinates of the minimum point of the curve?

e What are the values of x when $y = -1$?

4 **a** Copy and complete this table for $y = (x - 2)^2$.

x	−2	−1	0	1	2	3	4	5	6
$x - 2$	−4								
$y = (x - 2)^2$	16								

b Draw axes with values of x from −2 to 6 and values of y from 0 to 16. Draw the curve.

c What is the equation of the line of symmetry?

d What are the co-ordinates of the minimum point of the curve?

e What are the values of x when $y = 7$?

5 **a** Copy and complete this table for $y = 2x^2 + 3$.

x	−4	−3	−2	−1	0	1	2	3	4
$2x^2$	32								
+ 3	+3								
$y = 2x^2 + 3$	35								

b Draw axes with values of x from −4 to 4 and values of y from 0 to 40. Draw the curve.

c Describe the main features of the curve.

6 **a** Copy and complete this table for $y = x^3 + 2$.

x	−3	−2	−1	0	1	2	3
x^3	−27						
+ 2	+2						
$y = x^3 + 2$	−25						

b Draw axes with values of x from −3 to 3 and values of y from −25 to 30. Draw the curve.

c Use your graph to find the co-ordinates of the points where the curve crosses the x axis and the y axis.

d Describe the symmetry of the graph.

Developing fluency

Reasoning

1 **a** Draw the graph of $y = x^2 - 4x + 3$, taking values of x from −1 to 5.

b Use your graph to solve the equation $x^2 - 4x + 3 = 0$.

c Use your graph to estimate the values of x when $y = 6$.

d Explain why it is not possible to find a value of x for which $y = -2$.

Exam-style

2 **a** Draw the graphs of $y = x^3 - 8$ and $y = -x^3 + 8$ on the same sheet of paper, taking values of x from −2 to 3.

b Use your graphs to solve the equation $x^3 - 8 = 0$.

c For what values of x is $x^3 - 8$ greater than $-x^3 + 8$?

d Describe the relationship between the two curves.

Extension

3 **a** Copy and complete this table of values for the curve $y = x^3 - 6x^2 + 11x - 6$.

x	0	1	2	3	4
x^3					
$-6x^2$					
$11x$					
-6					
$y = x^3 - 6x^2 + 11x - 6$					

b Draw the curve.

c Use your graph to solve the equation $x^3 - 6x^2 + 11x - 6 = 0$.

d For how many values of x is $y = 1$? Use your graph to find them to 1 decimal place.

e Find the values of x (again to 1 decimal place) when $y = 4$.

Exam-style

4 Mo stands on top of a cliff 60 m high. He throws a stone into the air.
The equation for the height, y metres, of the stone above the beach is $y = 60 + 20t - 5t^2$ where t is the time in seconds.

a Draw the graph of y against t. Take values of t from 0 to 6 and y from 0 to 100.

b What is the greatest height of the stone above the beach?

c Find how long the stone is in the air.

Extension

5 Use a co-ordinate grid with values of x from –3 to +3 with 2 cm intervals and values of y from –15 to 15 with 5 units every 2 cm.

a Draw the graphs of $y = (x - 1)^2 - 4$ and $y = 5x - x^3$.

b Find the values of x where the two graphs cross. Give your answers to 1 decimal place.

Reasoning

6 Alex stands at the edge of a 10 m high cliff and throws a pebble into the sea.
The height above sea level of the pebble at time t seconds is modelled by the equation $h = 10 + 8t - 5t^2$.

a Copy and complete this table of values.

t	0	0.5	1	1.5	2	2.5
10						
$+8t$						
$-5t^2$						
$h = 10 + 8t - 5t^2$						

b Draw the graph of h against t.

c What is the maximum height reached by the pebble?

d How long is the pebble above the height of the cliff?

e **i** When does the pebble hit the sea?

ii Does your graph have any meaning after this time?

Problem solving

1 Here are the graphs of $y = (x - 4)(x + 1)$ and $y = (x + 4)(1 - x)$:

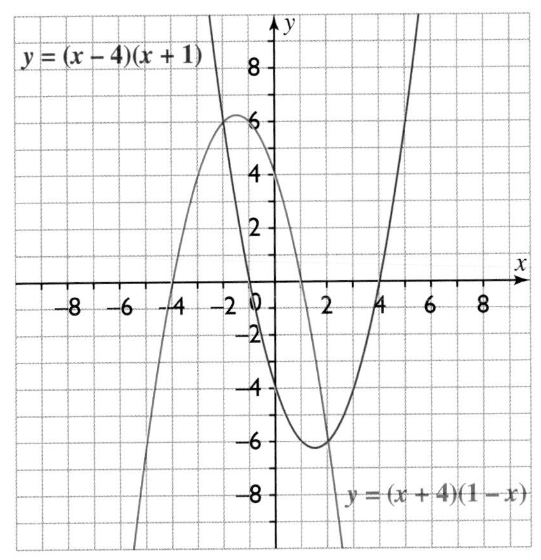

a For which values of x is $y = (x + 4)(1 - x)$ always positive?

b For which values of x is $y = (x - 4)(x + 1)$ always negative?

c Find the values of x when the blue line is below the red line.

d What is the minimum point of $y = (x - 4)(x + 1)$?

e What is the maximum point of $y = (x + 4)(1 - x)$?

2 Here are the graphs of $y = x^3 - 4x$ and $y = 4x - x^3$:

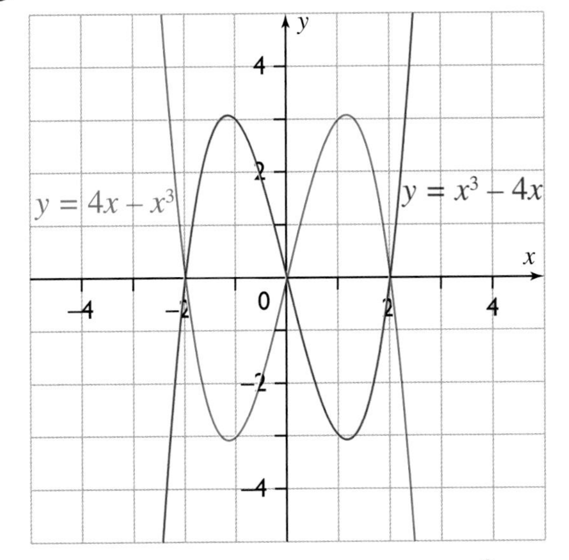

a For which values of x is $y = 4x - x^3$ positive?

b For which values of x is $y = x^3 - 4x$ negative?

c Find the values of x when $4x - x^3$ is greater than $x^3 - 4x$.

d Describe the relationship between the two curves.

e Describe the symmetry of $y = x^3 - 4x$.

3 Janey's company manufactures golf balls. Janey and Pete tested one of the balls by throwing it into the air from the top of the cliff. Pete made a video.

Back at work, they found the height of the ball at 2 second intervals.

Time (s) t	2	4	6	8
Height (m) h	40	40	0	−80

a Show that the path of the ball is consistent with the quadratic function $h = 30t - 5t^2$.

b Draw the graph of h against t, for values of t from 0 to 8.

c For how long was the ball more than 25 m above the cliff?

d The ball landed 8 seconds after Janey threw it in the air. How high was the cliff?

4 Narinder makes rectangular table mats. The length of each mat is x cm and its area is A cm^2. Each mat he makes has a perimeter of 80 cm.

a Show that $A = x(40 - x)$.

b Draw the graph of A against x for values of x from 0 to 40.

c Find the length of a table mat with area 375 cm^2.

d Explain how your graph shows you that the table mat with the greatest area is square-shaped.

5 José makes garden ornaments out of concrete. They stand on a base of radius x cm. The volume of concrete needed, V cm^3, is given by the formula

$V = 2x^2(10 - x)$.

a Draw a graph of V against x, taking values of x from 0 to 10.

b A base needs 200 cm^3 of concrete. Use your graph to estimate its radius.

c One day, José has an order for 5 ornaments of radius 6.2 cm. Use your graph to estimate how much concrete he needs.

6 An oil tanker carries a cargo of oil at a speed of v km per hour. The cost, £C, per hour of carrying the oil is given by the formula $C = 20v + \frac{6000}{v}$, for values of v between 5 and the maximum speed of 40.

a Make a table of values.

b Draw the graph of C against v.

c Use your graph to estimate the speed at which the value of C is the least.

Reviewing skills

1. **a** Copy and complete this table for $y = 5x - x^2$.

x	−1	0	1	2	3	4	5	6
$5x$	−5							
$-x^2$	−1							
$y = 5x - x^2$	−6							

b Draw axes with values of x from −1 to 6 and values of y from −8 to 8. Draw the curve.

c Describe the main features of the curve.

d Use your graph to solve the equation $5x - x^2 = 0$.

2. Fauzia is a junior rugby player. She is learning to take place kicks. The ball must go over the posts; the bar is 3 m high.

When the ball has travelled x m horizontally its height is y m.

For Fauzia's kicks $y = 0.7x - 0.02x^2$

a Make a table of values of y, taking x to be 0, 10, 15, 20, 25 and 35.

b Draw the graph of y against x.

c How far from the posts can Fauzia make a successful kick?

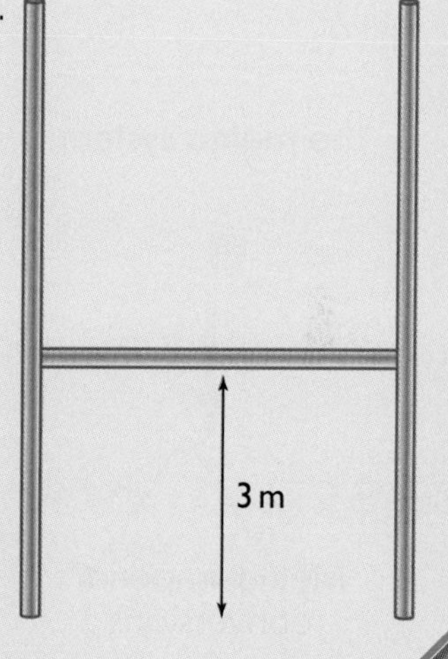

Geometry and Measures Strand 1 Units and scales

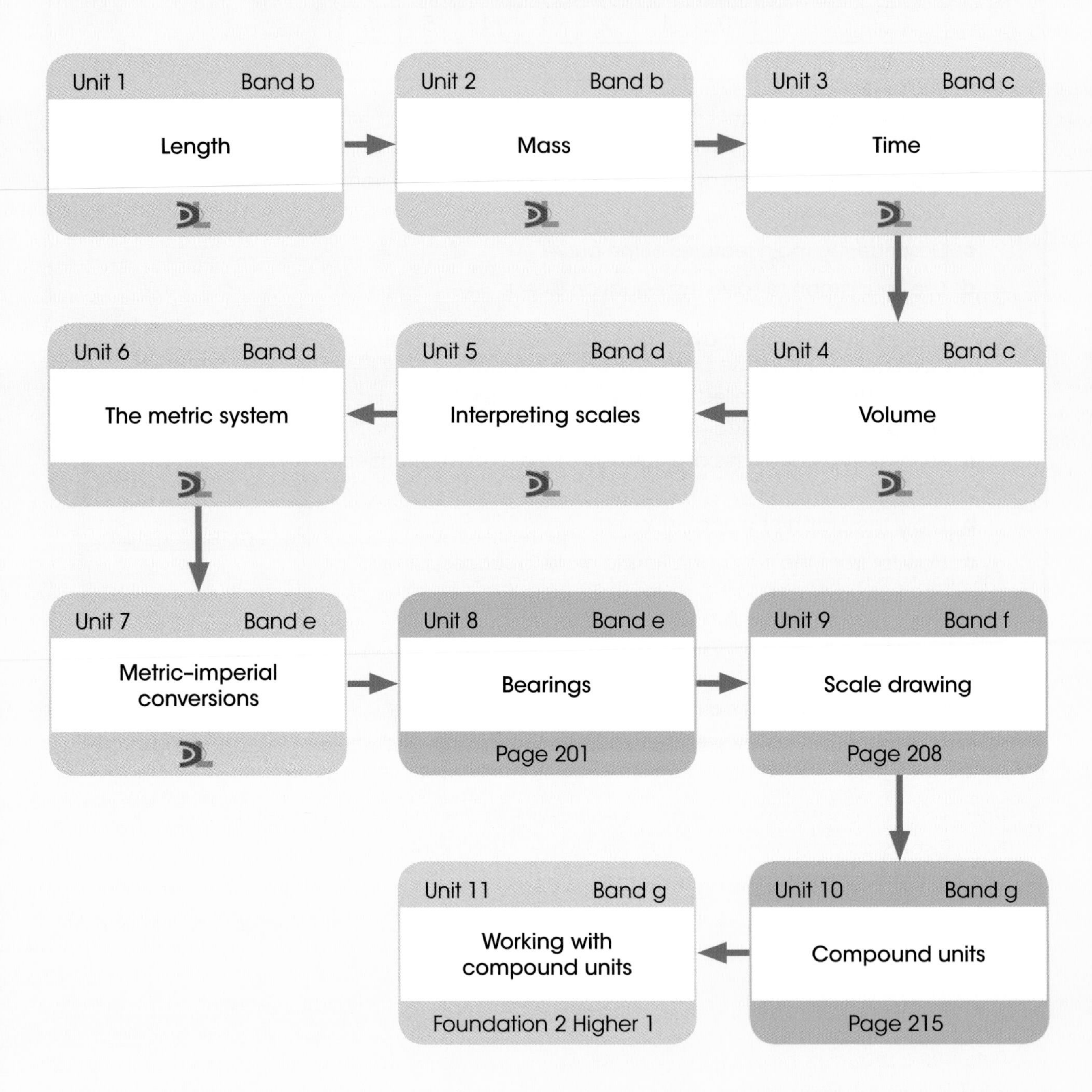

Units 1–7 are assumed knowledge for this book. They are reviewed and extended in the Moving on section on page 199.

Units 1–7 • Moving on

Exam-style

1 The diagram shows part of the fence that Jim is going to put along both sides of a path.

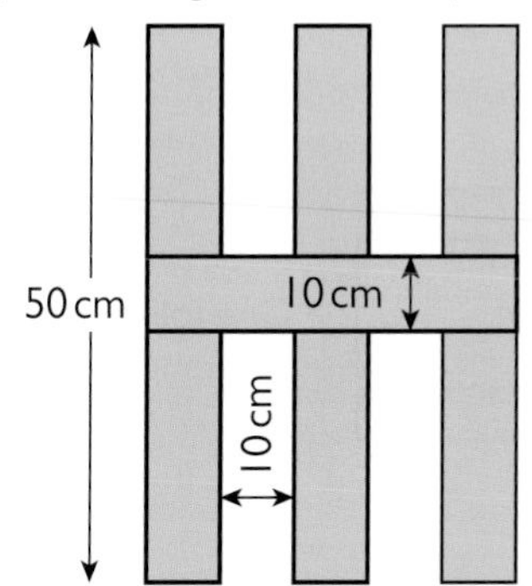

The fence is made of repeating blocks of three vertical planks of wood of height 50 cm and width 10 cm, with a space of 10 cm between each plank, together with one horizontal plank of width 10 cm and length 50 cm. These stretch along the path.

Jim buys the planks in packs of 10.

Each pack costs £6.99.

The length of the path is 8 m.

Work out the total cost of the packs Jim should buy.

Exam-style

2 Mo likes running.

He runs on five days each week.

On three of the days he runs 2500 metres each day.

On the other two days he runs 5 km each day.

Work out how far Mo runs altogether in a 4-week period.

Exam-style

3 Ali wants to send some homemade food to her friend.

She is going to pack this food in a box.

Fruit cake	1.5 kg
Jar of jam	340 g
2 bars of chocolate	145 g each bar
Box of sweets	200 g

Ali does not want the packed box to have a weight of more than $2\frac{3}{4}$ kg.

What is the greatest weight of packed box she can make from the food is this list?

4 Jim is organising a 10 km fun run.

There will be water stations at 4 km, 7 km and at the end of the fun run.

Jim assumes that each runner will drink 300 ml of water at each water station and 500 ml at the end.

There are 120 runners.

Work out an estimate for the total amount of water Jim will need to provide.

Give your answer in litres.

Exam-style

5 Helen uses this rule for cooking meat.

Allow 25 minutes for each 500 grams plus 15 minutes

The weight of the meat is 2 kg.
She needs to finish cooking the meat at 7.30 p.m.
What is the latest time she can start cooking the meat?

Exam-style

6 Leon puts sheets of paper into folders.
The thickness of a sheet of paper is 0.1 mm.
The thickness of the folder covers is 0.4 mm.
Each folder has 200 sheets of paper in it.
Leon has 10 folders.
Can he stack 10 folders into a gap with a height of 30 cm?

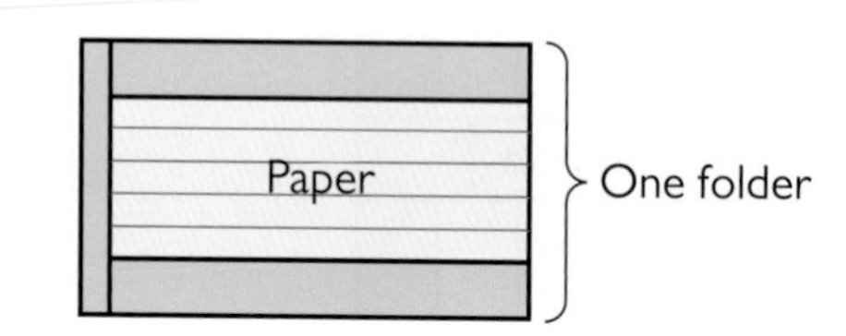

Exam-style

7 Ian is a keen swimmer.
Every weekday he swims 20 lengths of the pool.
On each day of the weekend he swims 40 lengths of the pool.
The length of the pool is 50 m.
His training target for the next four weeks is to swim 8 km.
Will he meet his target?

Exam-style

8 Sam is 5 feet 3 inches tall.
Penny is 163 cm tall.
Who is taller and by how much?
Use 2.54 cm = 1 inch and 12 inches = 1 foot.

9 The BMI of a person is calculated from this rule.

BMI = mass in kg ÷ (height in m)^2

A person's ideal BMI lies between 18 and 25.
Bill weighs 154 pounds and has a height of 183 cm.
Does Bill have an ideal BMI?
Give a reason for your answer.

10 Harry is an angler.
His largest catch in the UK weighed 7 pounds 6 ounces.
His largest catch in France weighed 4.2 kg.
Which was his largest catch?
Use 16 ounces = 1 pound and 2.2 pounds = 1 kg.

11 Jamal has an empty suitcase which weighs 7 kg.
The weight of his clothes is 12 pounds.
He has two pairs of shoes, each pair weighing 1.2 pounds.
Jamal also has a laptop which weighs $2\frac{1}{2}$ kg.
He wants to pack his clothes, shoes and laptop in the suitcase.
His luggage allowance is 23 kg.
Will he have to pay for excess luggage?
Give a reason for your answer.

Unit 8 • Bearings • Band e

Outside the Maths classroom

Navigating

What special problems are there when you navigate at sea?

Toolbox

A **bearing** is given as an **angle** direction.
It is the **angle** measured clockwise from North.
Compass bearings are always given using three figures, so there can be no mistakes.
A **back bearing** is the direction of the return journey.
If a bearing is less than 180°, the back bearing is 180° more than the bearing.
If a bearing is more than 180°, the back bearing is the bearing less 180°.
North is 000°, East is 090°, South is 180° and West is 270°.

Example – Drawing a bearing

B is 3 cm from A on a bearing of 120°.
Draw the bearing of B from A.

Solution

a Draw a North line at A.

b Measure an angle of 120° clockwise from North at A.

c Draw a line at this angle from A.
Use a ruler to mark point B 3 cm along this line.

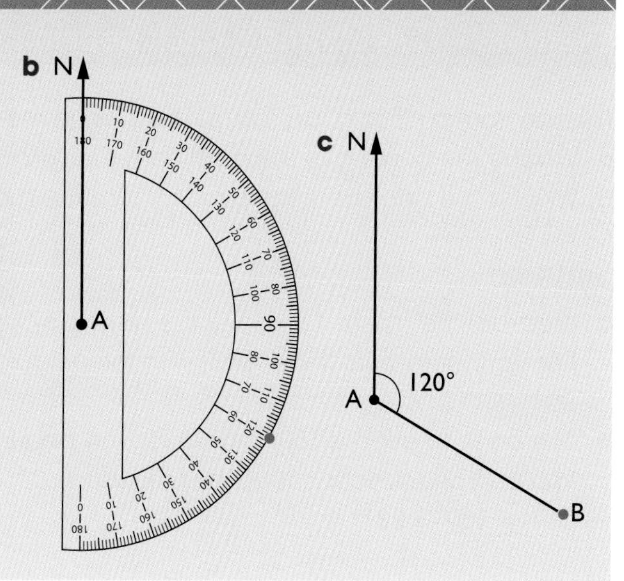

Example – Measuring a bearing

Here is a map of Great Britain.

Find the bearing of

a London from Penzance

b Birmingham from London.

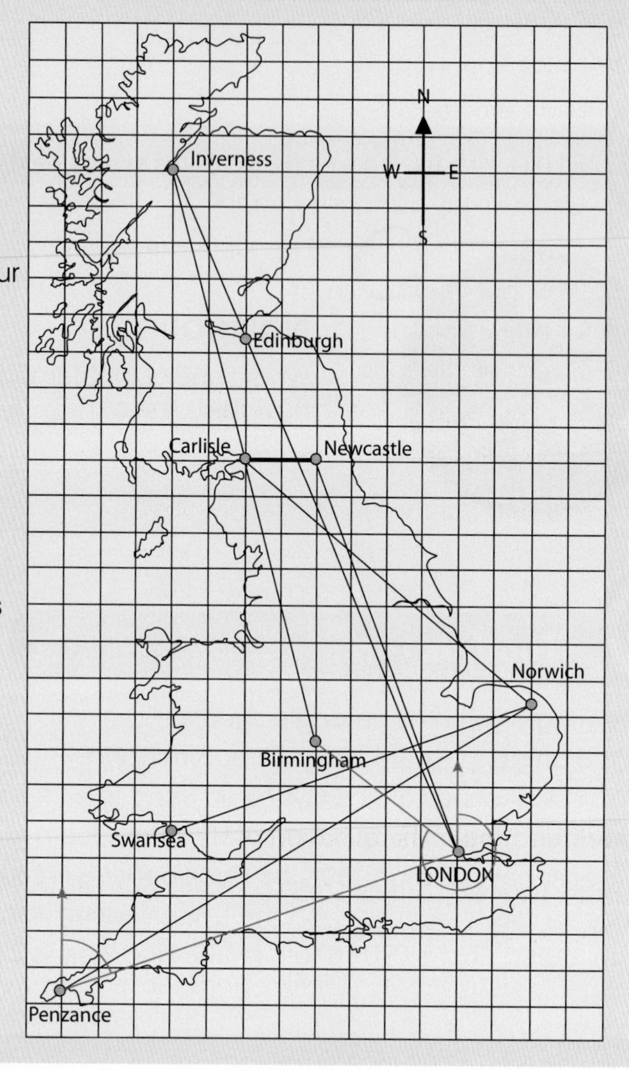

Solution

a The bearing is from Penzance so place your protractor on the map with the centre at Penzance and the zero line vertically up.

Use the grid lines to help you.

Measure the angle clockwise from North.

As 70° only has two digits, write your answer with a zero in front.

070°

b The bearing of Birmingham from London is a reflex angle.

Place your protractor on the map with the centre at London and the zero line vertically up.

If you have a 360° protractor you can measure the angle directly.

If you have a 180° protractor, measure the angle anticlockwise from North.

Subtract this angle from 360°.

$360° - 53° = 307°$

Example – Calculating a back bearing

The line from a hill to a radio mast is shown in the diagram.

a What is the bearing of the mast from the hill?

b What is the back bearing to return from the mast to the hill?

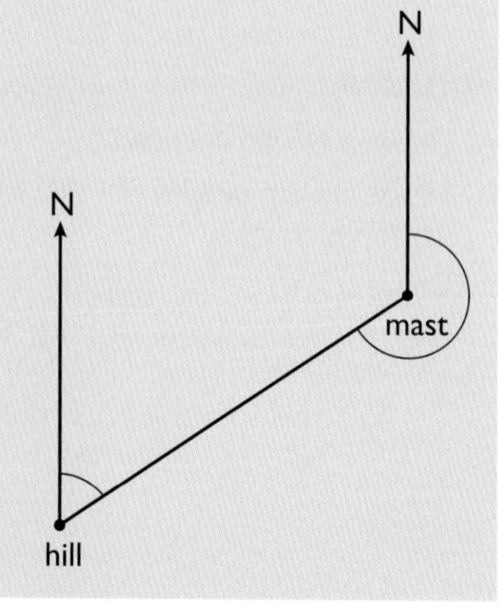

Solution

a Measure the marked angle at the hill. It is 57°.

Remember to write the answer with three figures.

057°

b The back bearing is 180° more than the bearing.

You can check your answer by measuring.

$057° + 180° = 237°$

Practising skills

1 Write down the three-figure bearings of A, B, C and D from O.

a

N
A
40°
O

b

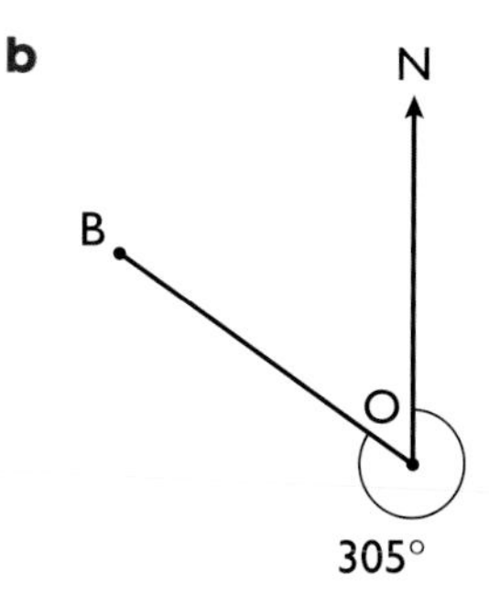

c

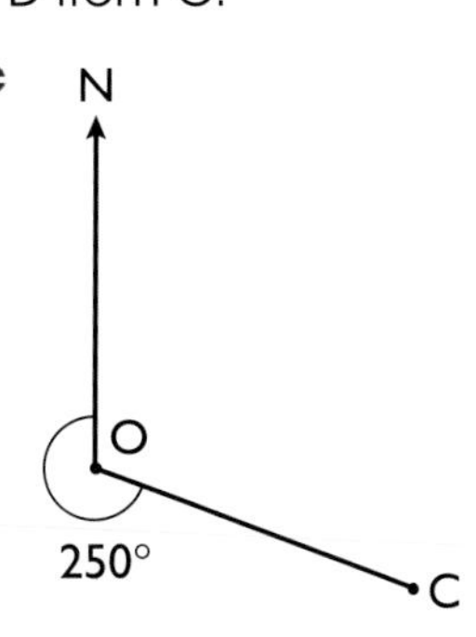

d

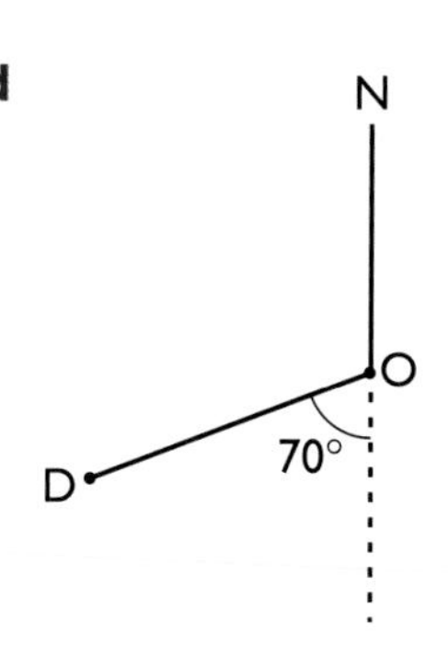

2 Measure the bearings of each of the following towns from Avonford.

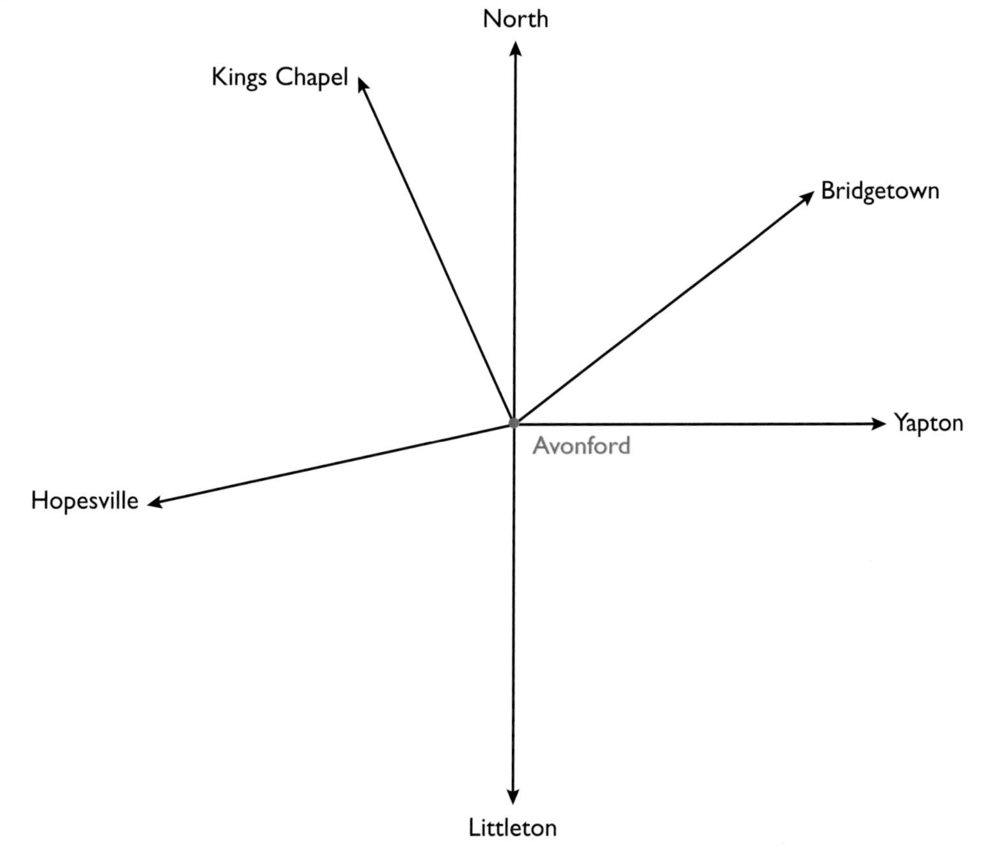

3 Write down the bearing for each of these compass directions.

a South **b** East **c** North-East **d** South-West

4 Write down the compass points with these bearings.

a 000° **b** 135° **c** 315° **d** 270°

5) For each of these
   i draw an accurate diagram
   ii calculate the angles x and y
   ii give the bearing from A to B
   iv give the back bearing from B to A.

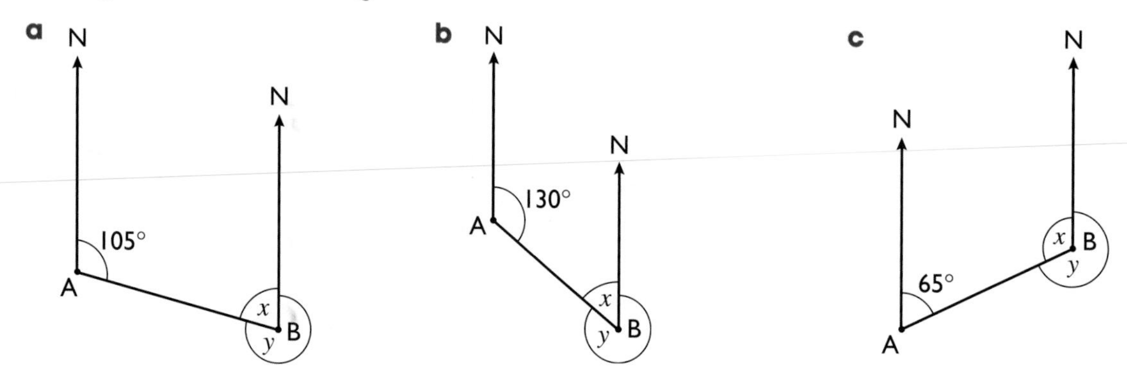

6) Draw accurate diagrams to show these bearings.
   a The bearing from A to B is 060°. B is 6 cm from A.
   b The bearing from C to D is 200°. D is 5 cm from C.
   c The bearing from E to F is 330°. F is 4.2 cm from E.
   d The bearing from G to H is 105°. H is 57 mm from G.

Reasoning

7) A helicopter and an aeroplane are flying at the same altitude.
The bearing of the aeroplane from the helicopter is 070°.
Jack says the bearing of the helicopter from the aeroplane is 110°.
Is Jack right?
Explain your answer fully.

Developing fluency

1) The diagram shows the position of two ships A and B.

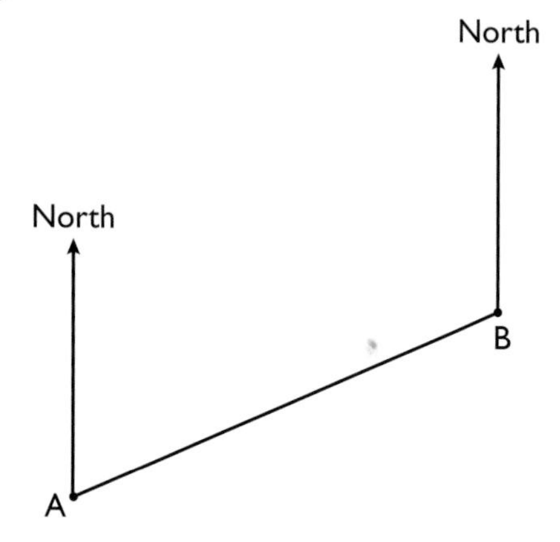

   a Measure the bearing of ship B from ship A.
   b Calculate the back bearing to travel from ship B to ship A.

2. Applebury is South-East of Newtown.
   a What is the bearing of Applebury from Newtown?
   b What is the bearing of Newtown from Applebury?

Exam-style

3. **a** A pilot flies from London to Manchester on a bearing of 312°. Calculate the bearing of London form Manchester.
   b The flight path from Plymouth to Edinburgh is 017°. Calculate the bearing of Plymouth from Edinburgh.

Reasoning

4. The bearing of Sunnyside from Avonford is 133°.
   The bearing of Westbrook from Avonford is also 133°.
   a What can you say about the positions of these three towns?

   b Who is right? Explain your answer fully.

Exam-style

5. Alex is organising an orienteering competition.
   Checkpoint B is 4 km from checkpoint A on a bearing of 155°.
   Checkpoint C is 7 km from B on a bearing of 220°.
   a Draw an accurate diagram showing the three checkpoints.
   Alex walks from checkpoint C back to checkpoint A.
   b How far does she walk?
   c What is the bearing of checkpoint A from checkpoint C?
   d What is the bearing of checkpoint C from checkpoint A?

6. The diagram shows the position of two ships P and Q and a nearby lighthouse, L.
   Find the bearing of
   a P from Q **b** Q from P
   c Q from L **d** L from Q
   e L from P **f** P from L.

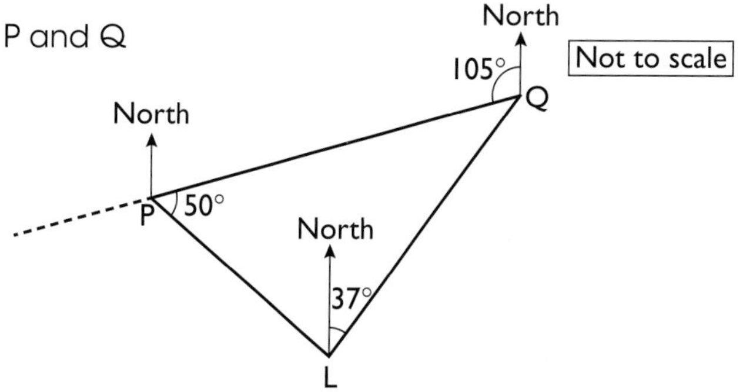

Problem solving

Exam-style

1. Alton, Broxham and Colton are three villages.
   The distance of Alton from Colton is 15 km.
   The distance of Broxham from Colton is 15 km.
   Alton is due East of Broxham.
   The bearing of Alton from Colton is 140°.

   Colton

   Not to scale

   Broxham   Alton

   a Draw an accurate diagram showing the three villages.

   b Find the bearing of Colton from Broxham.

Exam-style | Extension

2. An aeroplane flies round a triangular route.
   It starts from point X and flies 50 km on a bearing of 130° to a point Y.
   At Y, it turns and flies to a point Z.
   From Z, it turns and flies back to X.
   The points X, Y and Z form an equilateral triangle.

   a Draw a diagram showing X, Y and Z.

   b Work out the bearing of Z from Y.

   c Work out the bearing that the aeroplane flies from Z to X.

Exam-style | Extension

3. Dalton is a town 4 miles due North of Fran's home.
   Foxham is a village 8 miles from Fran's home.
   The bearing of Foxham from Dalton is 210°.

   a Draw a suitable accurate diagram to find the bearing of Foxham from Fran's home.

   b State one assumption you made to answer part **a**.

Exam-style | Extension

4. A, B and C are 3 markers in the sea.
   The bearing of A from B is 040°.
   The bearing of C from B is 100°.
   The bearing of A from C is 340°.
   Prove that A, B and C form the corners of an equilateral triangle.

Exam-style | Extension

5. A boat is travelling due East.
   It travels a distance of 6 miles every hour.
   At 1 p.m. the bearing of an island from the boat is 030°.
   At 2.30 p.m. the bearing of the island from the boat is 305°.
   Find, by measurement, the closest distance the boat gets to the island between 1 p.m. and 2.30 p.m.

Exam-style | Extension

6. In a race, boats sail clockwise around four markers A, B, C and D in the sea.
   The markers form a square.
   A boat starts from A and sails towards B on a bearing of 110°.
   Work out the other three bearings that the boat must travel on to finish the course.

Reviewing skills

1. A boat sails from port A on a bearing of 045°.
   The boat sails 5 miles every hour.
   B is a point 7 miles due East of A.
   A ship leaves B at the same time and sails on a bearing of 030°.
   The ship sails 6 miles every hour.
   Find, by measurement, the bearing of the boat from the ship after 30 minutes.

Unit 9 • Scale drawing • Band f

Outside the Maths classroom

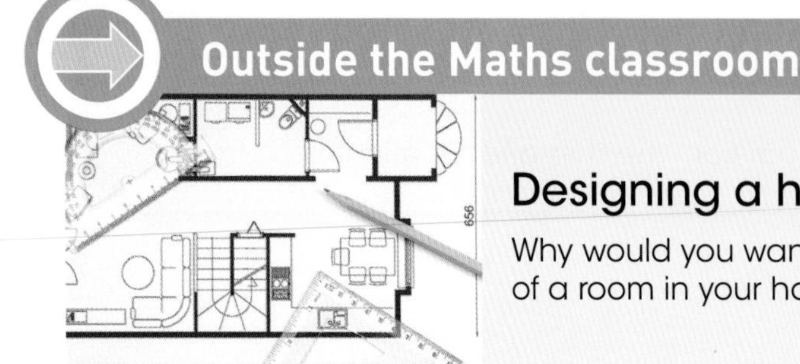

Designing a house

Why would you want to make a plan of a room in your home?

Toolbox

A scale drawing is the **same shape** as the original but a **different size**.

All the lengths are in the same **ratio**.

On a scale drawing where 1 cm on the scale drawing represents 2 m on the actual object, the scale can be written as

$\frac{1}{200}$ or

1 cm = 2 m or

1 : 200

$\frac{1}{200}$ is sometimes referred to as the **scale factor**.

Warning: Be careful with areas and volumes.

For example, 1 cm = 10 mm but 1 cm^2 = 10^2 = 100 mm^2 and 1 cm^3 = 10^3 = 1000 mm^3.

Example – Using a scale

Kitty has a toy car on a scale of $\frac{1}{50}$th.

The toy car is 6 cm long.

The real car is 1.7 m wide.

a How long is the real car?

b How wide is the toy car?

Solution

The toy car is $\frac{1}{50}$ the size of the real car.

So the real car is 50 times the size of the toy car.

a Length of real car = 6 × 50 cm ← To find the length of the real car, multiply by 50.

= 300 cm

The real car is 300 cm or 3 m long.

b 1.7 m = 170 cm ← It is easier to convert to cm first.

Width of toy car = 170 ÷ 50 cm ← To find the width of the toy car, divide by 50.

= 3.4 cm

The toy car is 3.4 cm wide.

Example – Reading a map

Adebola is working out distances between places on a map.
The map has a scale of 1 : 50 000.
On the map, the bowling alley is 4.7 cm from the aqueduct.

a How far is the bowling alley from the aqueduct in kilometres?

Adebola's house is 3.8 km from her school.

b How far is her house from her school on the map?

Solution

a 4.7 × 50 000 = 235 000 cm ← To find a distance on the ground, multiply by the scale factor of 50 000.

235 000 ÷ 100 000 = 2.35 km ← There are 100 000 cm in 1 km.

b 3.8 × 100 000 = 380 000 cm ← Convert to centimetres first.

380 000 ÷ 50 000 = 7.6 cm ← To find a distance on the map, divide by the scale factor of 50 000.

Practising skills

1 Look at these rectangles.
Find these scales.

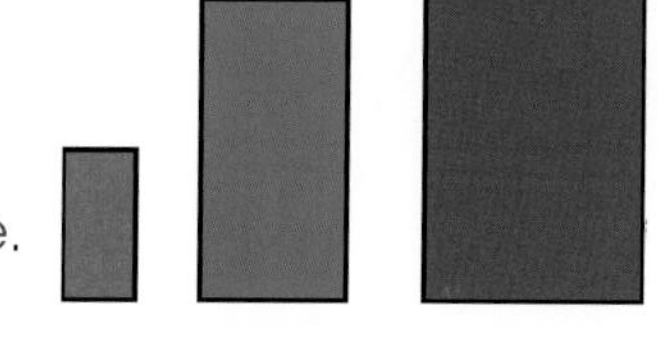

a The width of red rectangle : the width of green rectangle.

b The width of red rectangle : the width of blue rectangle.

c The width of green rectangle : the width of blue rectangle.

2 A map has a scale of 1 : 50 000.
The distance between two villages on the map is 7 cm.

a What is the actual distance between the villages?

Littleton is 10.4 km from Avonford.

b How far apart are the two towns on the map?

3 Find the scale of these maps in the form 1 : n.

a A street map where 3 cm represents 600 m.

b An ordnance survey map where 20 cm represents 500 m.

c A road map where 7 cm represents 17.5 km.

d A map of the world where 8 cm represents 800 km.

4 A model train engine is made to a scale of 1 : 50.
The length of the model is 40 cm.

a What is the length of the actual train engine?

The height of the actual train engine is 4 m.

b What is the height of the model?

5. A town map is drawn to a scale of 1 : 20000.
Copy and complete this table showing the distance between various places.

Places	Distance on map (cm)	Distance in real life (km)
Library to Sports centre		1.2
School to park	2.5	
Cinema to supermarket	7.5	
Café to cinema		2
Bowling alley to river		1.8

Developing fluency

Problem solving

1. This is the plan of a flat. It is not drawn to scale.
Using the measurements given on the plan, make an accurate scale drawing.
Use the scale 1 cm = 2 m.

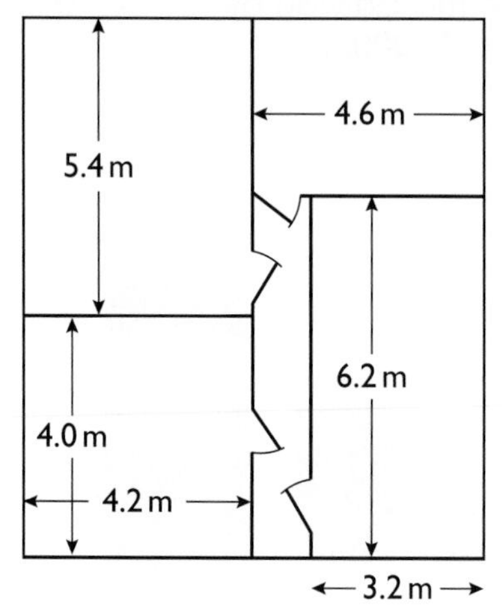

Exam-style

2. Zac is orienteering.
He starts at point A and then walks 6 km East followed by 7 km North-East to reach a lake at point B.
   - **a** Make a scale drawing of Zac's journey.
   - **b** How far is point B from Zac's starting point?
   - **c** What is the bearing from the lake to point A?
   - **d** What is the bearing of the lake from point A?

3. The diagram shows the plan of a garden.
The scale is 1 cm = 2 m.
Copy and complete the table to show the true measurements of the garden.

Item	Plan measurement	True measurement
Length of patio		
Width of patio		
Length of lawn		
Length of vegetable patch		
Width of pond		
Length of pond		
Width of house		
Length of shed		
Width of shed		
Length of path		

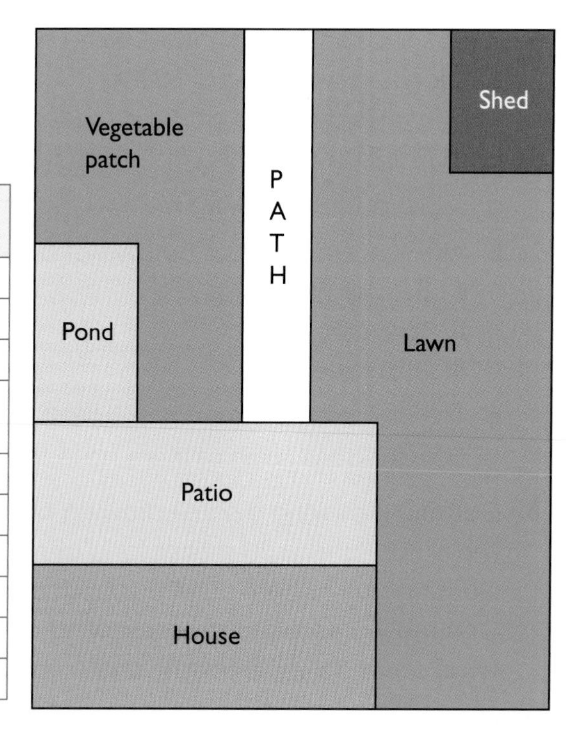

4. A shop sells toy boat sets for 3- to 5-year-olds.
There are different sizes.
The small boat is exactly half the size of the standard boat.
The scale of the small boat : standard boat is 1 : 2.

a The mast in the small set is a cylinder with a height of 20 cm.

i What shape is the mast in the standard set?

ii What is the height of the mast in the standard set?

b This is an accurate scale drawing of the boat from the small set.

i How long is the boat?

ii What are the dimensions of the sails?

iii Make an accurate scale drawing of the boat from the standard set.

c The shop also sells a super-sized boat.
It is 20 times bigger than the standard set.

i How long is the super-sized boat?

ii How tall is it?

iii Can children climb on it?

Reasoning

5. Three friends set off on a map-reading activity. They start together at point A and travel for half an hour.
Tom jogs on a bearing of 300° at 10 kilometres per hour.
Molly cycles due West at 20 kilometres per hour.
Evan walks on a bearing of 120° at 6 kilometres per hour.
   a Make a scale drawing to show their travels.
   b At the end of the half hour, what is the distance between
      i Evan and Tom
      ii Tom and Molly
      iii Evan and Molly?
   c The friends then meet at point B. It is 5 km from A on a bearing of 225°.
   How far does each of them have to travel?

Reasoning

6. Gemma is making a scale drawing of a swimming pool.
The actual pool is 30 metres long and 15 metres wide.
Gemma has a sheet of A4 paper and needs at least a 1 cm border around her diagram.
Gemma wants to make her drawing as large as possible.
What scale do you think she should use?

7. On a map a forest is shown as a rectangle with an area of 13.6 cm^2. It is 4 cm long.
The scale of the map is 1 : 50 000.
   a What is the area of the real forest?
   b **i** What distance on the ground is represented by 1 cm?
      ii What area on the ground is represented by 1 cm^2?
   c Use your answer to check the area you found in part **a ii**.
   A lake has a surface area of 0.7 km^2.
   d What is the area of the lake on the map?

Problem solving

Exam-style

1. Here is a sketch of the side view of the roof space of a building.

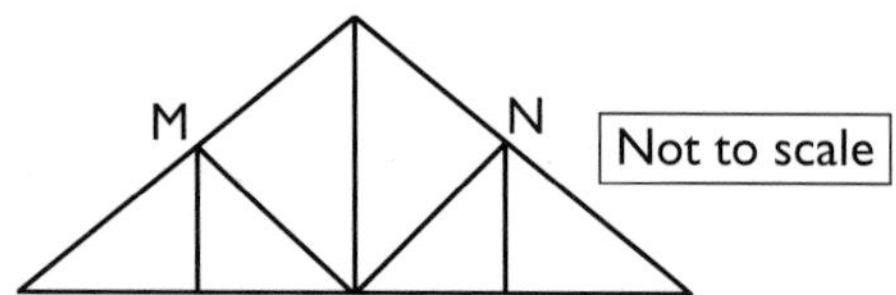

The lines in the sketch show the wooden frame which supports the roof.
The width of the frame is 8 m and the height of the frame is 3 m.
M and N are the midpoints of the sloping sides.
The frame has a single line of symmetry.
   a Make an accurate scale drawing.
   b Find, by measurement, the total length of wood used to make the frame.

Exam-style

2 A hall is 8 m square.
Desks 50 cm square are to be placed in the hall.
The desks are to be at least 1 m apart and at least 1 m from any edge of the hall.
Show how to fit the maximum number of desks in the hall.

Exam-style

3 Hannah wants to make a bird box from wood.
She makes a rough sketch of the front of the bird box.

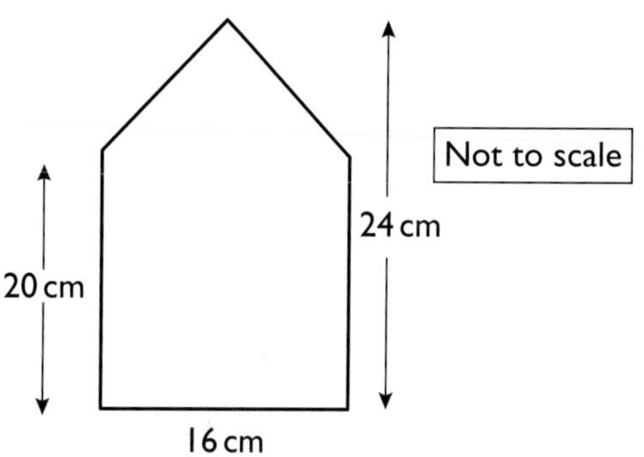

Not to scale

Hannah wants to drill a circular hole of radius 2 cm in the front.
The centre of the hole must be 12 cm from each of the bottom corners.

a Make an accurate scale drawing of the front of the bird box.

b Find, by measurement, the distance of the centre of the circular hole from the top corner.

Exam-style

4 The diagram is a sketch of the gable end of a house.
The end of the house has a vertical line of symmetry.
Building regulations state that the top of the house must be less than 4 m above the level AB.

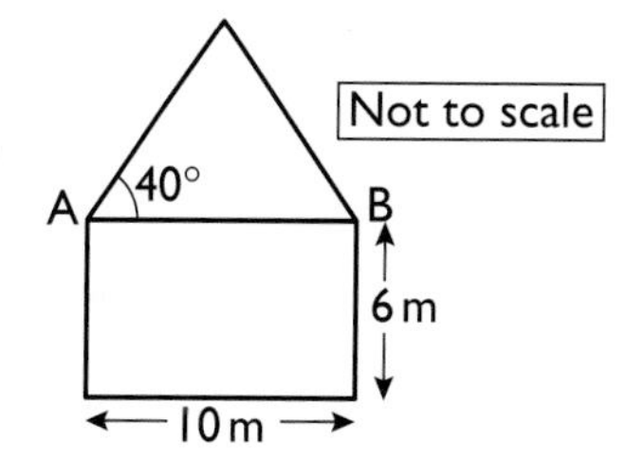

Not to scale

a Make an accurate scale drawing of the end of the house.

b Find, by measurement, if the top of the house is less than 4 m from the level AB.

Exam-style

5 Rosie sails her boat from port A on a bearing of 070° for 20 miles to a point B.
From B she sails on a bearing of 100° for 15 miles to a point C.

a Make an accurate drawing of Rosie's journey.

b Find, by measurement, the bearing of the course that Rosie must set to get back to port A directly, and how far she has to go.

Exam-style

Extension

6 Model trains are made for a 00 gauge track.
The scale is 1 : 76.
Pete has a model of the Mallard locomotive.
The Mallard is 70 feet long.
Find the length in centimetres of Pete's model.
Use 2.54 cm = 1 inch and 12 inches = 1 foot.

Reviewing skills

1. The diagram shows a sketch of a tunnel for a railway. It is a circle of radius 4 m.

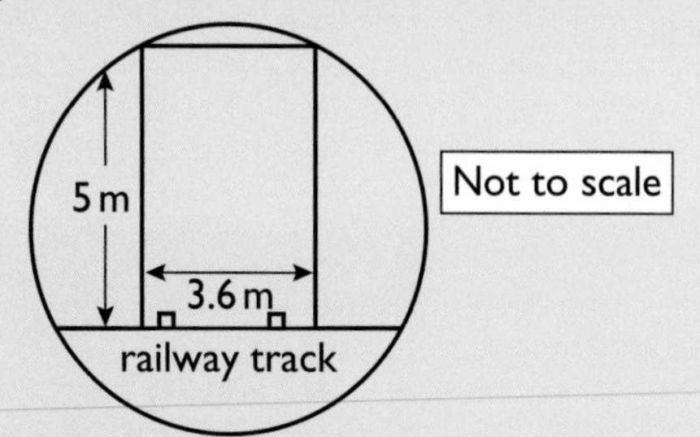

The diagram includes a horizontal chord. This shows where the railway track will be laid.

The diagram also includes a rectangle. It shows the space that must be allowed to let trains through.

a Make a scale drawing to show this.

b Find, by measurement, the height of the railway track above the lowest point of the tunnel.

Unit 10 • Compound units • Band g

Outside the Maths classroom

Planning a journey

How do journey planning websites provide an estimated travel time?

Toolbox

A **compound measure** is a measure involving two quantities.

An example is **speed**, which involves distance and time.

$$\text{speed} = \frac{\text{distance}}{\text{time}}$$

The unit of distance might be km and the unit of time might be hours. In this case the compound unit for speed will be kilometres per hour (km/hr or km hr^{-1}).

Other examples are grams per cubic centimetre for density and Newtons per square metre for pressure.

Example – Working with density

A metal block has a volume of 1000 cm^3.

It has a mass of 8.5 kg.

Using the formula

$$\text{density} = \frac{\text{mass}}{\text{volume}}$$

find the density of the block in kilograms per cm^3.

Solution

$$\text{density} = \frac{\text{mass}}{\text{volume}}$$

$= 8.5 \div 1000$

$= 0.0085$ kilograms per cm^3 ← This unit can also be written as kg cm^{-3} or kg/cm^3.

Example – Working with speed

Jason leaves London on a motorbike at 06:00 and travels to his parents' house 432 km away. He arrives at 10:00.

a Find his average speed
- **i** in kilometres per hour
- **ii** in kilometres per minute
- **iii** in metres per second

b Write the unit 'metres per second' in two other ways.

Solution

a His journey takes from 6 o'clock to 10 o'clock so it lasts 4 hours.

$$\text{Average speed} = \frac{\text{distance}}{\text{time}}$$

i $\frac{\text{distance}}{\text{time}} = \frac{432 \text{ kilometres}}{4 \text{ hours}}$

$= 108$ kilometres per hour

ii $\frac{\text{distance}}{\text{time}} = \frac{432 \text{ kilometres}}{240 \text{ minutes}}$ ← 4 hours = 4 × 60 = 240 minutes

$= 1.8$ kilometres per minute

iii $\frac{\text{distance}}{\text{time}} = \frac{432\,000 \text{ metres}}{14\,400 \text{ seconds}}$ ← 432 kilometres = 432 × 1000 = 432 000 metres; 4 hours = 4 × 60 × 60 = 14 400 seconds

$= 30$ metres per second

b Other ways of writing 'metres per second' include m s^{-1} and m/s.

Practising skills

1 Work out the average speed in km/h of each of these.

- **a** An aeroplane travelling a distance of 900 km in 2 hours.
- **b** A ferry travelling 8 km in 15 minutes.
- **c** A car travelling 60 km in $1\frac{1}{2}$ hours.

2 Copy and complete the conversion diagram to convert m/s to km/h.

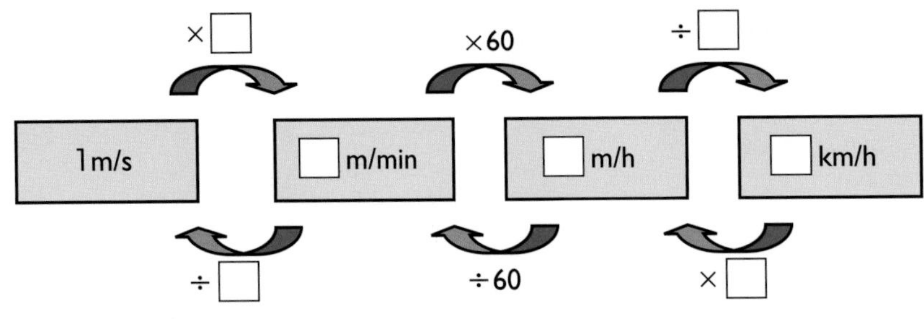

(3) Usain sprints 200 m in 20 seconds.
Work out Usain's average speed in

a metres per second

b metres per hour

c kilometres per hour.

(4) **a** Tom cycles for 3 hours. He travels a distance of 120 km.
Work out his average speed.

b A tortoise travels for 20 minutes. It travels 40 metres.
Work out its speed in

i metres per minute

ii cm per second.

c Andrew swims for $1\frac{1}{2}$ hours. He covers 60 lengths, each of 20 metres.
What is his speed in metres per second?

(5) Jamie and Sarah have part-time jobs.
In one week, Jamie gets £51 for 6 hours work.
Sarah works 9 hours and gets £70.20.
Find their rates of pay.
Who is the better paid?

Developing fluency

Remember 5 miles is approximately 8 km.

(1) Bob's train departs at 12:38 and arrives at its destination 210 miles away at 14:18.
What is the train's average speed?

Reasoning

(2) Jake and Amy go to different garages to fill their cars with petrol.
Jake pays £34.50 for 25 litres.
Amy pays £63.90 for 45 litres.
Who gets the cheaper petrol? Show all of your working.

Exam-style

(3) Jo's lawn is rectangular.
It is 16 m long and 11.5 m wide.
She wants to feed it with fertiliser.
She buys a 2.5 kg box which covers 500 m^2.
How many grams should she use?

(4) **a** Harry cycles at 8.5 km/h.
What is his speed in m/s?

b A cheetah can sprint at speeds of up to 30 m/s.
What is this speed in miles per hour?

Exam-style

5) Maya drives her car for 30 minutes at a steady speed of 70 mph.
She then drives for a further 48 miles taking $\frac{3}{4}$ hour.

a Find the total distance she travels.

b Work out Maya's average speed for the whole journey.

6) This conversion graph can be used to change between gallons and litres.

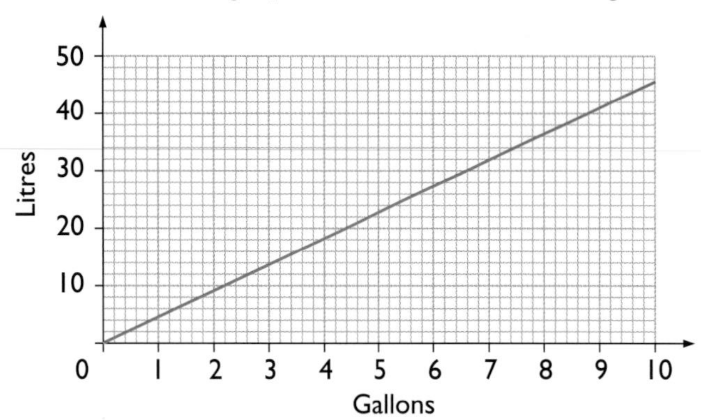

Art has oil-fired central heating.
The tank for the oil holds 280 gallons of oil when full.
Art's tank is only $\frac{1}{4}$ full.
Oil costs 50p per litre.
Work out an estimate for the cost of filling the oil tank.

7) **a** Find the distance, in km, travelled by a satellite travelling at 8600 metres per second for 1 year.

b What is the speed of the satellite in miles per hour?

8) A company makes metal files for use in engineering and metalwork.
The files have rows of sharp teeth for cleaning surfaces.
A smooth file has 55 teeth per inch. A rough file has 20 teeth per inch.
Both files are 12 cm long.
Work out the difference in the number of teeth on the two files.
You may use 2.5 cm = 1 inch.

Problem solving

Exam-style

1) The speed limit on some main roads in France is 110 kilometres per hour.
The speed limit on motorways in the UK is 70 miles per hour.
Which speed limit is higher and by how much?

2) A factory refines precious metal.
It refines 15 kg of precious metal per hour for 8 hours a day, 5 days a week.
Work out the amount of precious metal it refines in a year.

Exam-style

3) Hassan drives his car at a constant speed of 60 mph for 1 hour and then for 2 hours at a constant speed of 48 mph.

a Find the total distance he travels.

b Work out Hassan's average speed for the whole journey.

Exam-style

4 Jenny has an empty pool.

She is going to pump water into the pool at the rate of 20 litres per second.

The pool is in the shape of a cuboid 10 m long by 5 m wide.

Work out the length of time it will take to fill the pool to a depth of 150 cm.
$1\ m^3 = 1000$ litres.

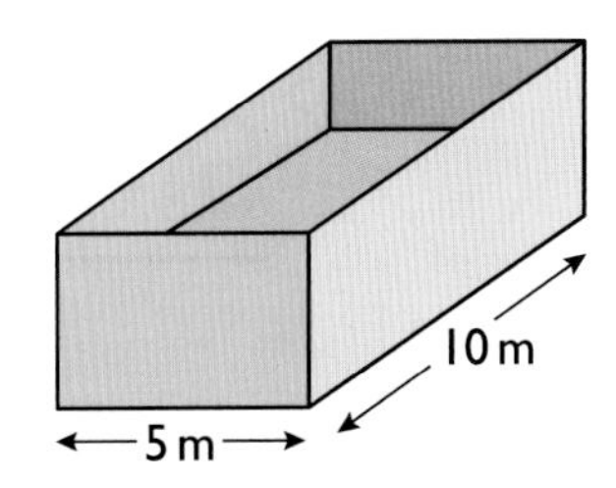

Exam-style

5 A farmer wants to spread fertiliser on a field.

The field is rectangular and is 600 m long by 90 m wide.

a Find the area of the field in hectares. Use 1 hectare = 10 000 m^2.

b The farmer puts on the fertiliser at 184 kg per hectare.
Show that 1000 kg of fertiliser will be enough.

Exam-style

6 A balloon takes off from the ground.

It ascends at a steady speed of 6.5 metres per second for 10 minutes.

It then descends at 8.4 metres per second for 6 minutes.

It then ascends at 6.5 metres per second for a further 3 minutes.

Work out how far the balloon is now above the ground.

Reviewing skills

1 Harry drives 130 km in 2 hours.
Find his average speed in

a km/h

b m/s.

2 Mel works a 40-hour week and earns £316 per week.
Toby earns £287 for a 35-hour week.
Who has the better hourly rate, and by how much?

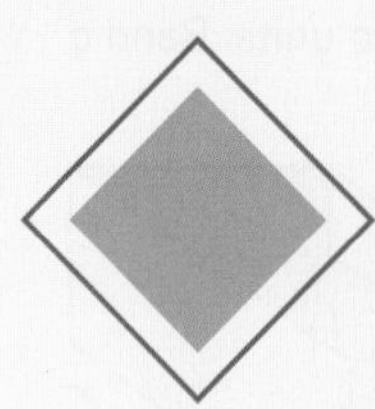

Geometry and Measures
Strand 2 Properties of shapes

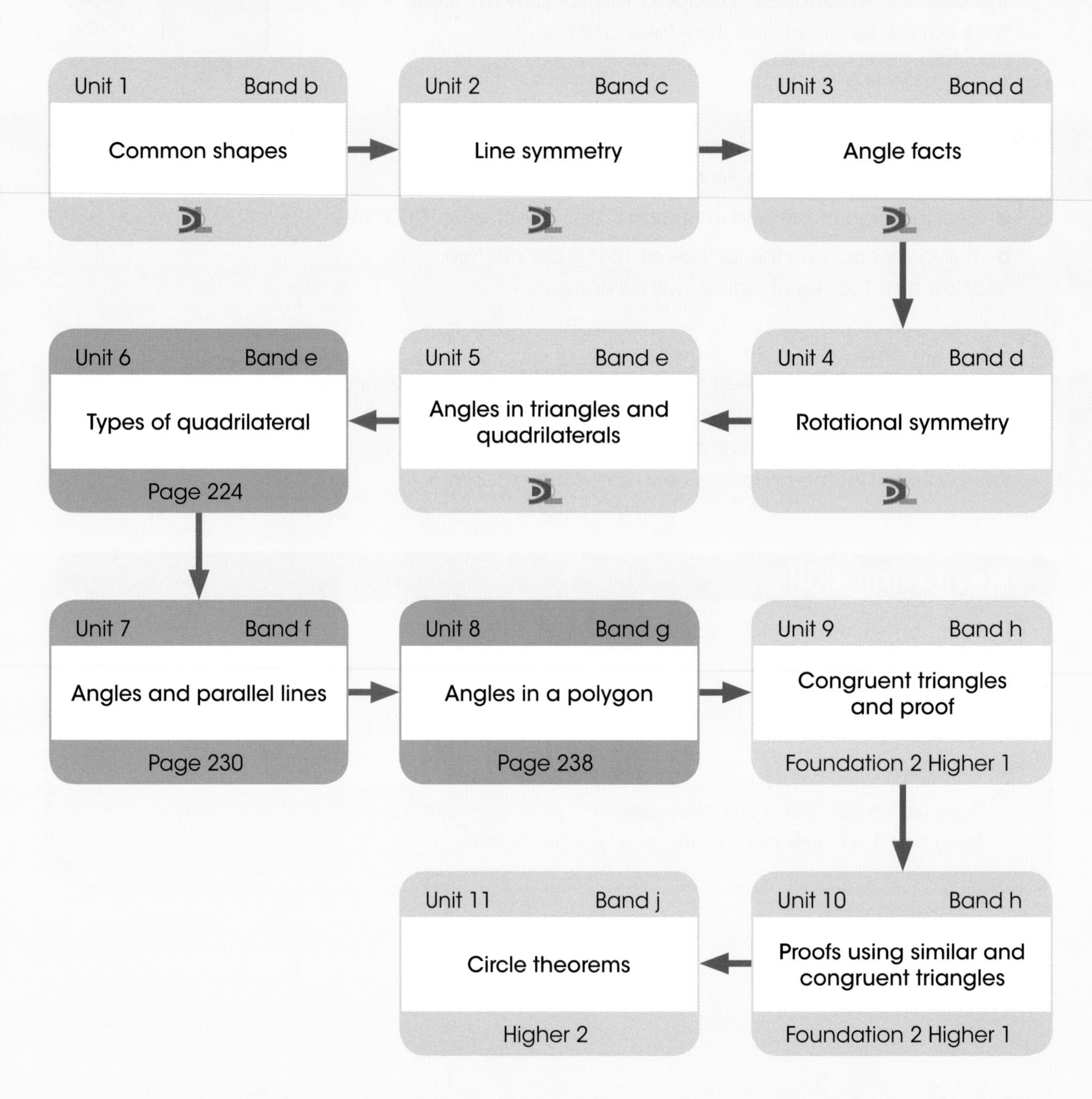

Units 1–5 are assumed knowledge for this book. They are reviewed and extended in the Moving on section on page 221.

Units 1–5 • Moving on

1. Nimer is a designer.
He wants to design a logo with rotational symmetry of order 4.
He is going to use this shape as a part of the logo.

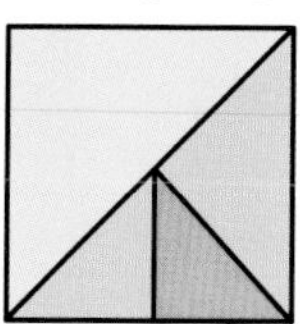

Draw a possible finished design.

2. The pie chart gives some information about the membership of a club.
$\frac{1}{4}$ of the membership are children.
The number of men is four times the number of women.
Work out the angles for children, for women and for men.

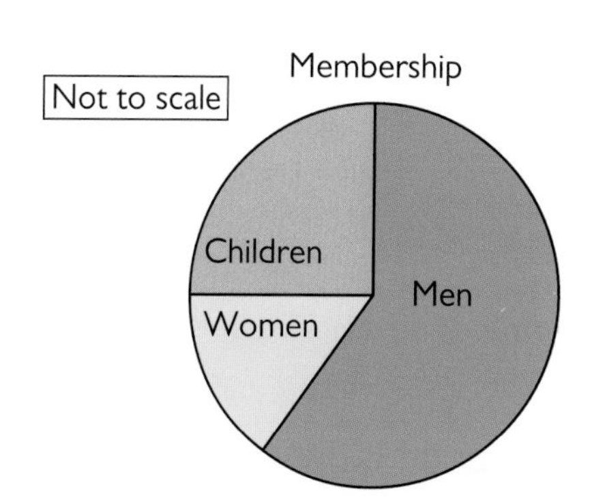

3. ABC is a straight line.
Angle ABD is three times the size of angle DBC.
Work out the size of angle DBC.

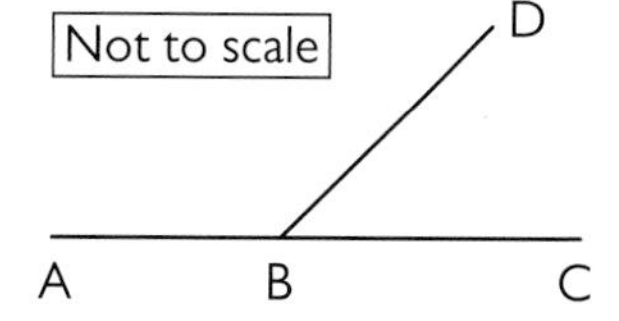

4. Work out the angle between the hour hand and the minute hand of a clock at 03:30.

5. The reflex angle AOC is three times the size of angle AOB.

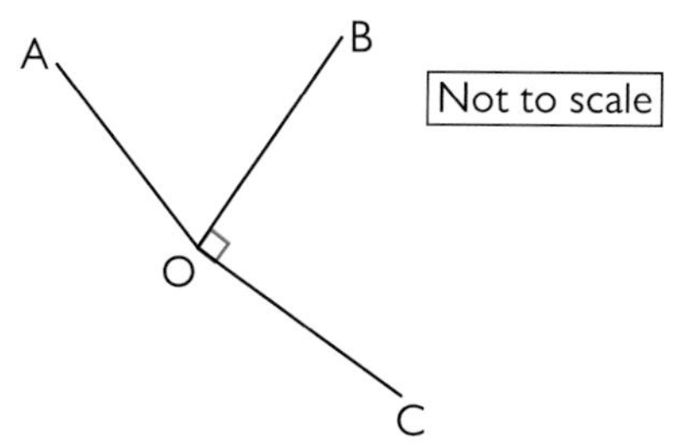

Find the size of the reflex angle AOC.

Exam-style

6 Leo thinks that angle COD is twice angle AOB.

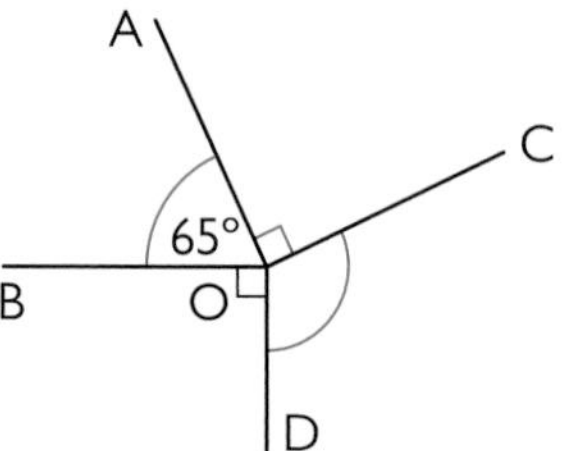

Show that Leo is wrong.

Exam-style

7 You can fit rectangles around a point without leaving any gaps.

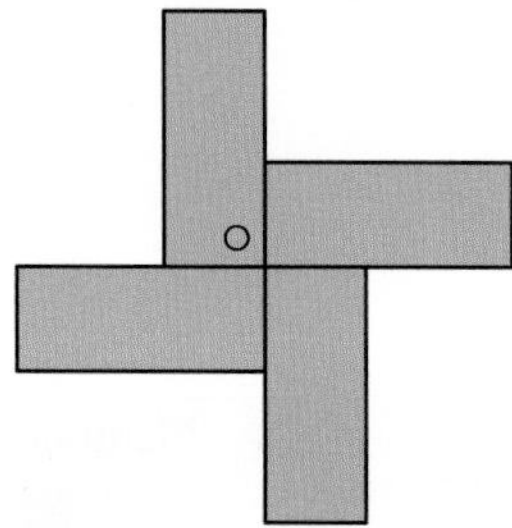

The shape below has two lines of symmetry.

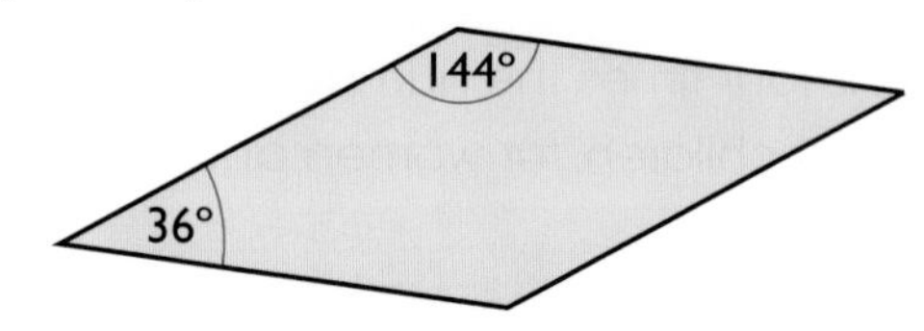

a Show that four of these shapes can be fitted around a point without leaving any gaps.

b Show how more than four of these shapes can be fitted around a point without leaving any gap.

Exam-style

8 The hands of a clock both point in the same direction at 12 o'clock.
Work out the angle between the hour hand and the minute hand after 30 minutes have passed.

9 The diagram shows a piece of card in the shape of a triangle.
The card is placed on a flat table.

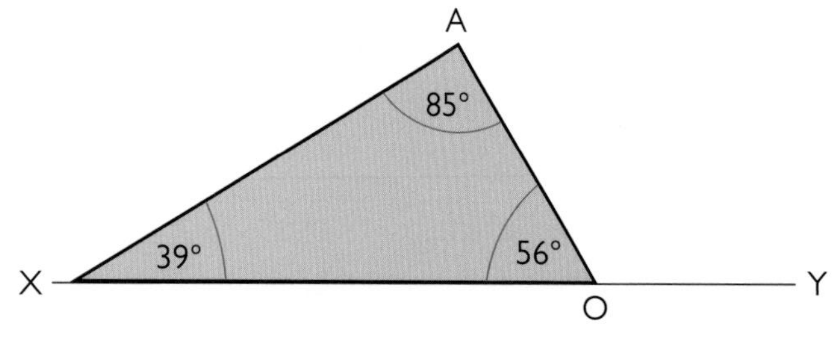

The triangle is rotated clockwise through an angle x about the point O until its side OA rests on the table.

Work out the size of angle x.

Exam-style

10 ABC is a straight line.
Angle ABD = 60°.
Angle CBE is one third the size of angle DBC.

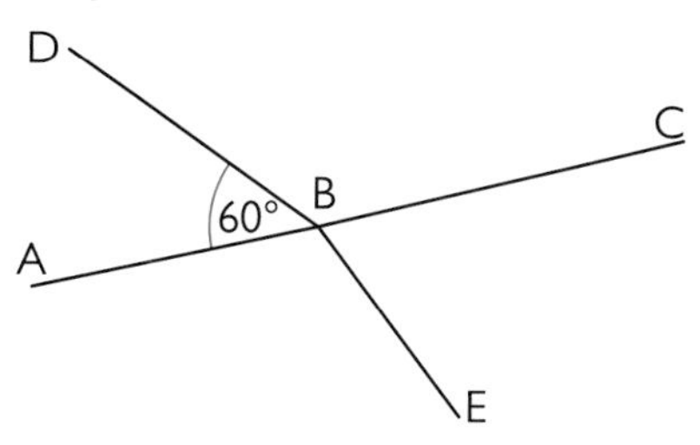

Find the size of angle ABE.

Exam-style

11 A, B, C, D and E are points.
ABC is a straight line.
Angle EBD is twice the size of angle DBA.
Angle EBC is three times the size of angle EBD.
Work out the size of angle EBC.

12 Here is a triangle PQR.

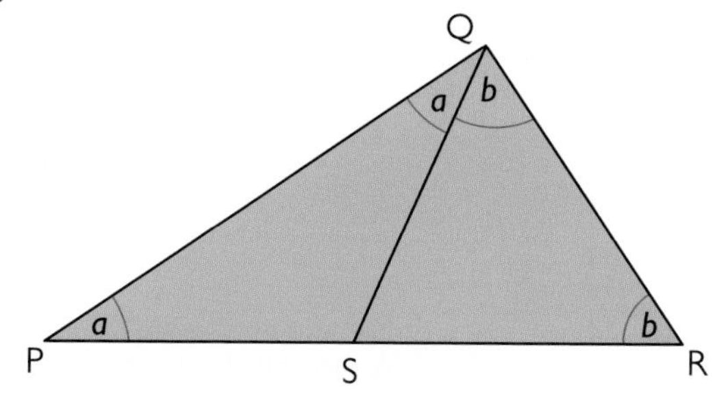

S is a point on the side PR.
Show that angle PQR is a right angle.

Exam-style

13 In triangle ABC, angle ABC = 90°.
Angle BAC is twice the size of angle BCA.
Find the size of angle BAC.

14 This figure has rotational symmetry of order 4.

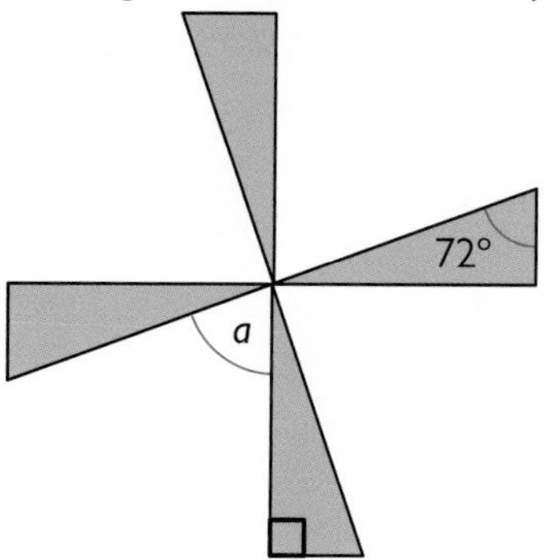

Work out the size of angle a.

Unit 6 • Types of quadrilateral • Band e

Outside the Maths classroom

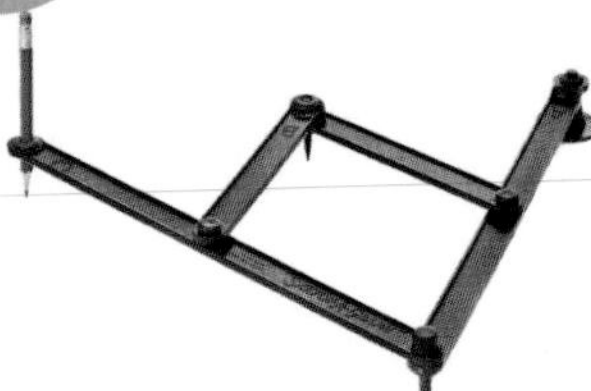

Pantograph

Is the shape formed by the arms of a pantograph always a parallelogram?

Toolbox

When classifying shapes the key things to look for are

- the number of sides
- the lengths of the sides
- the lines of symmetry
- whether opposite sides are parallel
- the sizes of the angles.

There are seven special types of quadrilateral to remember.

A **square**

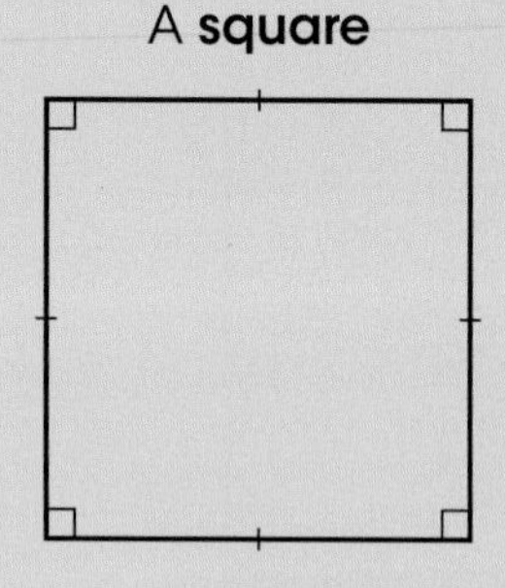

A **rectangle**

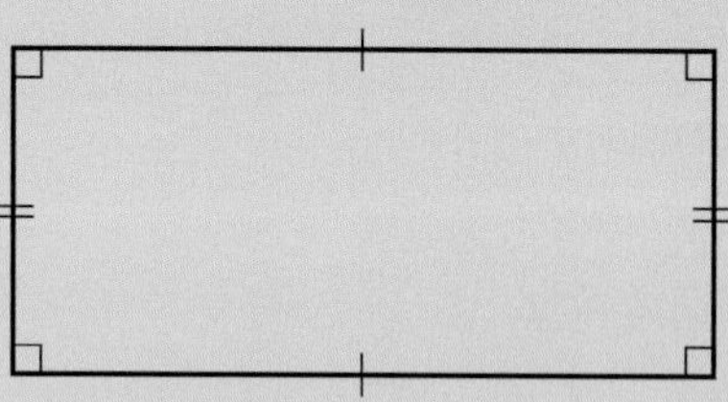

A **parallelogram**

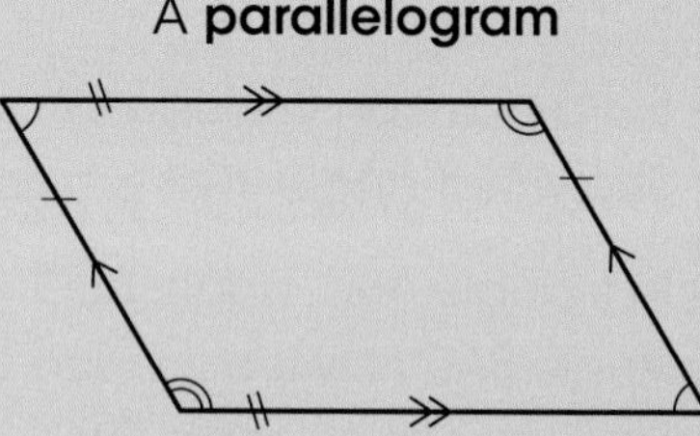

A **rhombus**

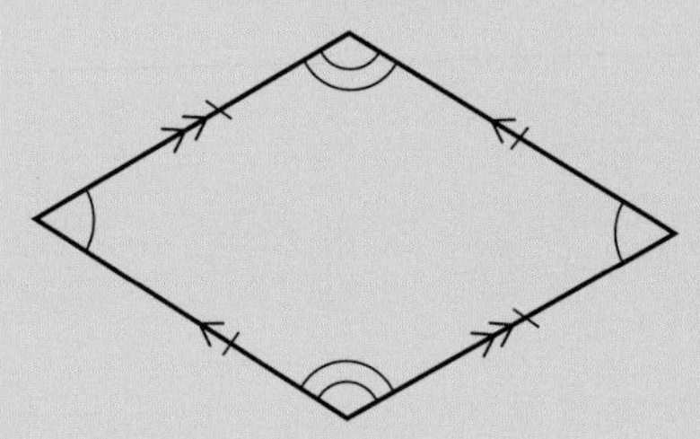

A **kite**

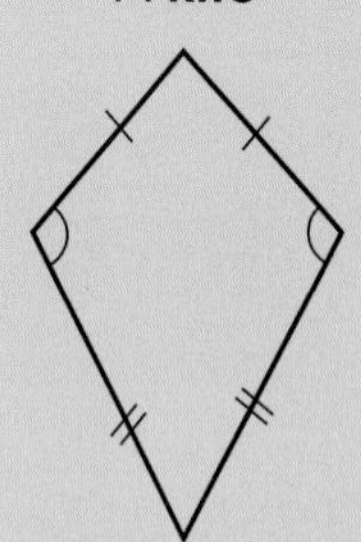

An **arrowhead (deltoid)**

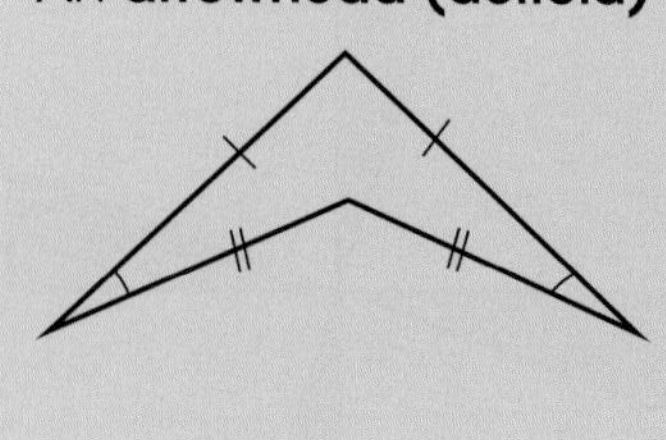

A **trapezium**

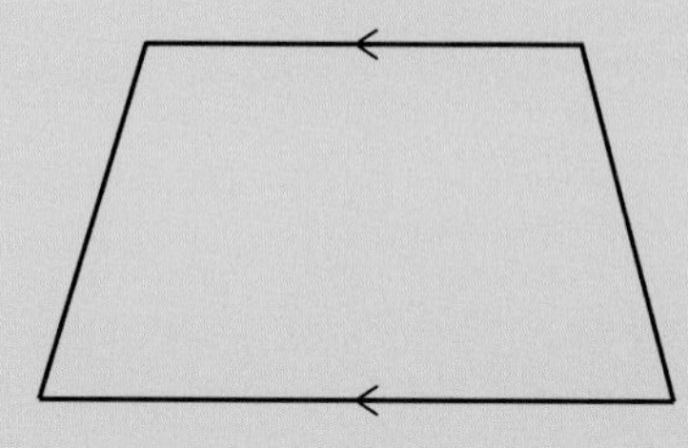

Equal sides are marked with the same number of dashes.
Parallel sides are marked with the same number of arrows.
Equal angles are marked with the same number of arcs.

Example – Classifying quadrilaterals

What type of quadrilateral is Suzanne thinking of?

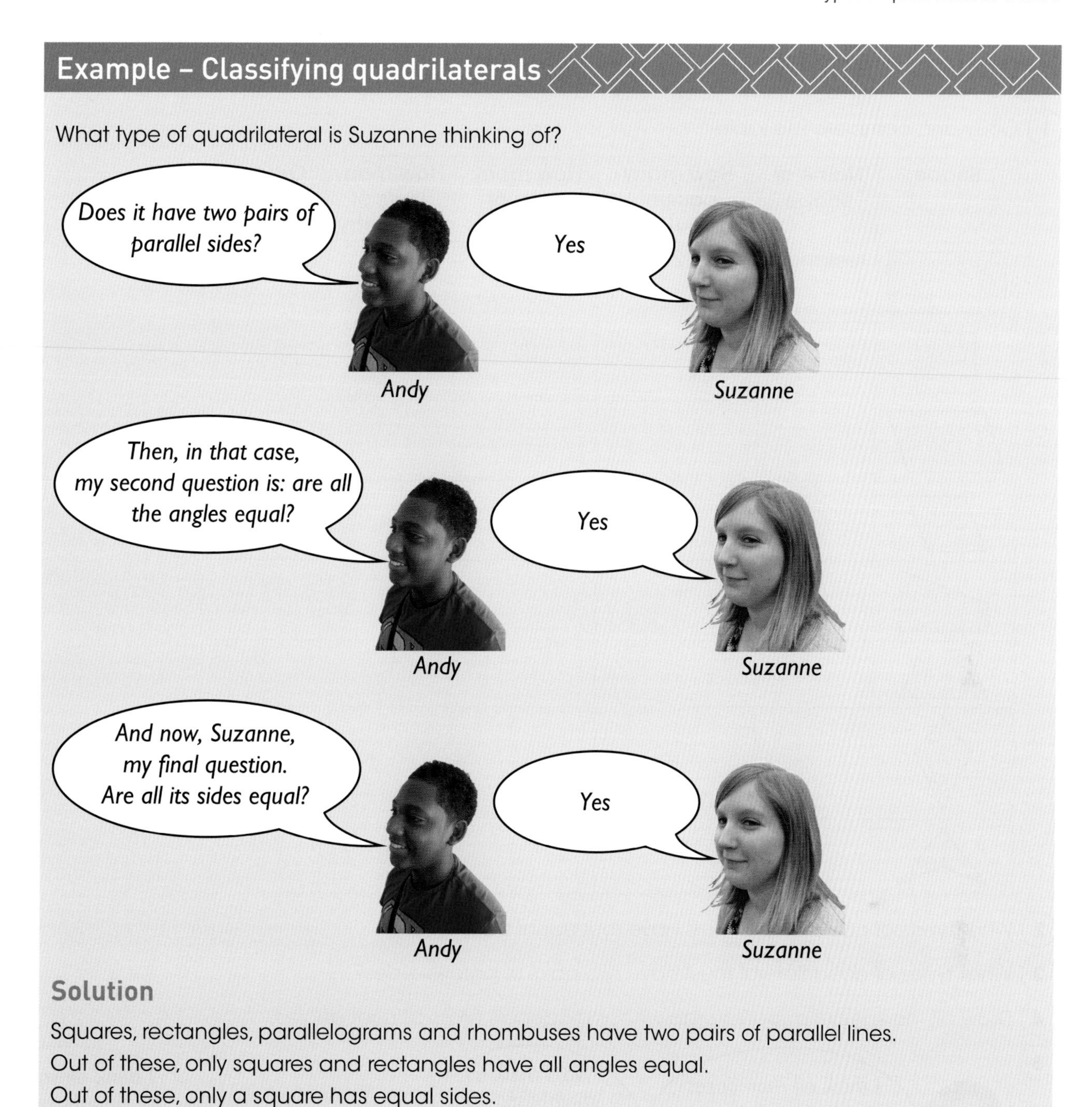

Solution

Squares, rectangles, parallelograms and rhombuses have two pairs of parallel lines.
Out of these, only squares and rectangles have all angles equal.
Out of these, only a square has equal sides.
Suzanne's quadrilateral must be a square.

Practising skills

1 Copy and complete this table.

Shape	Name of shape	How many pairs of parallel sides	How many pairs of equal sides	How many lines of symmetry	Order of rotational symmetry
	rectangle	2			
			2		
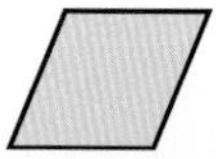					
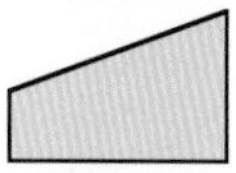					
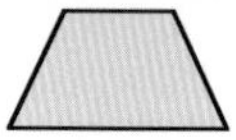					0
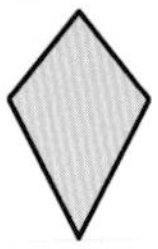					
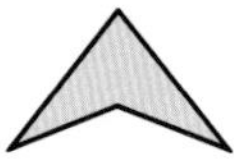				1	

2 What types of quadrilateral must have four equal sides?

3 What types of quadrilateral must have two pairs of parallel sides?

4 Draw all the special quadrilaterals whose diagonals cross at right angles.

5

Who is right?

Give a reason for your answer.

6 **a** Draw a kite with one right angle.

b Draw a kite with two right angles.

c Is it possible to draw a kite with exactly three right angles?

d Is it possible for an arrowhead to have a right angle? If so, draw an example.

Reasoning

Reasoning

Developing fluency

Exam-style

1. Draw x and y axes from 0 to 8.
   Plot the points A(2, 2), B(8, 2) and C(6, 5).
   Plot the point D, and write down its co-ordinates, when ABCD is
   a an isosceles trapezium
   b a parallelogram.

2. Draw x and y axes from 0 to 8.
   Plot the points A(1, 2), B(6, 0) and C(4, 5).
   Plot the point D, and write down its co-ordinates, when ABCD is a kite with a right angle.

3. **a** Which special quadrilaterals can be formed by joining two congruent triangles?
   Give their names and draw sketches of them.
   b Which special quadrilaterals can be formed by joining two congruent isosceles triangles?

Reasoning

4. **a** How many different quadrilaterals can you make on a 3 by 3 pinboard?

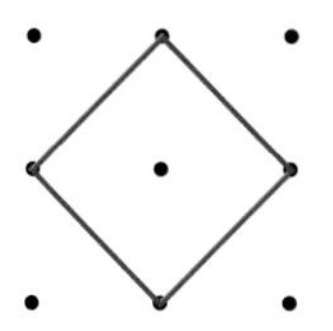

   b How many different quadrilaterals can you make on a 4 by 4 pinboard?

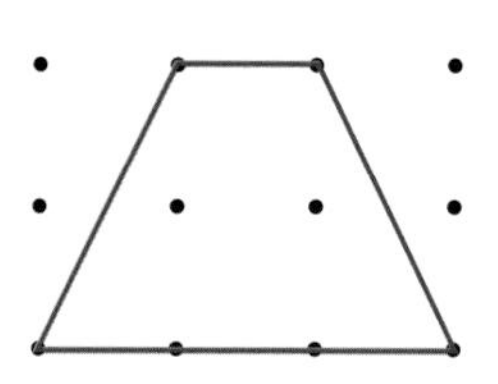

5. The 12 points A to L are equally spaced round a circle.
   a Find which quadrilaterals you can draw by joining some of them.
   Find also which quadrilaterals you cannot draw in this way.
   b Now include the centre O.
   Can you now draw all the quadrilaterals?

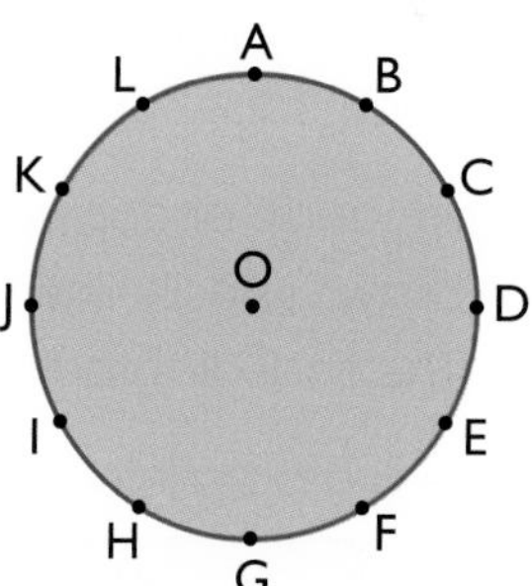

Exam-style

Extension

6. Here is the trapezium PQRS. The angles a and b are shown in the diagram.

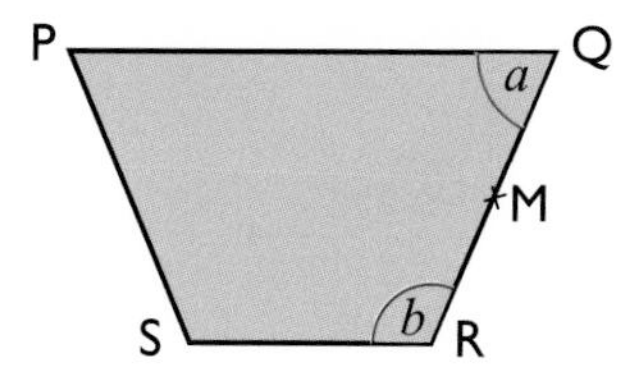

M is the midpoint of QR.
   a PQRS is rotated through 180° about M to get P′Q′R′S′.
   Draw a diagram showing PQRS and P′Q′R′S′ together.
   b Show that $a + b = 180°$.

Problem solving

1 The diagram shows a pattern called a **tangram**.

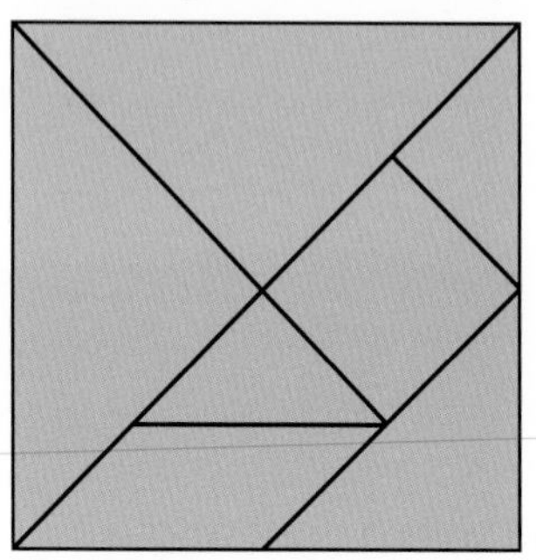

The diagram has been drawn accurately.

Show how three of the parts of the tangram can be used to make a rectangle.

Exam-style

2 The diagram shows a square inside a rhombus. One of the angles of the rhombus is 68°.

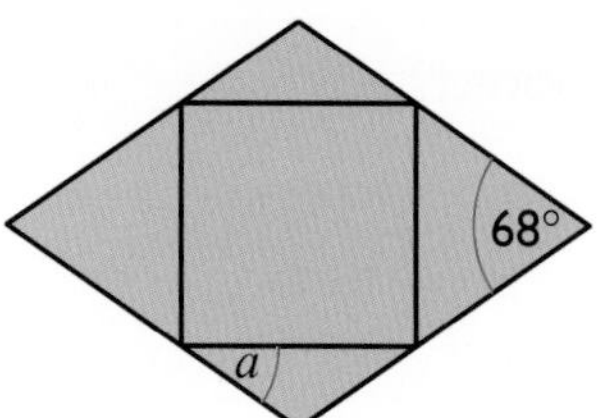

The corners of the square are the midpoints of the sides of the rhombus.

Find the size of the angle a.

3 Here is a parallelogram.

Make a copy of the parallelogram.

a Draw a single line which divides the parallelogram into two identical trapeziums. Each trapezium must have two right angles.

Make another copy of the parallelogram.

b Draw a single line which divides the parallelogram into two identical trapeziums. The final diagram must have the same order of rotational symmetry as the parallelogram.

Exam-style

4 Here is a parallelogram.

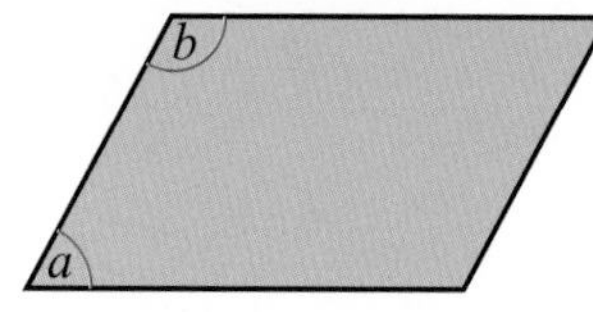

Use the properties of the angles of a parallelogram to show that $a + b = 180°$.

Exam-style

5) ABCD is a parallelogram. Angle CBF = 68°.
E is a point on the line ABF. Angle ADE = angle EDC = x°.
Find the value of x.

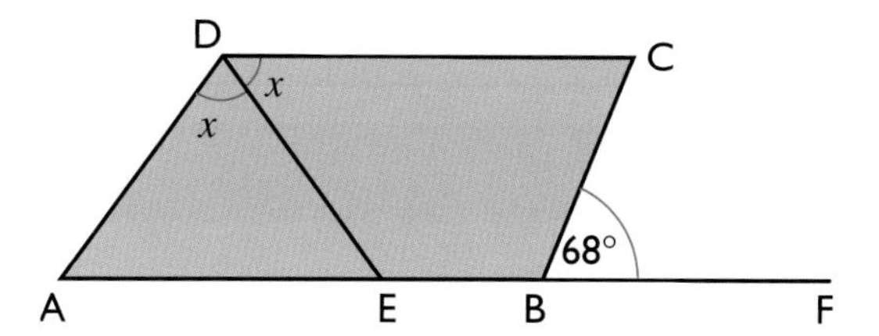

Exam-style

6) The diagram shows a rhombus PQRS. Angle QPR = 42°.

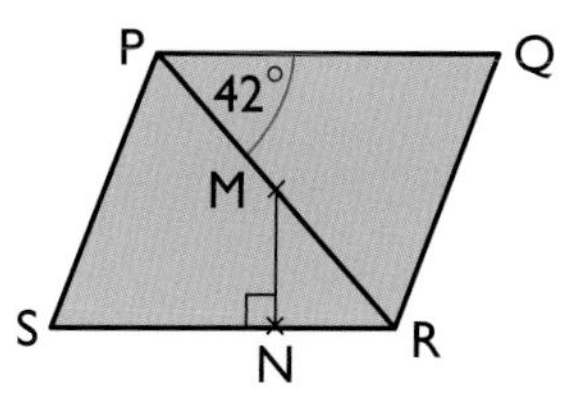

M is the midpoint of PR.
Find the size of the angle RMN.

Extension

Reviewing skills

1) Describe each of these shapes as fully and accurately as you can.

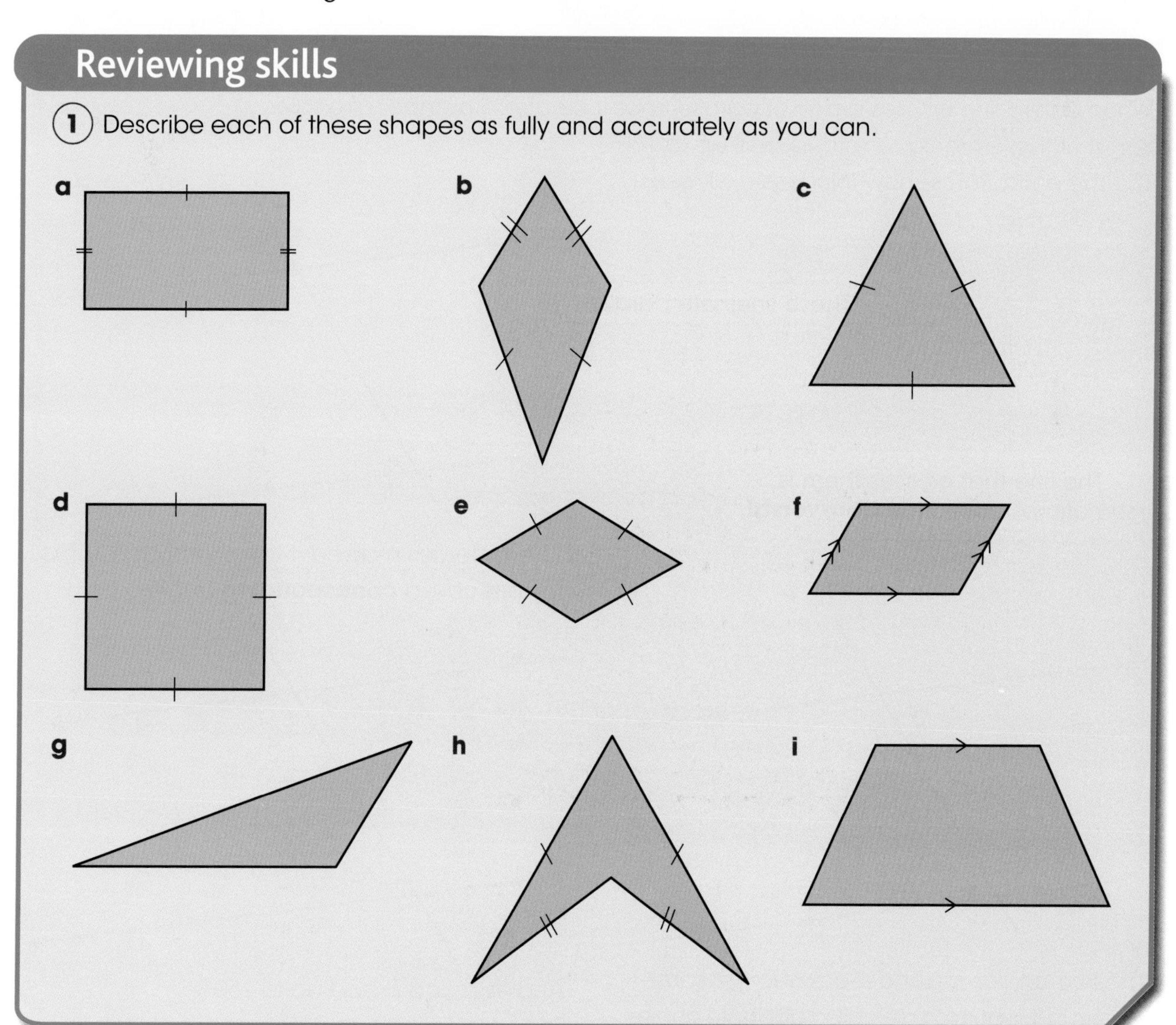

Unit 7 • Angles and parallel lines • Band f

Outside the Maths classroom

Chart navigation

Why would a navigator need to use a parallel ruler?

Toolbox

Parallel lines go in the same direction. They never meet, no matter how far they are extended.
When a third line crosses a pair of parallel lines it creates a number of angles.
Some of these are equal; others add up to 180°.
The three diagrams show things you will see.

These lines are parallel.

The line that crosses them is sometimes called the **transversal**.

The two angles marked a are exactly the same.
They are called **corresponding** angles.

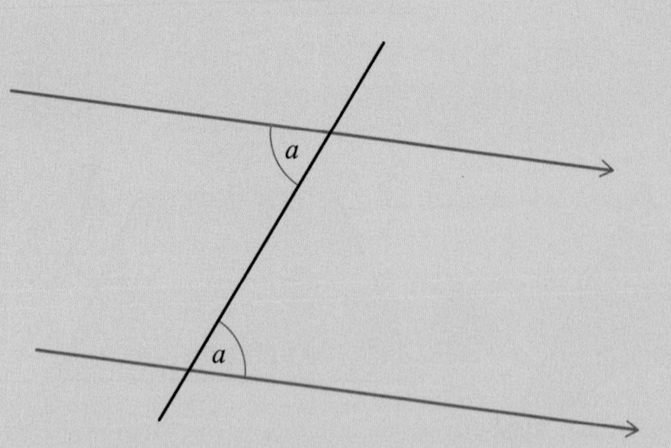

The two angles marked a are also the same.
Angles in this position are called **alternate** angles.

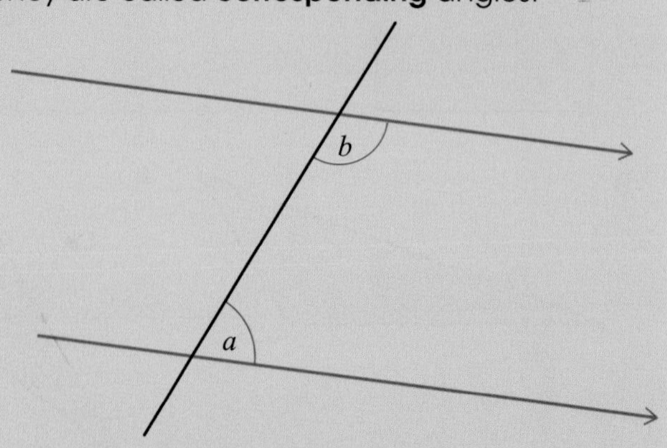

Angles a and b add up to 180°.
They are called **supplementary** angles.

Example – Using alternate angles

The diagram shows a pair of parallel lines and an intersecting line.

Work out the size of the lettered angles.
Give a reason for each of your answers.

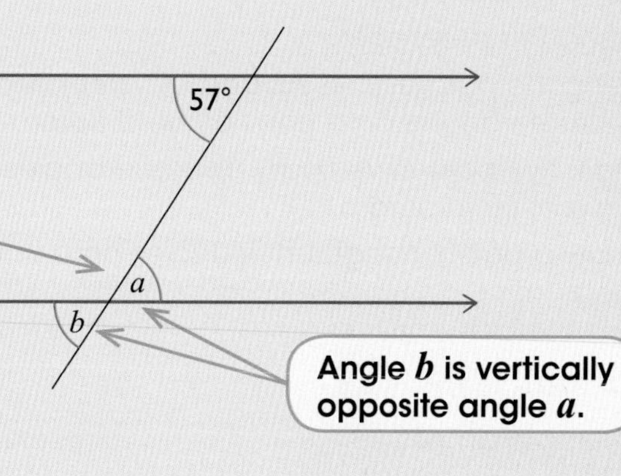

Solution

Alternate angles are equal.

$a = 57°$

Opposite angles are equal.

$b = 57°$

Example – Using supplementary and corresponding angles

The diagram shows a pair of parallel lines and an intersecting line.

Work out the size of the lettered angles.
Give a reason for each of your answers.

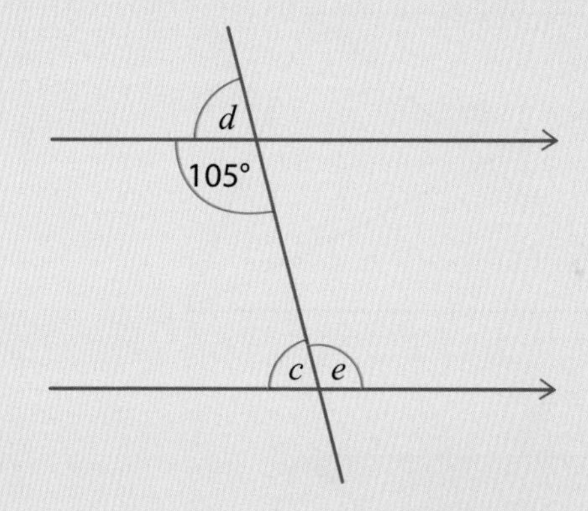

Solution

Supplementary angles add up to 180°.

$c = 180 - 105 = 75°$

c and d are corresponding angles.

Corresponding angles are equal.

$d = 75°$

Alternate angles are equal.

e is an alternate angle to the given angle.

$e = 105°$

Example – Finding angles in a parallelogram

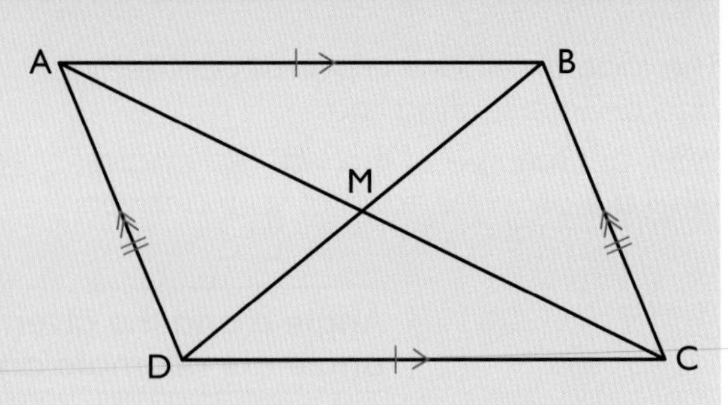

ABCD is a parallelogram.

Its diagonals meet at M.

a Using the parallel lines AB and DC, mark two pairs of alternate angles on a copy of the diagram.

b Mark two pairs of vertically opposite angles on a copy of the diagram.

c Show a pair of congruent triangles on a copy of the diagram.

Solution

a Alternate angles

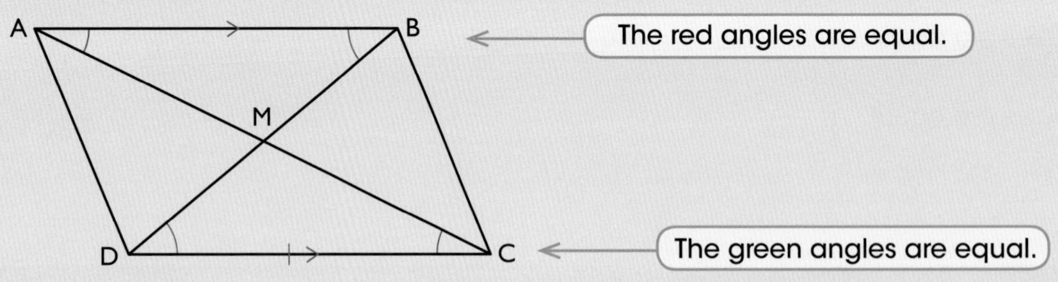

b Vertically opposite angles

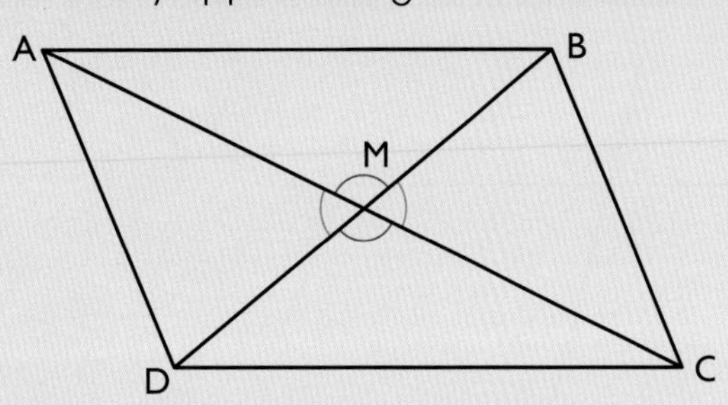

c Congruent triangles

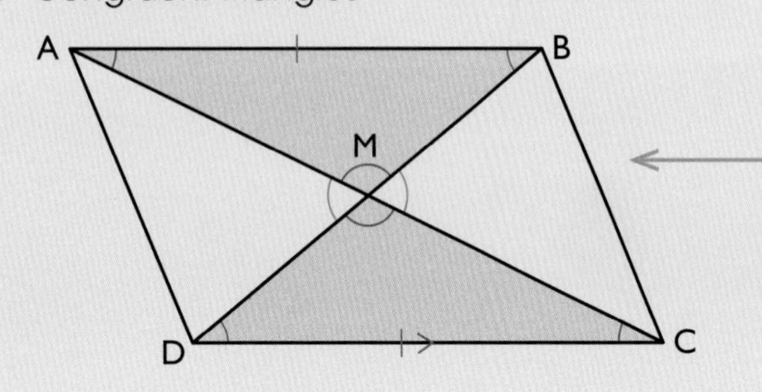

The three angles of these triangles are the same so they are the same shape.
Also AB and CD are the same length so they are the same size.

Practising skills

1) Find the size of each lettered angle in these diagrams.

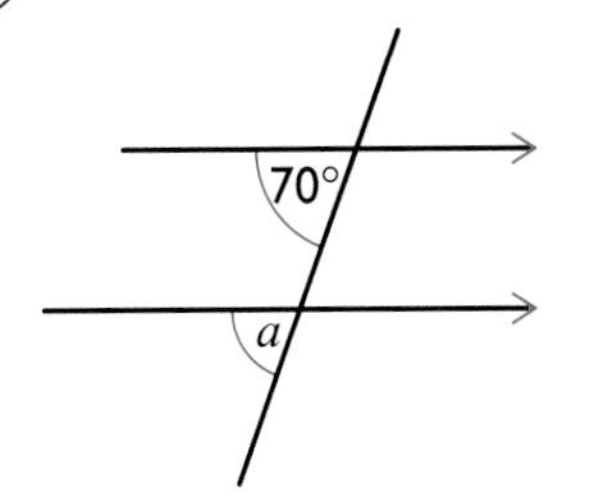

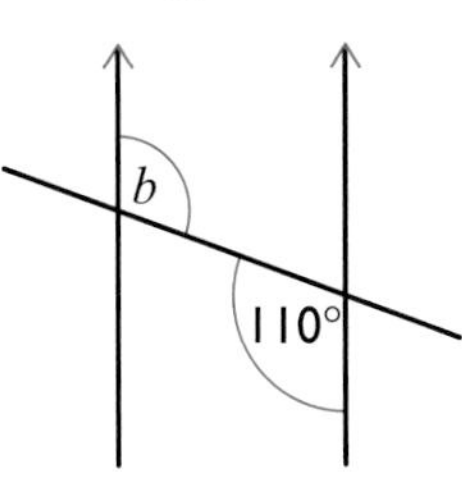

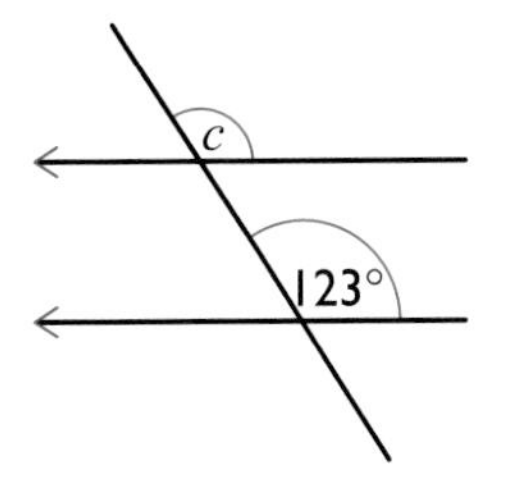

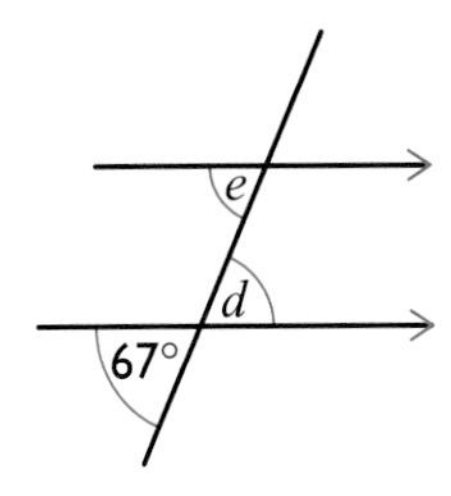

2) Find the size of each lettered angle in these diagrams.

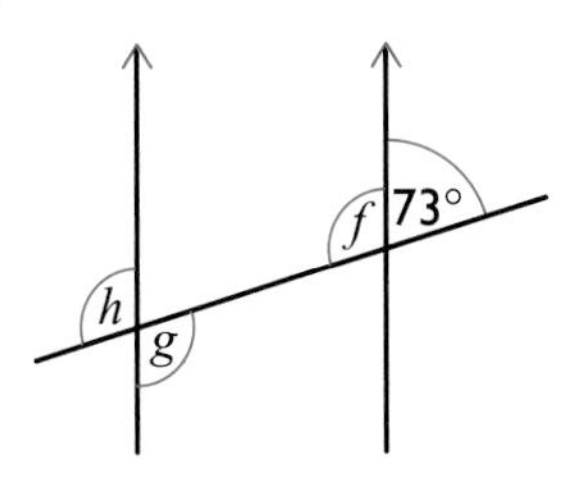

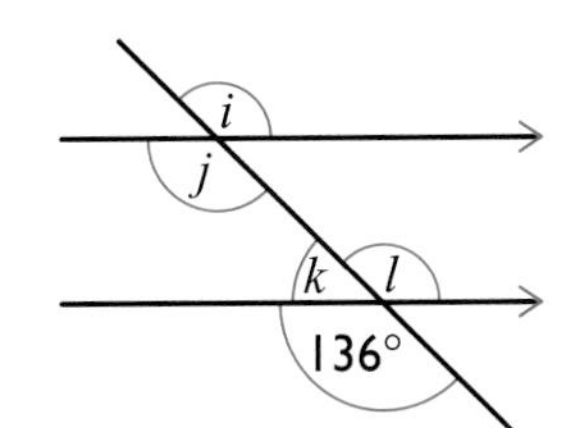

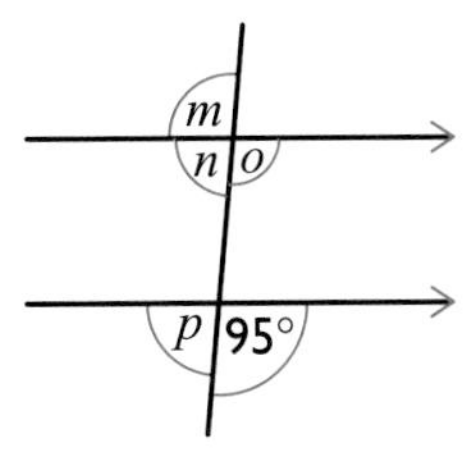

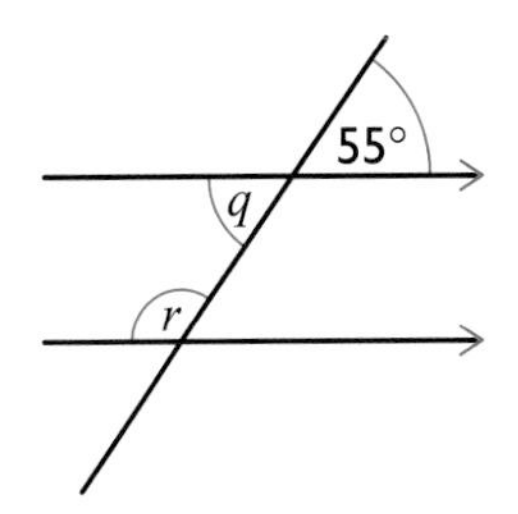

Reasoning

3) Joe says that in this diagram, there are only two different sizes of angle marked.
The angle a is 50°.

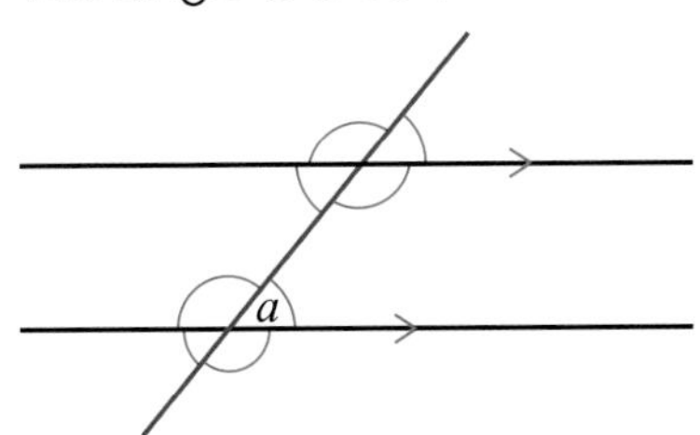

Find the values of the other angles in the diagram.
Is Joe right?

4) Write down the sizes of the lettered angles in this diagram.
For each one give your reason.

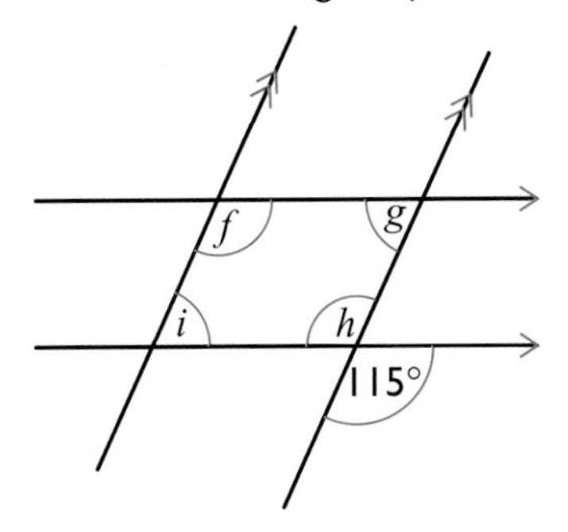

Developing fluency

1. This fence is going down a hill.
   The posts are vertical.
   The wires are parallel.
   a What is angle a?
   b What angle do the wires make with the horizontal?

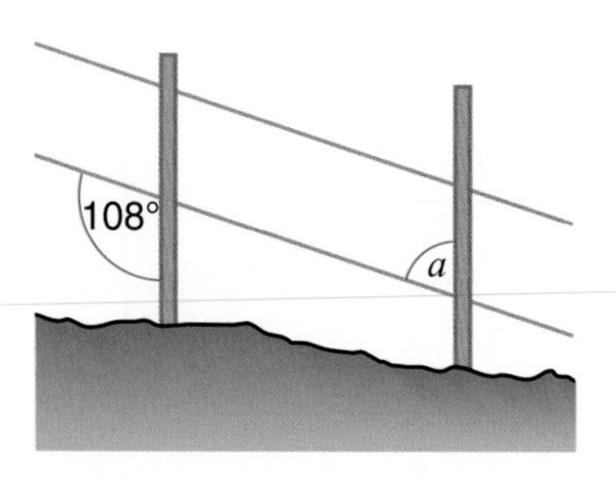

2. The diagram shows part of a fence panel.
   Find the angles a, b, c and d.

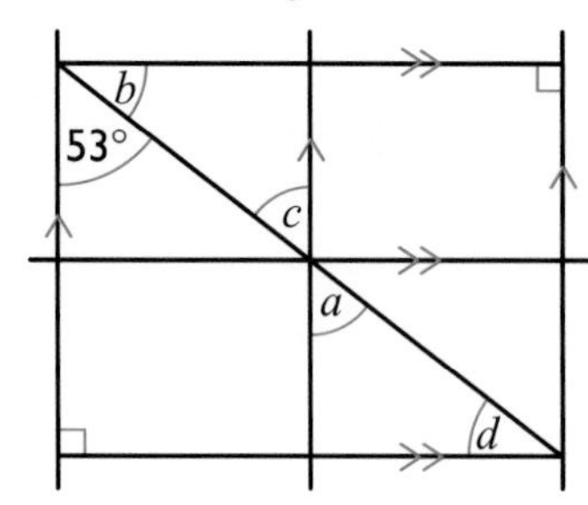

Exam-style

3. Write down the sizes of the lettered angles in these diagrams.
   For each one give your reason.

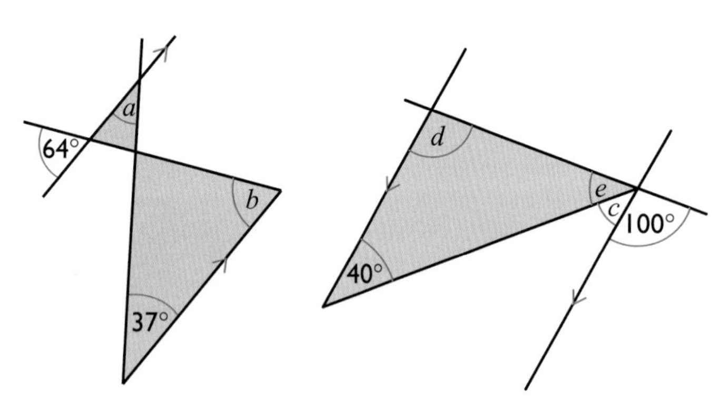

Exam-style

4.

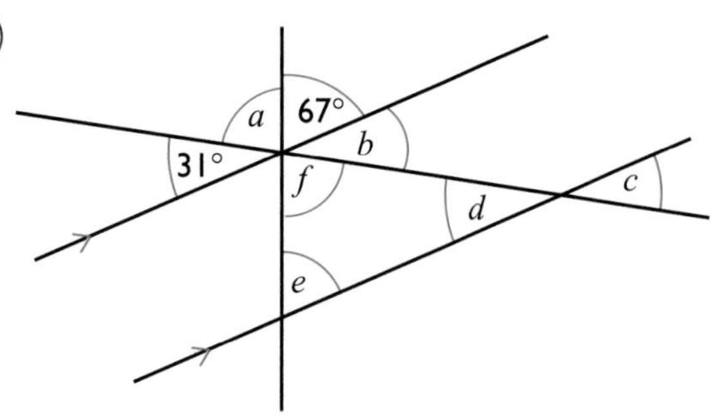

   a Write down the size of each lettered angle.
   For each one give your reason.
   b What is $f + d + e$?

Exam-style

5) This diagram represents an electricity pylon.
Find the sizes of the angles marked a, b and c. Give your reasons.

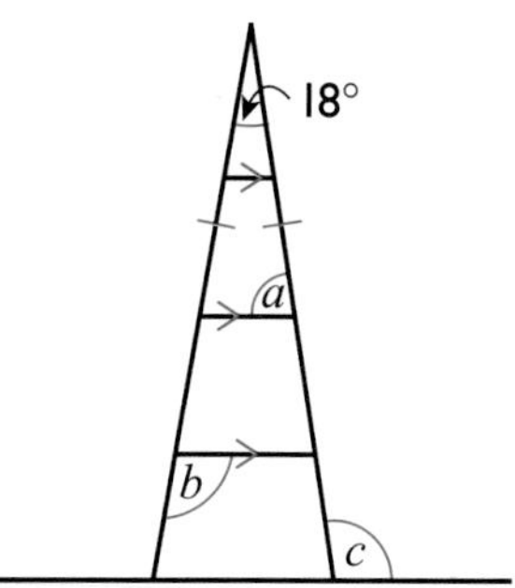

Exam-style

6) In the diagram, the straight lines ABC and DEF are parallel.

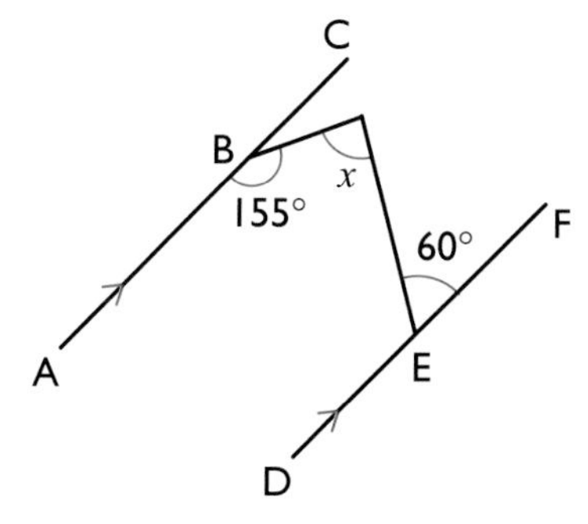

Find the size of angle x. Give your reasons.

Problem solving

Exam-style

1)

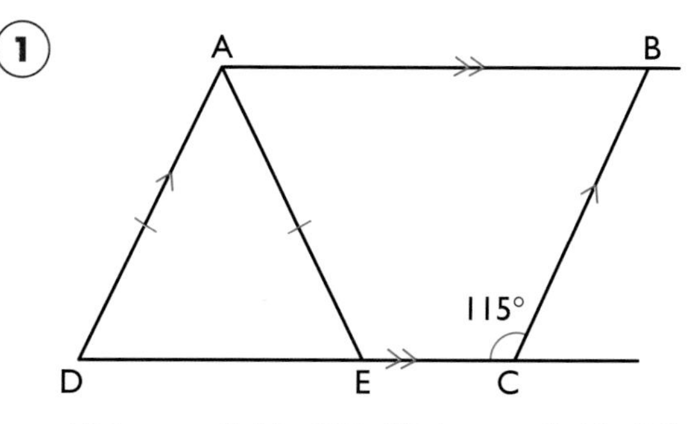

AB is parallel to DC. AD is parallel to BC.
E is a point on DC such that ADE is an isosceles triangle.
Find the size of angle DAE. Give your reasons.

Exam-style

2) This diagram has a single line of symmetry and a pair of parallel lines.

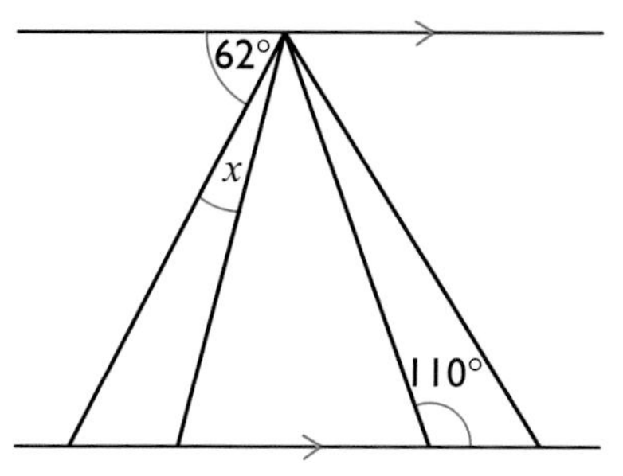

Find the size of angle x. Give your reasons.

Extension

Exam-style

3 The diagram shows a square and a pair of parallel lines through two of its vertices.

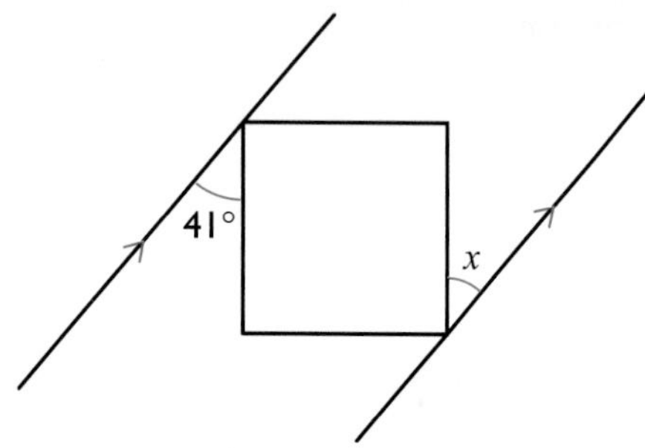

Jim thinks that $x = 41°$.
Is Jim correct?
Give a reason for your answer.

Extension

Exam-style

4 The diagram shows an equilateral triangle and a pair of parallel lines, through two of its vertices.
Find the value of x. Give your reasons.

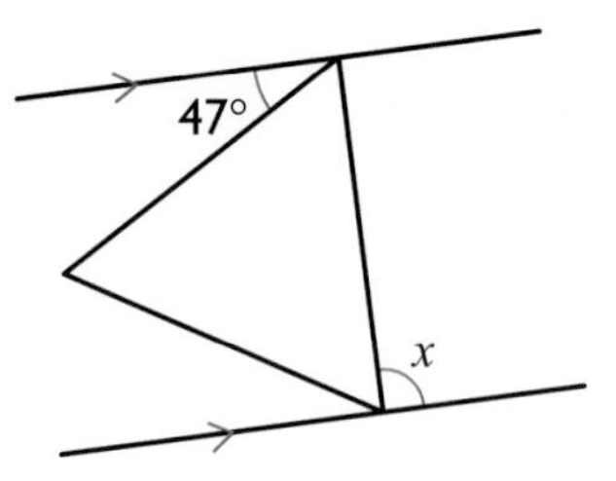

Exam-style

5 ABC and DEF are parallel lines.
Angle ABE is three times angle BED.
Work out the size of angle ABE.

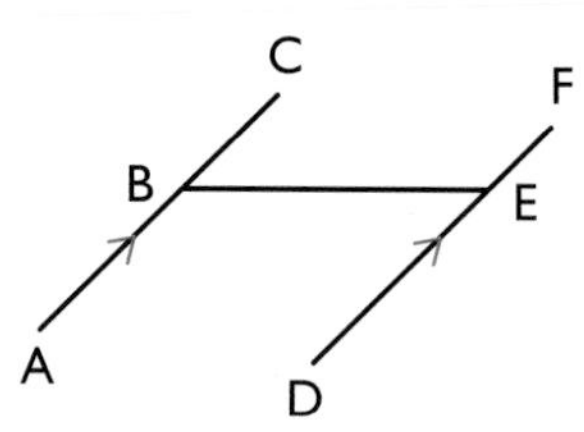

Extension

Exam-style

6

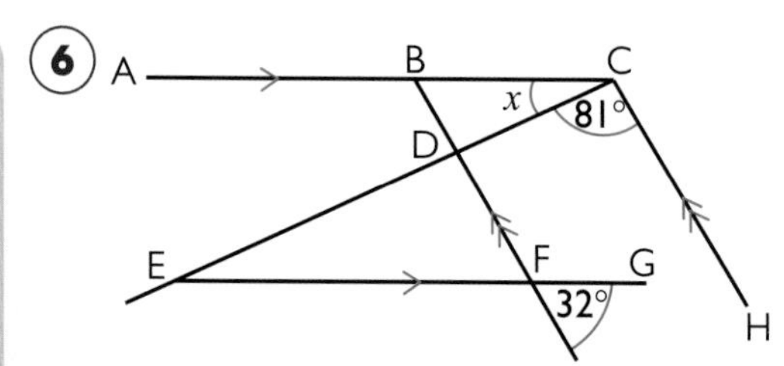

ABC, EDC, BDF and EFG are straight lines.
AC is parallel to EG.
BF is parallel to CH.
Find the value of x.

Extension

Reviewing skills

1. Find the size of each lettered angle in these diagrams.
For each one write down the angle fact(s) that you use.

a

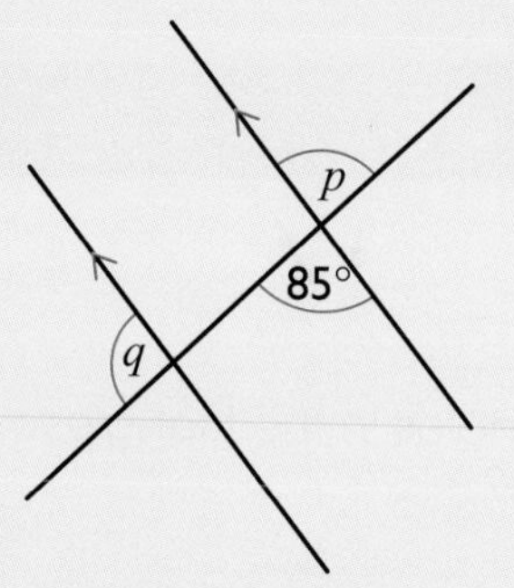

b

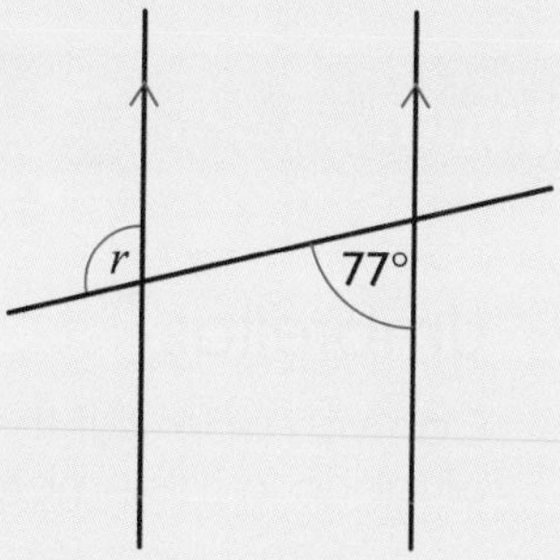

c

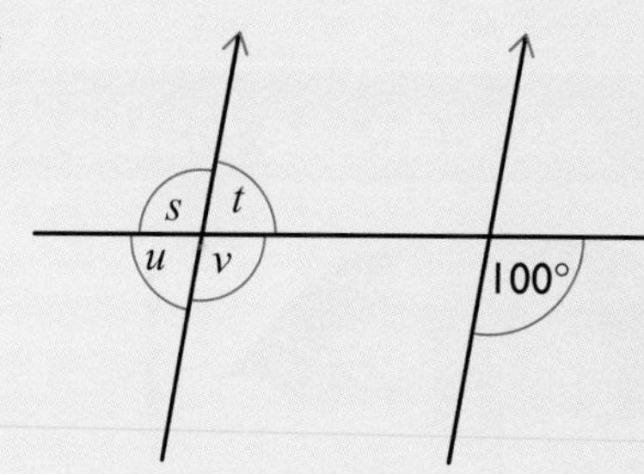

2. Jack makes a wire fence.

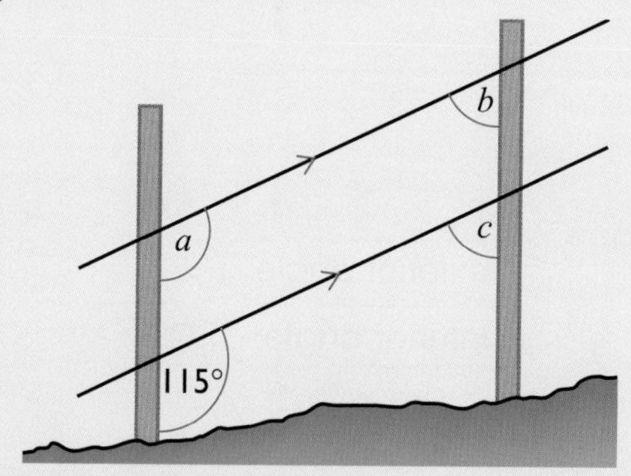

The posts are vertical.
The wires are parallel.
Work out the size of the angles a, b and c.

3. In the diagram, ABCD is a parallelogram.

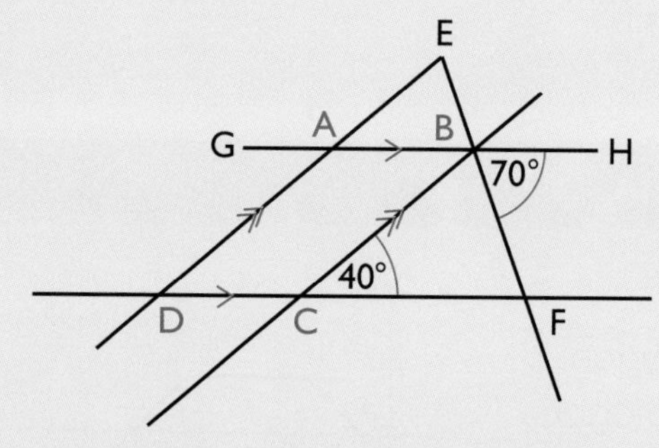

Angle BCF = 40° and angle HBF = 70°.
Show that triangle EDF is an isosceles triangle.

Unit 8 • Angles in a polygon • Band g

Outside the Maths classroom

Umbrellas

Can you work out the angles in the triangles of fabric in this umbrella?

Toolbox

An **interior angle** is the angle inside the corner of a shape.

An **exterior angle** is the angle that has to be turned through to move from one side to the next.

At each vertex,
interior angle + exterior angle = 180°.

In a **regular polygon** all of the interior angles are equal and all of the exterior angles are equal.

The exterior angles of a polygon always make one complete turn and so **add up to 360°**.

exterior angle

interior angle

The sum of the interior angles of a polygon $=180(n - 2)°$ where n is the number of sides.

For example, the sum of the interior angles of a hexagon is
$180 \times (6 - 2) = 720°$.

Example – Finding angles of regular polygons

The diagram shows a regular octagon.

a Work out the size of an exterior angle of a regular octagon.

b Hence find the size of an interior angle of a regular octagon.

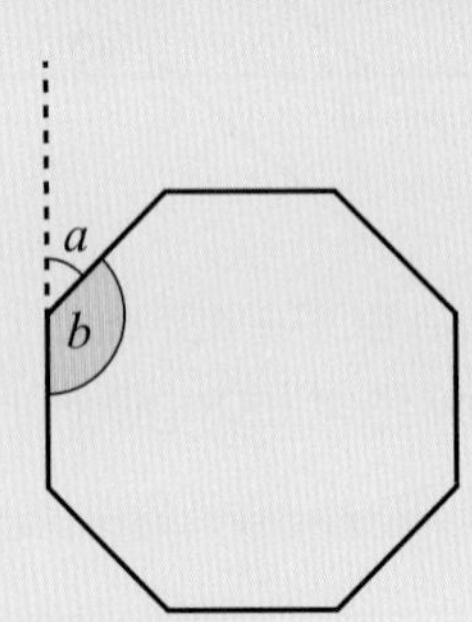

Solution

a The exterior angles of a polygon add up to 360°.
For a regular octagon,
each exterior angle = 360° ÷ 8 = 45° ← This is angle a in the diagram.

b Interior angle + exterior angle = 180°
For a regular octagon,
each interior angle = 180° – 45° ← This is angle b in the diagram.
= 135°

Example – Finding the number of sides of a regular polygon

A regular polygon has an interior angle of 162°.
How many sides does the polygon have?

Solution

First find the exterior angle.

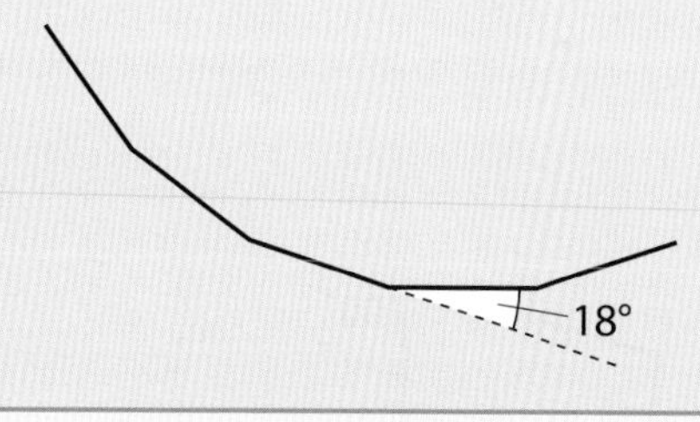

Interior angle + exterior angle = 180°

162° + exterior angle = 180°

exterior angle = 180° – 162° = 18°

The exterior angles add up to 360°.

Number of exterior angles = 360 ÷ 18 = 20

The polygon has 20 sides. ← **The number of sides is the same as the number of exterior angles.**

Practising skills

1. **a** Calculate the exterior angle, a, of a regular hexagon.
   b Calculate the interior angle, b, of a regular hexagon.

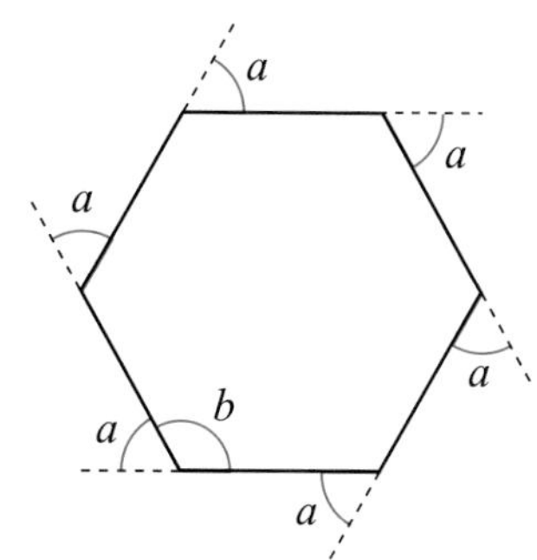

2. **a** Calculate the exterior angle of a regular octagon.
   b Calculate the interior angle of a regular octagon.

3. Copy and complete the following table:

Regular polygon	Number of sides	Size of each exterior angle	Size of each interior angle	Sum of interior angles
Equilateral triangle				
Square				
Pentagon				
Hexagon				
Octagon				
Decagon				
Dodecagon				
Pendedecagon	15			
Icosagon	20			

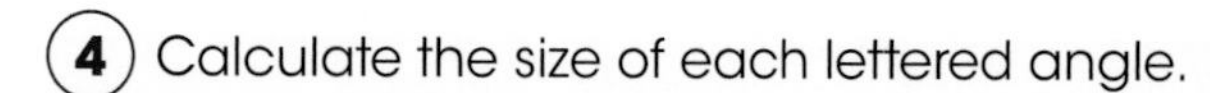

4 Calculate the size of each lettered angle.

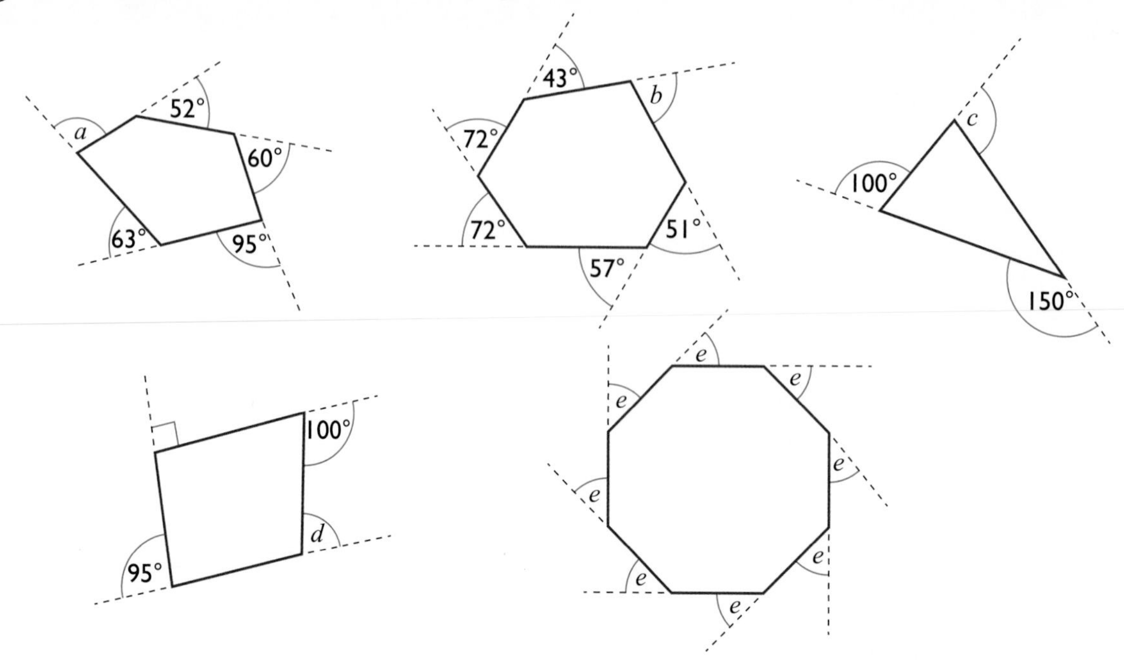

Exam-style

5 The external angle of a regular polygon is 10°.

a How many sides does the polygon have?

b What is the size of one interior angle?

c What is the sum of the interior angles of the polygon?

6 Seb draws a pentagon.

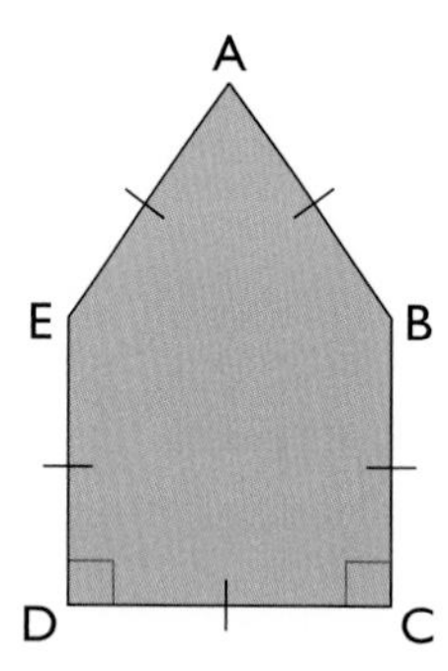

a Is Seb's pentagon regular?
Give a reason for your answer.

b What is the sum of the interior angles of a pentagon?

c Find each of the interior angles in Seb's pentagon.

Developing fluency

1. The diagram shows a pentagon.

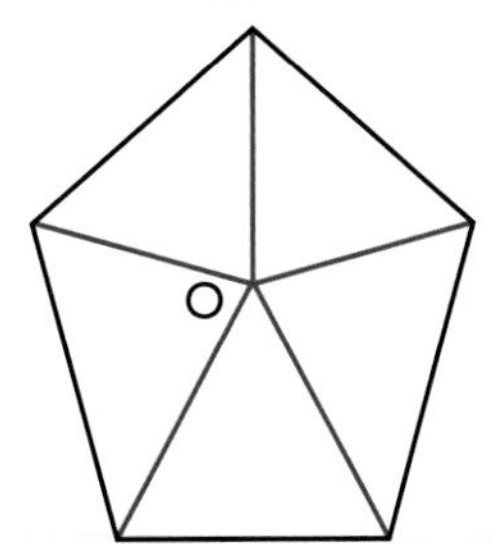

The vertices are all joined to a point O in the middle, making five triangles.

a Work out the total of all the angles in the five triangles.

b What is the total of the angles at the point O?

c Use your answers to parts **a** and **b** to find the total of the interior angles of the pentagon.

d Find the size of each interior angle if the pentagon is regular.

2. This diagram shows a hexagon.
One of the vertices, D, is joined to all the other vertices.

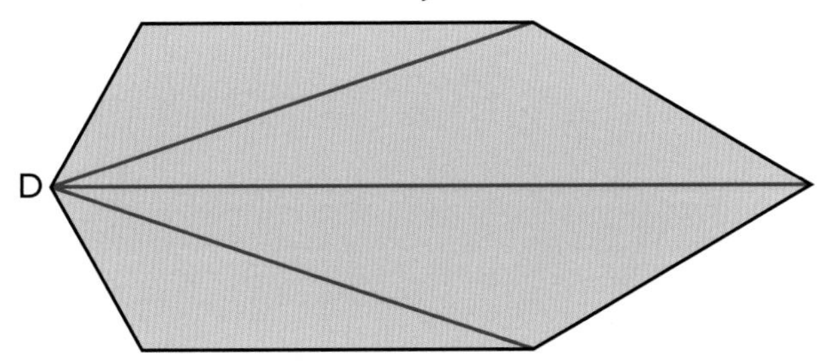

a The lines from D divide the hexagon into triangles. How many triangles are there?

b What is the total of all the angles in the triangles?

c What is the total of all the interior angles of the hexagon?

d Find the size of each interior angle if the hexagon is regular.

3. The diagram shows a hexagon split into six triangles.

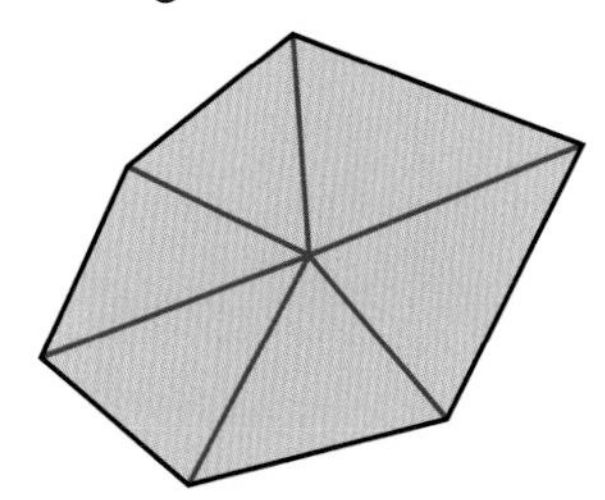

a What is the sum of

  i all the angles in all the triangles

  ii the angles at the centre

  iii the interior angles of the hexagon?

b Use the diagram to explain why the sum of the interior angles of a polygon with n sides is
$n \times 180° - 360°$

c Show that the formula in part **b** is the same as $(n - 2) \times 180°$.

Exam-style

4 A regular polygon has exterior angles of 30°.

a What is the sum of the exterior angles?

b How many sides does the polygon have?

c What is the size of each interior angle?

d What is the sum of the interior angles of the polygon?

Exam-style

5 **a** A regular polygon has an interior angle of 174°.
How many sides does the polygon have?

b The sum of the interior angles of a regular polygon is 1980°.
How many sides does the polygon have?

c A regular polygon has an exterior angle of 10°.
What is the sum of its interior angles?

Reasoning

6 Mark is tiling a floor using polygons for tiles.

a Mark tiles part of the floor using equilateral triangles. How many tiles meet at each vertex?

b Can Mark tile the floor using regular hexagons?
Explain your answer fully.

c Can Mark tile the floor using regular pentagons?
Explain your answer fully.

7

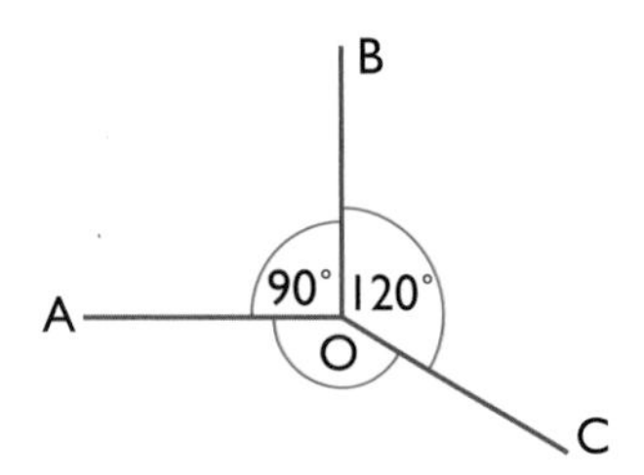

a In the diagram, the lines AO and BO are two sides of a regular polygon, BO and OC are two sides of another regular polygon and OC and OA are two sides of a third regular polygon.
How many sides does each of the three polygons have?
What are their names?

b The line CO is rotated by 15° clockwise about O.
How would you answer part **a** when it is in this position?

Exam-style

8

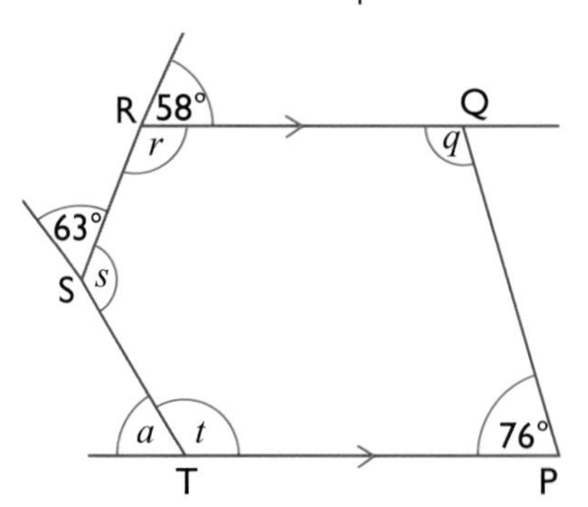

PQRST is a pentagon.
RQ is parallel to TP.

a Find the size of angle q.

b Find the sizes of the other interior angles r, s and t of the pentagon.

c Find the size of angle a.

Problem solving

Exam-style

1. A, B and C are three vertices of a regular 10-sided polygon.
   O is the centre of the polygon.

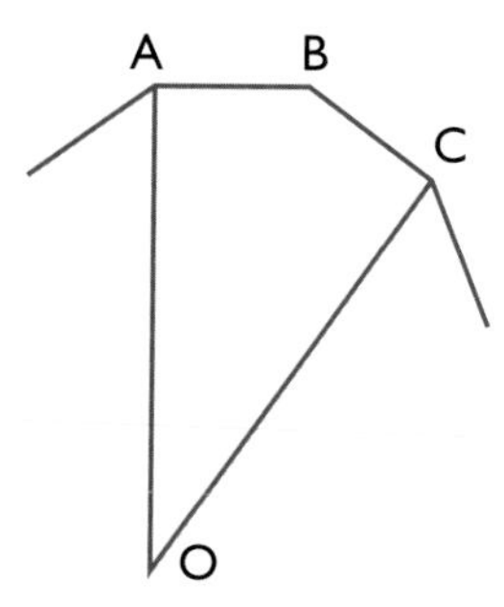

a Calculate the size of angle AOC.

b Show that OABC is a kite.

Exam-style

2. The diagram shows a regular hexagon and a square.

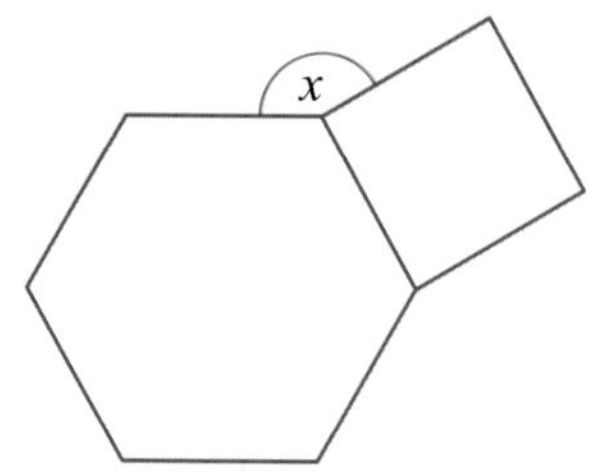

Work out the size of angle x.

Exam-style

3. The diagram shows a regular pentagon.

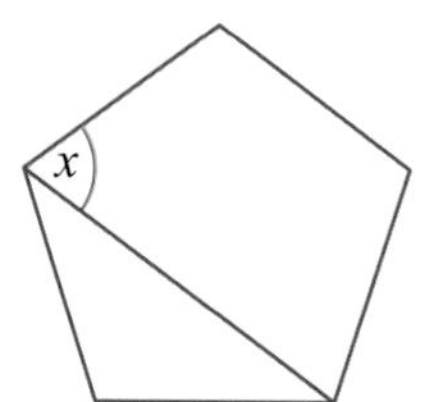

Find the size of the angle marked x.

Exam-style

4. The diagram shows a regular hexagon and a regular pentagon on the same base.

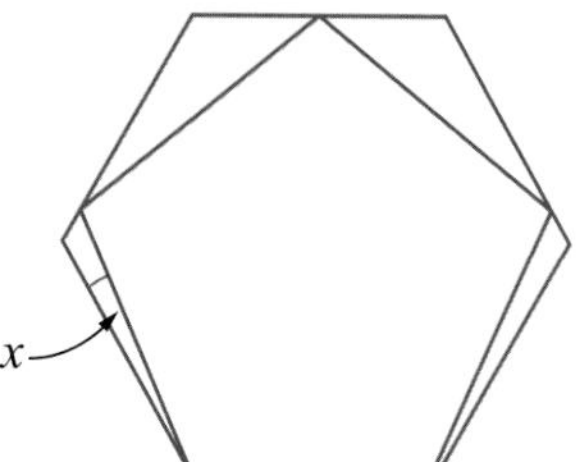

Work out the size of angle x.

Extension

Extension

5 The diagram shows two identical pentagons.

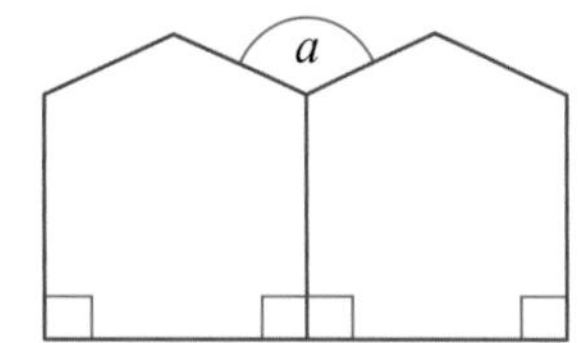

For each pentagon, two of the angles are right angles and the other three angles are the same.
Work out the size of angle a.

6 Here is a regular octagon.

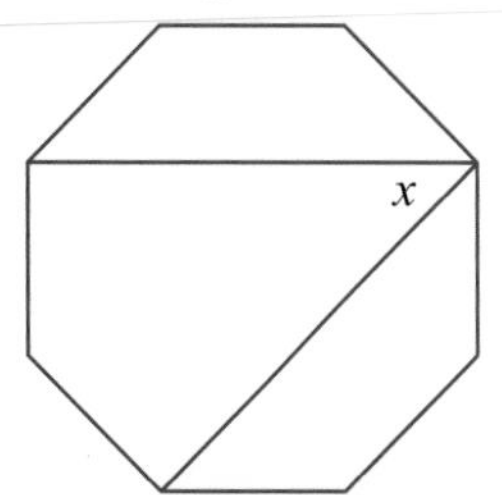

Work out the size of angle x.

Reviewing skills

1 Calculate the size of each lettered angle.

a

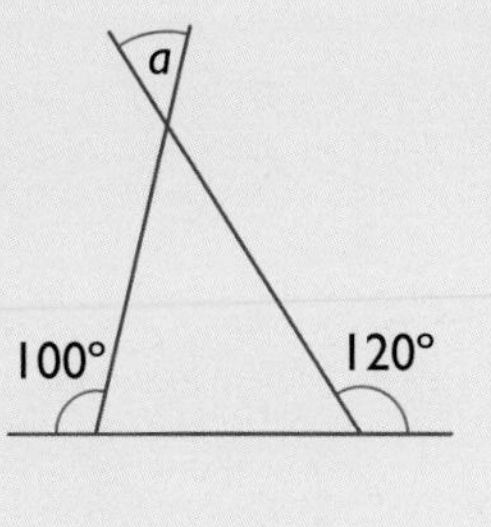

b

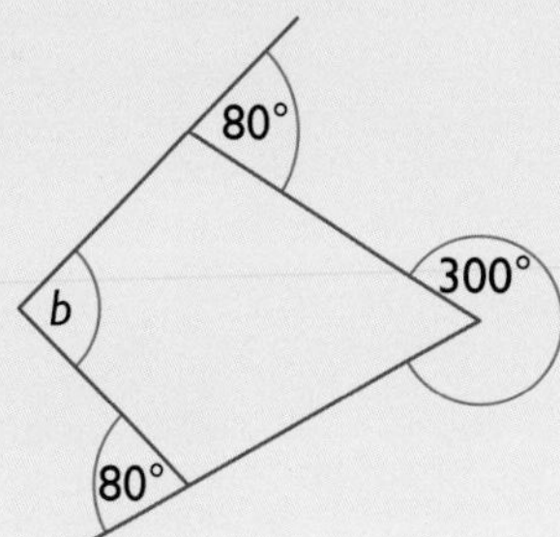

c

2 **a** The exterior angle of a regular polygon is 5°.
How many sides does the polygon have?

b Is there a regular polygon with an exterior angle of 7°?
Explain your answer.

3 **a** Show that the interior angle of a regular dodecagon (12 sides) is 144°.

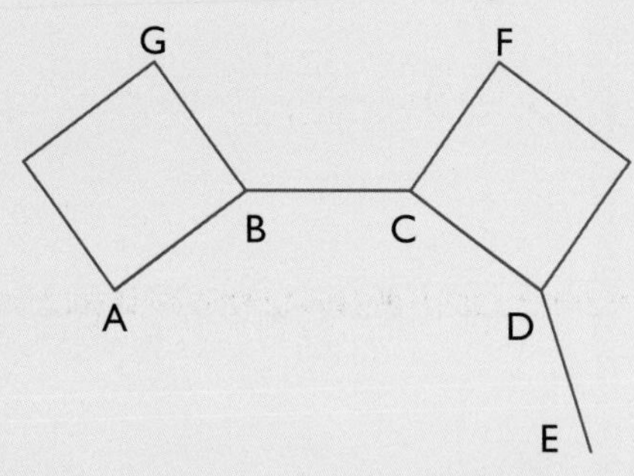

ABCDE is part of a regular dodecagon.
The two quadrilaterals are squares with sides the same length as AB.

b Find the size of the angle GBC.

c Show that angles GBC and FCB are each equal to the interior angle of a regular hexagon.

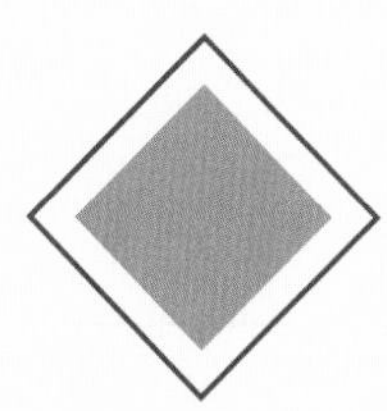

Geometry and Measures
Strand 3 Measuring shapes

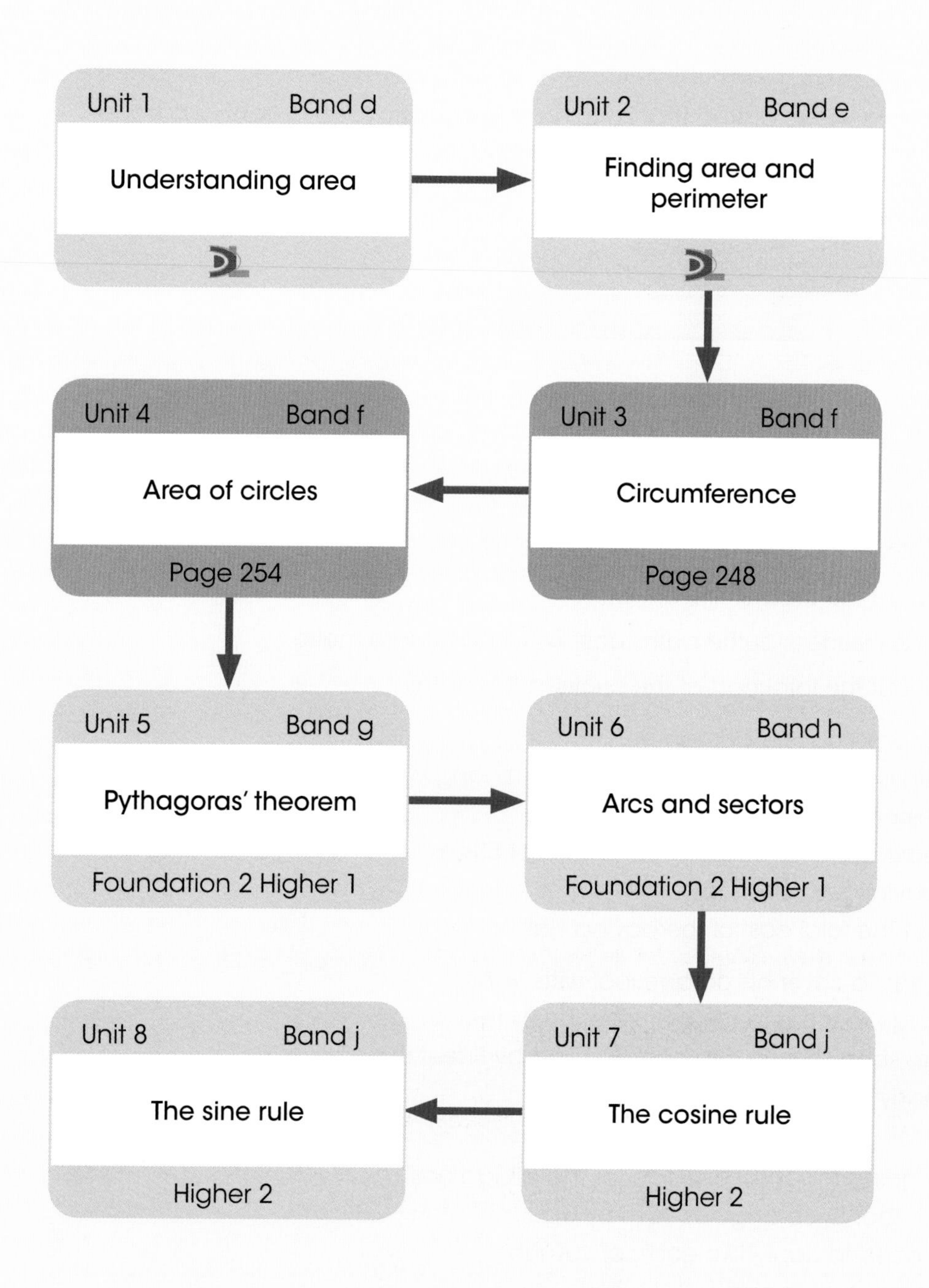

Units 1–2 are assumed knowledge for this book. They are reviewed and extended in the Moving on section on page 246.

Units 1–2 • Moving on

Exam-style

1. The diagram shows a drive that is going to be constructed in front of a house.

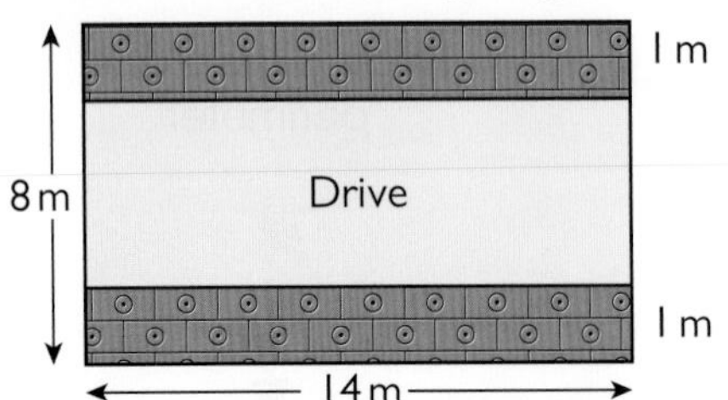

There is a brick path on each side of the drive.
The drive and the paths are rectangles.
The total width of the drive and paths is 8 m.
The width of each path is 1 m.
The length of the drive and of the paths is 14 m.

a Work out the area of the drive.

The bricks needed for the paths cost £48 for a square metre.

b Work out the total cost of the bricks in the paths.

Exam-style

2. Jim lays a path in his garden.
The path is in the shape of a rectangle 8 m long by 1 m wide.
Jim uses paving slabs to make the path.
Each paving slab is a square with sides of 50 cm.
Each paving slab costs £2.
Work out the total cost of the paving slabs.

Exam-style

3. Simon has to cover his garage roof with board.
His garage roof is a rectangle 8 feet by 24 feet.
His local shop sells square boards 4 feet by 4 feet in packs of 5.
How many packs will he need?

Exam-style

4. Alan is going to varnish the floor of the village hall.
The hall floor is rectangular, 12 m by 8 m.
He will need to apply two coats of varnish.
He can buy the varnish in 2 litre tins.
One tin is enough for 15 m^2.
How many tins of varnish will Alan need?

5. Janine has some square mosaic tiles.
Each tile has an area of 1 cm^2.
She can make a rectangle 2 cm by 27 cm using all the tiles.
What other rectangles could Janine make with her tiles?

6 Here is a design for a community hall.

The community hall is in the shape of a rectangle 4 m by 12 m.

The manager wants the kitchen to measure 3 m by 4 m.

The manager claims that the kitchen will take up less than $\frac{1}{6}$ of the area of the community hall.

Is the manager correct?

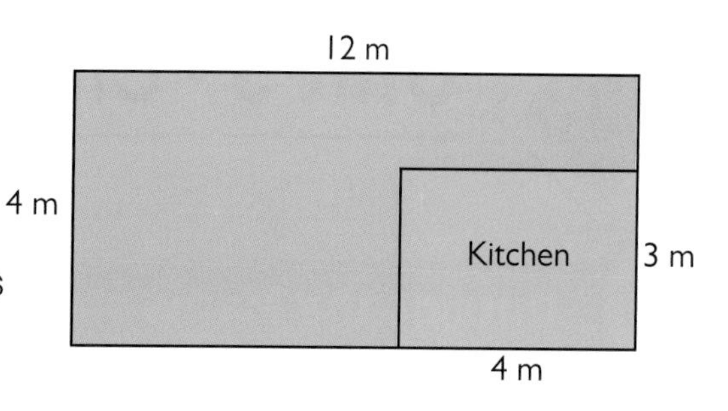

7 The diagram shows the plan of a room.

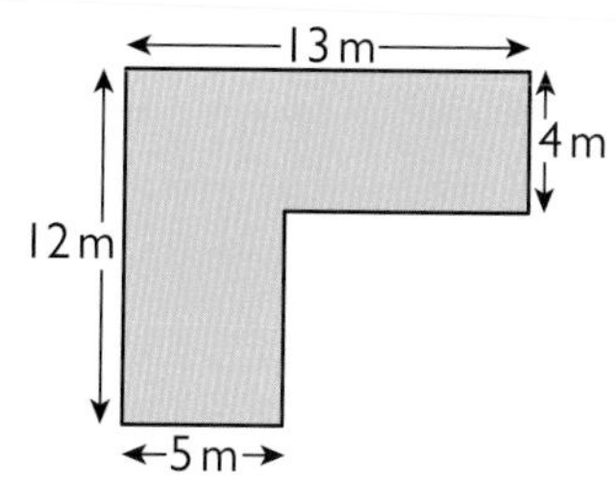

Andy wants to put skirting boards along all the edges of the room.

He will leave two gaps of 1 m each for the doors.

Each skirting board is 4 m long.

He can cut the skirting board to fit.

Andy buys the skirting board in packs of four.

How many packs will he need to buy?

8 Here is a diagram of a lawn.

The lawn is in the shape of a rectangle 9 m by 7 m.

A path is to be put all the way around the lawn using paving slabs.

Each paving slab is a square with sides of 50 cm.

The width of the path is to be 50 cm.

Work out the number of paving slabs that will be needed.

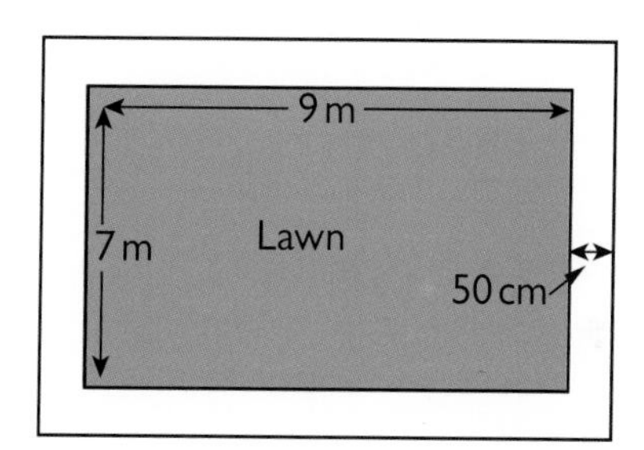

Unit 3 • Circumference • Band f

Outside the Maths classroom

Cogs and wheels

Where are cogs used? How do they work?

Toolbox

The **circumference** is the distance around the edge of a circle.

The **radius** of a circle is the distance from the centre to the circumference.

The **diameter** is the distance across the circle through the centre.

The circumference, C, of a circle of diameter d is

$$C = \pi d.$$

It is also given by

$$C = 2\pi r$$

where r is the radius.

π (pi) is a Greek letter which represents the number value 3.141 592 654… .

An approximate value of π is 3.14.

To be more accurate use π on your calculator.

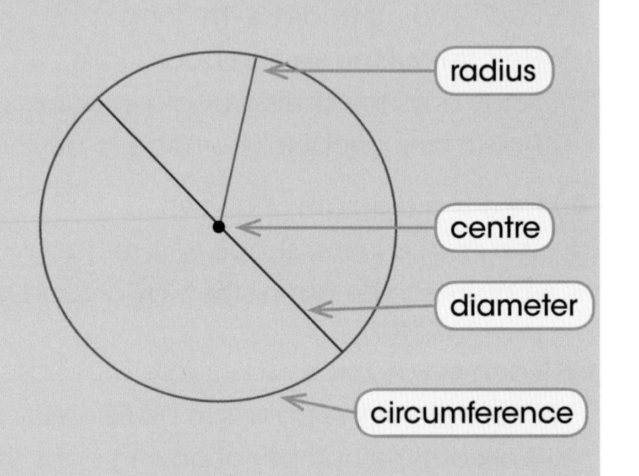

Example – Finding the circumference given the radius

Find the circumference of this circle.

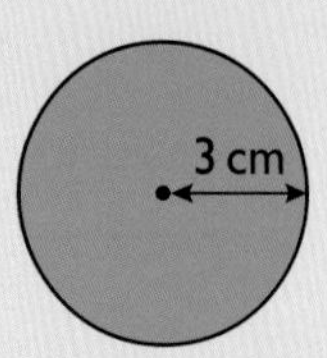

Solution

$C = 2\pi r$ ← The radius is given so use the formula $C = 2\pi r$.

$= 2 \times \pi \times 3$ ← $r = 3$ Use the value of π stored in your calculator or 3.14

$= 18.8$ to 1 d.p. ← Round your answer.

Example – Finding the diameter and radius given the circumference

A circle has a circumference of 5.76 m.

a Calculate the diameter.

b Calculate the radius.

5.76 m

Solution

a $C = \pi d$

$5.76 = \pi d$

$\frac{5.76}{\pi} = d$ ← Divide both sides by π.

$1.833... = d$

So d = 1.83 m to 2 d.p.

b $r = \frac{d}{2}$

$= \frac{1.833...}{2}$ ← Use the unrounded value of d from your calculator.

$= 0.916...$

So r = 0.92 m to 2 d.p.

Practising skills

1 Calculate the circumference of each of these circles. Give your answers to one decimal place.

a

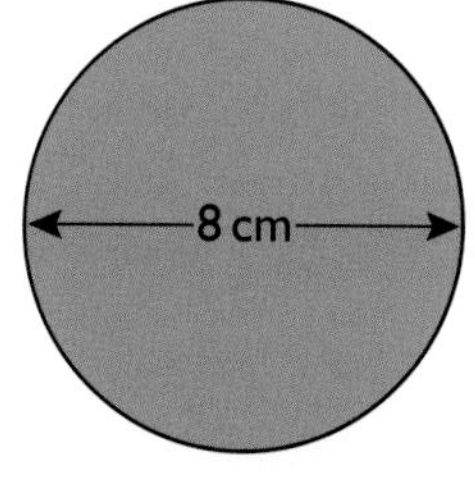

b

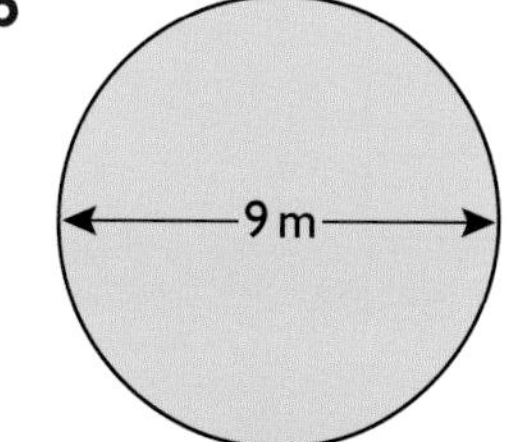

c

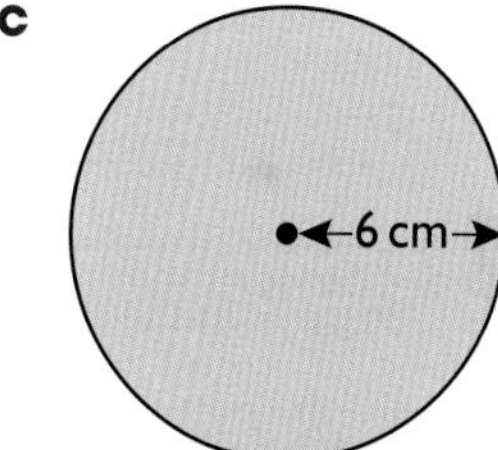

d

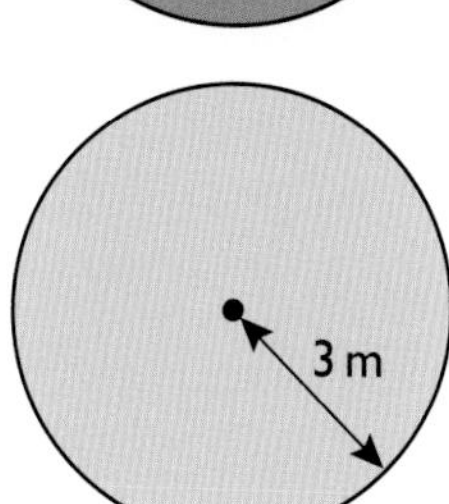

e

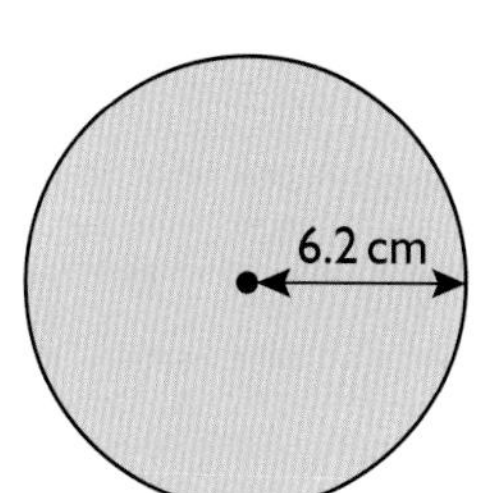

f

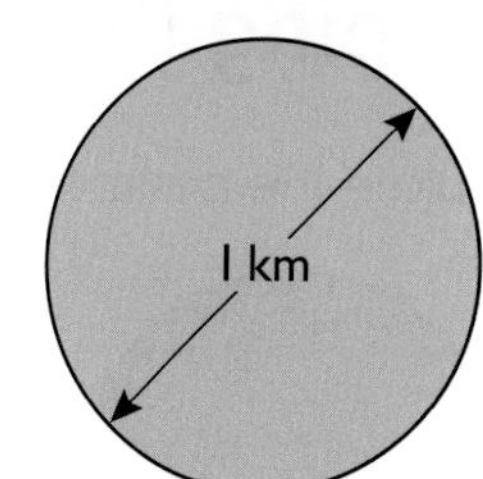

2 Calculate the circumference of each of these circles. Each circle is drawn accurately.

a

b

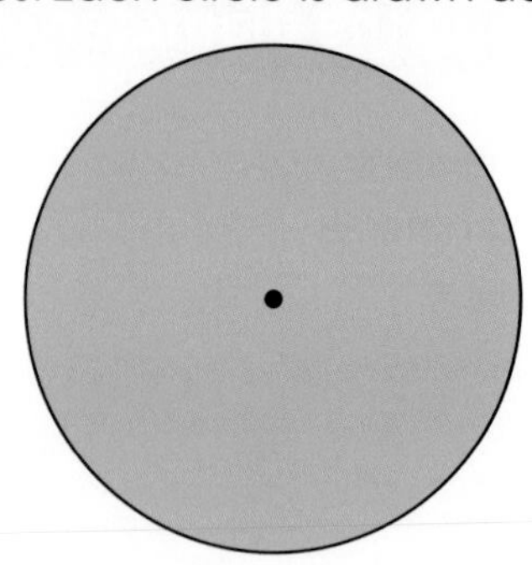

c

d

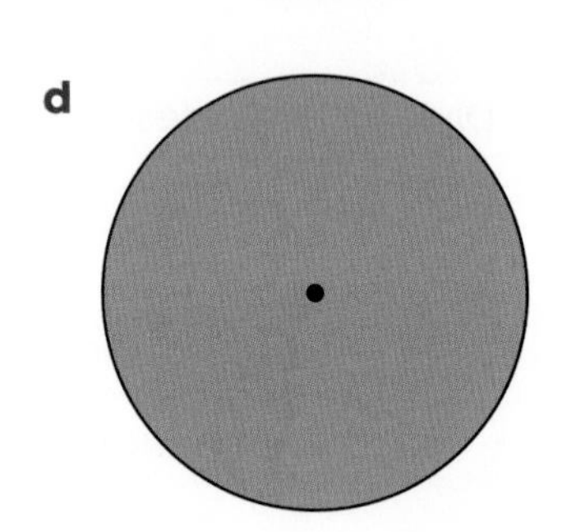

3 A circle has a circumference of 18 cm. Use the formula $C = \pi d$ to calculate its diameter to one decimal place.

4 Copy and complete this table which gives some measurements for circles.

	Radius	Diameter	Circumference
a	2 cm		
b		10 cm	
c		13 cm	
d	2.9 m		
e			57 cm
f			120 cm

Developing fluency

1 Calculate the perimeter of these shapes.

a

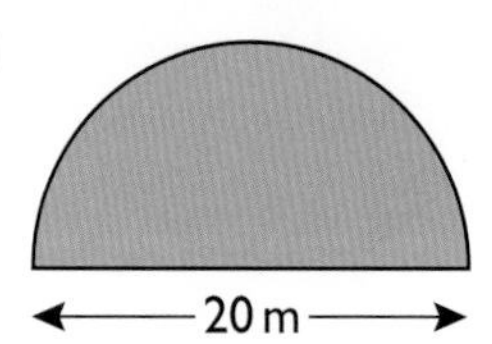

b

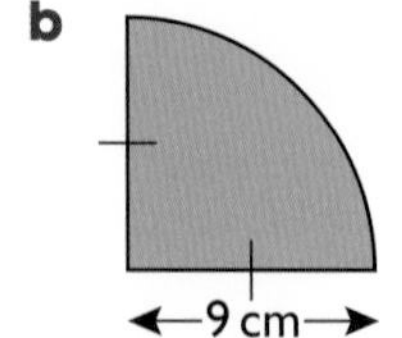

c

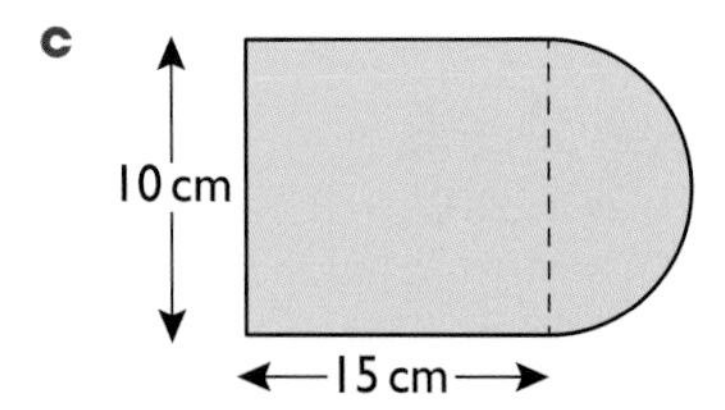

d

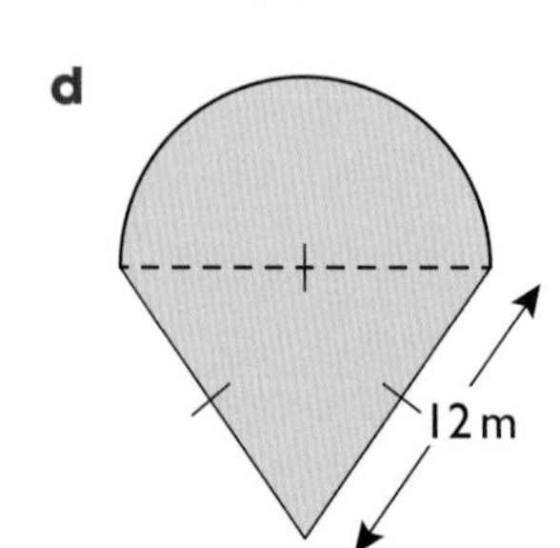

Exam-style

2. Casey is putting ribbon around her circular birthday cake which has a radius of 13 cm. There is an overlap of 2 cm. How long is the piece of ribbon?

3. A 400 m racetrack consists of two 90 m straights and two semicircles. Calculate the distance across the track, x metres.

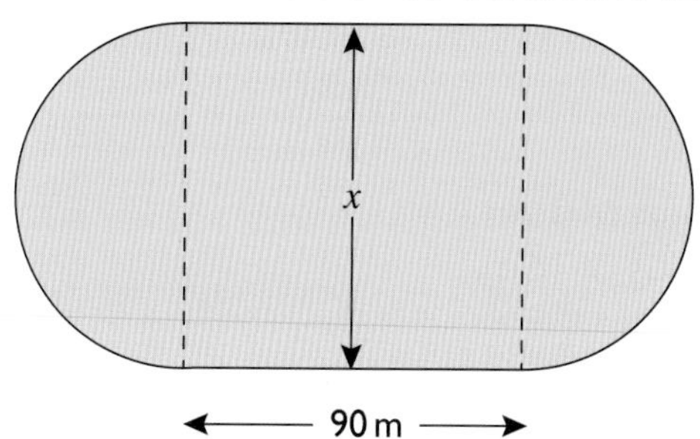

Exam-style

4. The large wheel of a tractor has a diameter of 160 cm. The wheel turns 60 times.
   a How far has the tractor moved?

   The small wheel of the tractor has a diameter of 1 m.

   b How many times has the small wheel turned during this journey?

Exam-style

Extension

5. The minute hand of a clock is 10 cm long and the hour hand is 7 cm long.
   a Calculate the difference in the distance travelled by the tip of each hand between 8 a.m. and 8.15 a.m.
   b Calculate the difference in the distance travelled by the tip of each hand between 6 a.m. and 6.05 a.m.

6. Alex makes a pond for his garden. Its shape is a semicircle.
   The diameter of the semicircle is 6 m.
   He puts tiles round the edge of the pond. Each tile is 10 cm wide.
   Estimate the number of tiles he needs.

Problem solving

Exam-style

1. A train has wheels of diameter 140 centimetres. The train runs 1 kilometre along a track.
   Work out how many complete turns one wheel makes.

2. A factory makes metal bands in the shape of circles to put round barrels.
   The bands at the top and bottom of a barrel each have a diameter of 40 cm.
   The band in the middle has a diameter of 50 cm.
   The factory makes enough metal bands for 800 barrels an hour.
   Work out the total length of the bands made in one hour.

Exam-style

3. The diagram shows a tin of peas and a label.
   The tin is in the shape of a cylinder.
   The diameter of the cylinder is 85 mm.
   The label is in the shape of a rectangle.
   The label must fit all the way around the tin with an overlap of 2 cm.
   Work out the width of the label, w mm.

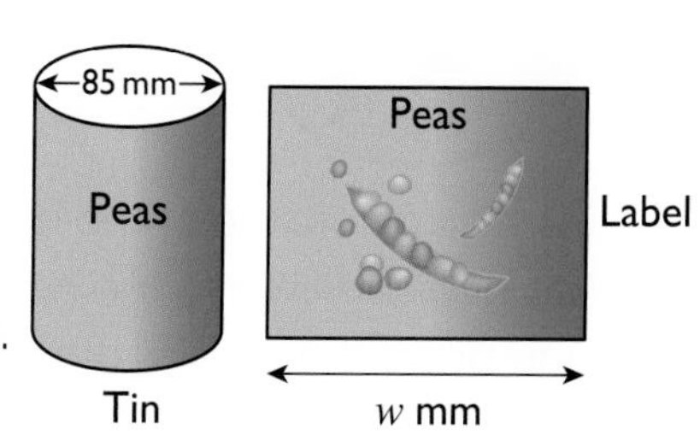

Exam-style

4 A road engineer is making a roundabout.

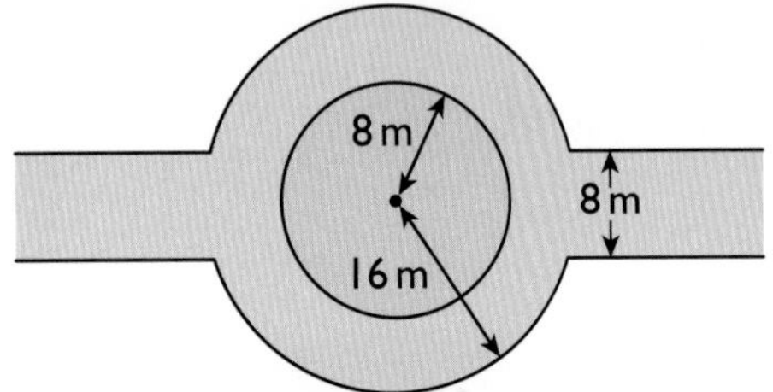

The inner edge of the roundabout is a circle with a radius of 8 m.
The outer edge of the roundabout is a circle with a radius of 16 m.
The circles have the same centre.
The engineer leaves two gaps each of width 8 m for roads to enter the roundabout.
He puts metal barriers around the inner and outer edges of the roundabout, but not in the gaps.
Approximately what length of metal barrier does the engineer use?

Extension

Exam-style

5 Alice is making a mat for her bathroom.
The mat has two ends which are semicircles.
Each semicircle has a diameter of 70 cm.
The mat will have a perimeter of 400 cm.
The length of the straight part of the mat is d metres.
Work out the value of d.

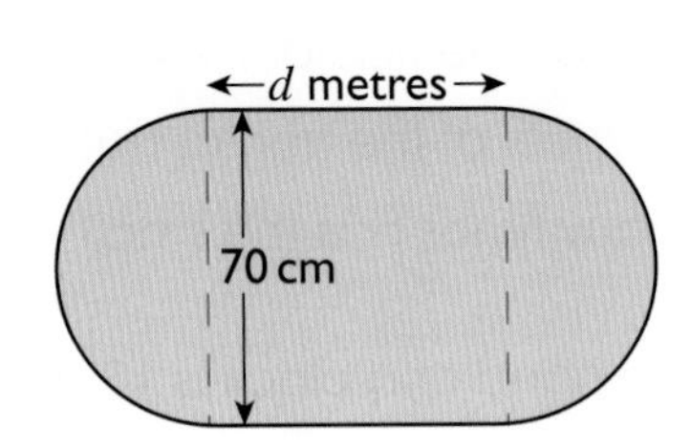

Extension

Exam-style

6 Here is a diagram of a window.

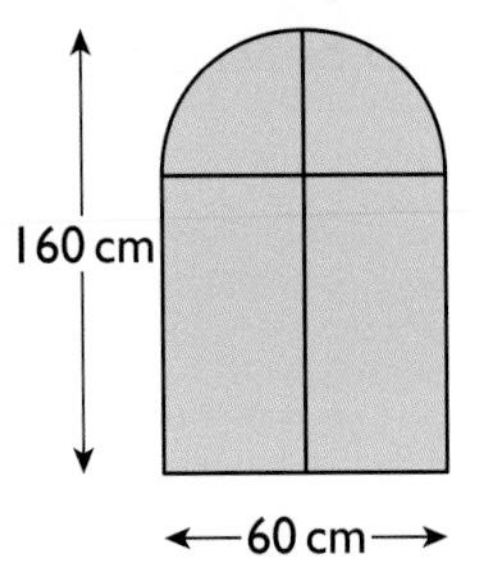

The window is made of two identical rectangles and two identical quarter circles.
The width of the window is 60 cm.
The height of the window is 160 cm.
Each pane of glass is surrounded by lead supports. These are shown by the lines in the diagram.
Work out the total length of the lead supports.

Extension

Exam-style

7 Here is a circular hoop.

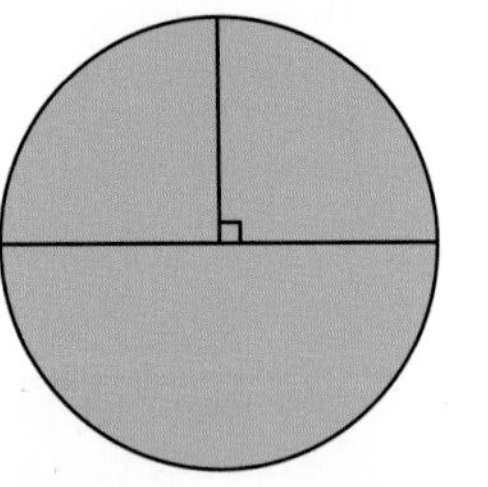

The circumference of the hoop is 2 metres.
Two bars are fixed to the hoop to strengthen it.
One bar is a diameter and the other is a radius of the hoop.
Work out the total length of the two bars.

Extension

Reviewing skills

1 Calculate the circumference of each of these circles. Give your answers to one decimal place.

a

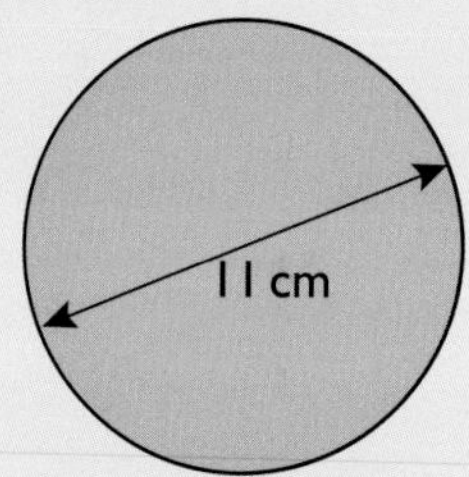

b

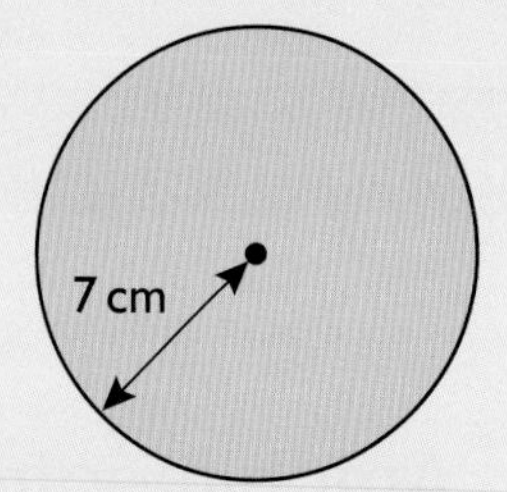

c

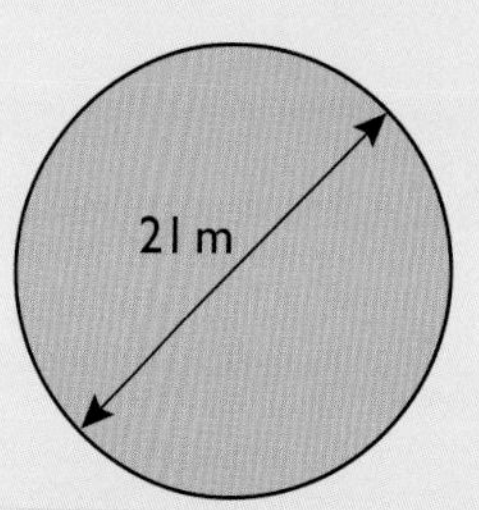

2 A circle has a circumference of 26 cm. Calculate to one decimal place

a the diameter of the circle

b the radius of the circle.

3 Calculate the perimeter of each of these shapes.

a

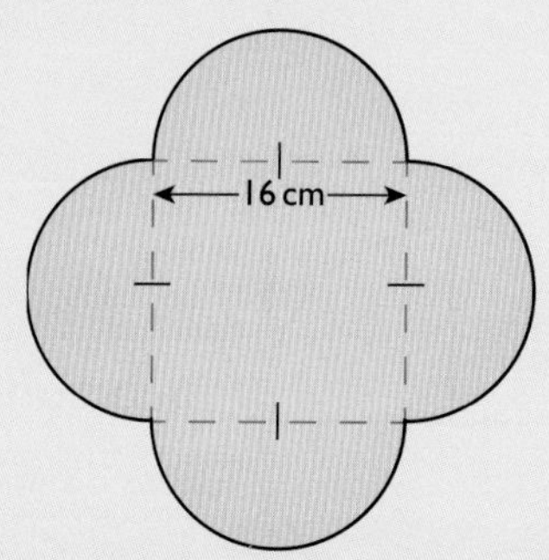

b

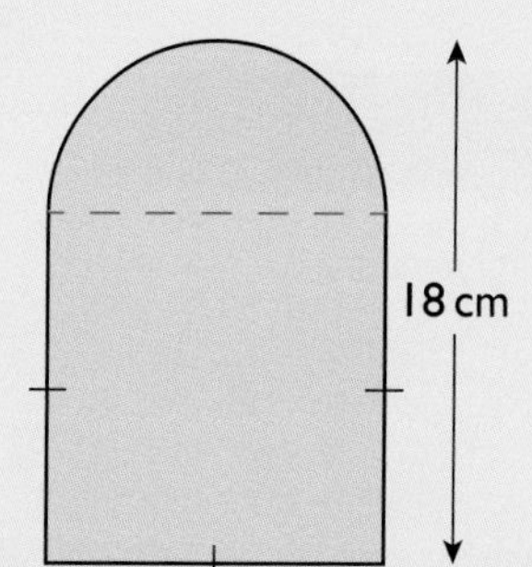

4 Cain and his dad go cycling. Cain's bicycle wheels have a diameter of 45 cm.
His dad's bicycle wheels have a diameter of 80 cm.
They both cycle 5 km.
Whose bicycle wheels turn more times? By how many turns?

Unit 4 • Area of circles • Band f

Outside the Maths classroom

Amphitheatres

How many seats are there in this amphitheatre?

Toolbox

The area, A, of a circle of radius r is given by

$$A = \pi r^2.$$

A **semicircle** is half a circle.

Its area is $\frac{1}{2}\pi r^2$.

A **quadrant** is a quarter of a circle.

Its area is $\frac{1}{4}\pi r^2$.

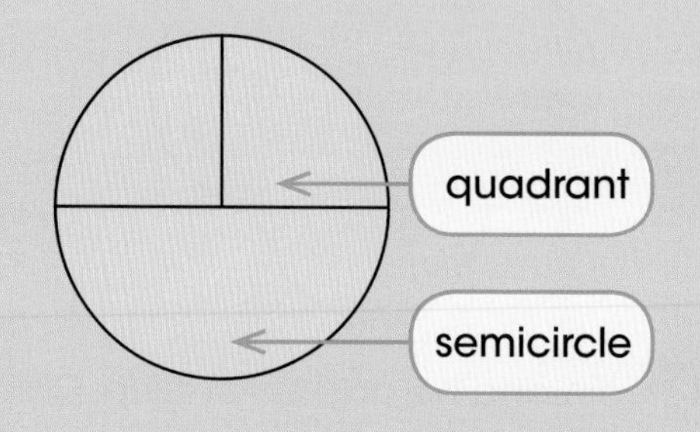

Example – Finding the area of a circle given the diameter

Find the area of this circle.

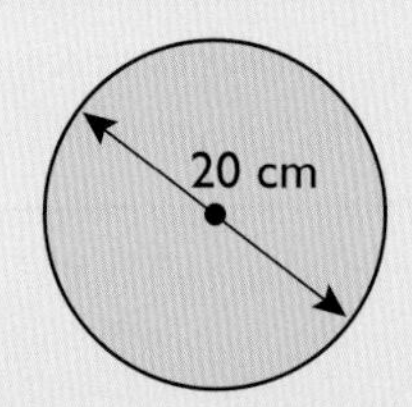

Solution

First find the radius.

$r = \frac{d}{2} = \frac{20}{2} = 10\text{cm}$

$A = \pi r^2$

$= \pi \times 10^2$

$= \pi \times 100$

$= 314.2\text{cm}^2$ (to 1 d.p.)

Example – Finding the radius and diameter of a circle given the area

The area of this circle is 50 cm^2.

Calculate

a the radius

b the diameter.

Give your answers to the nearest centimetre.

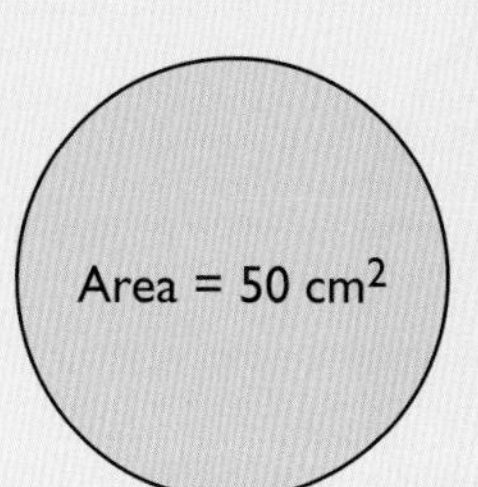

Solution

a $A = \pi r^2$

$50 = \pi r^2$

$\frac{50}{\pi} = r^2$ ← Divide both sides by π

$\sqrt{\frac{50}{\pi}} = r$ ← Take the square root of both sides.

$r = \sqrt{\frac{50}{\pi}}$

$r = 3.989...$ ← Your calculator often gives you a long number. Write it like this, using ... for the later digits. That is better than writing them all down.

The radius is 4 cm (to the nearest centimetre).

b Diameter = 2 × radius

= 2 × 3.989... ← Use the most accurate value. Keep the number on your calculator until you get your final answer.

= 7.979...

The diameter is 8 cm (to the nearest centimetre).

Practising skills

1 Calculate the area of each of these circles. Give your answers to one decimal place.

a

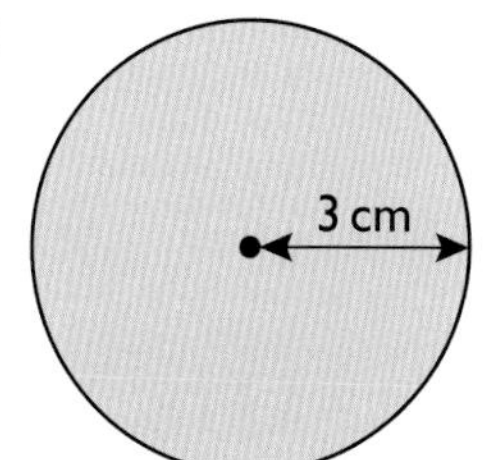

b

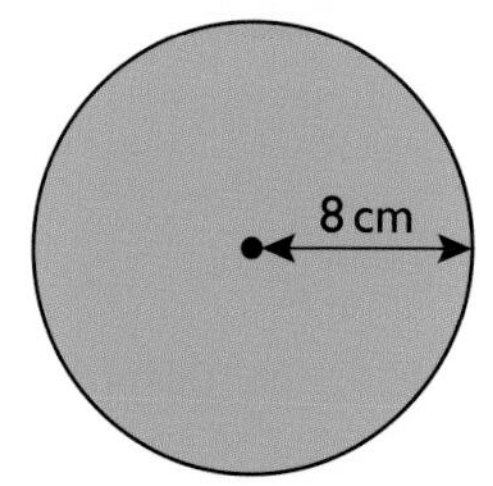

c

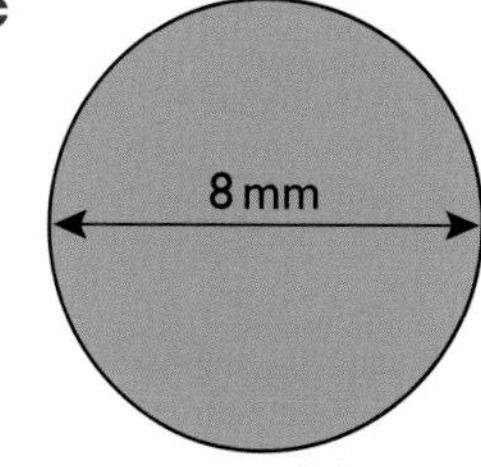

d

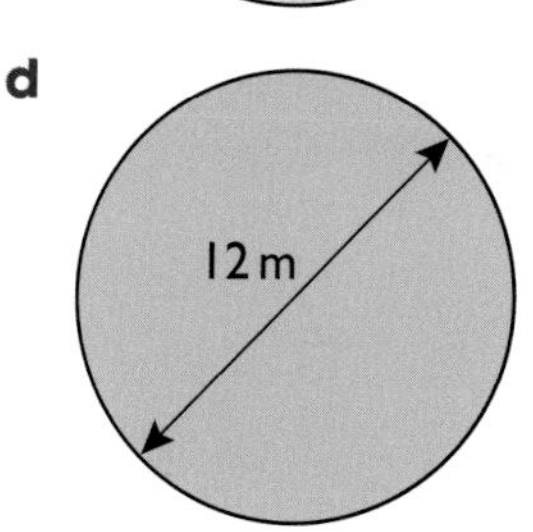

e

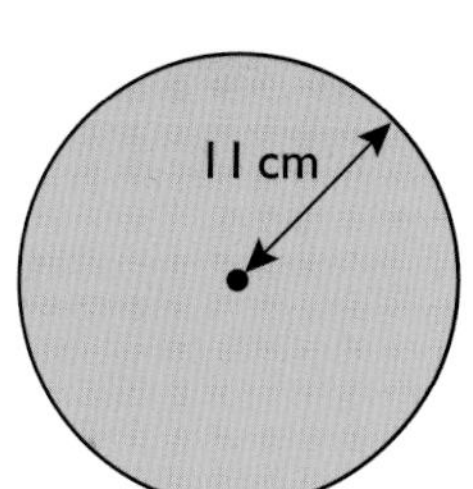

f

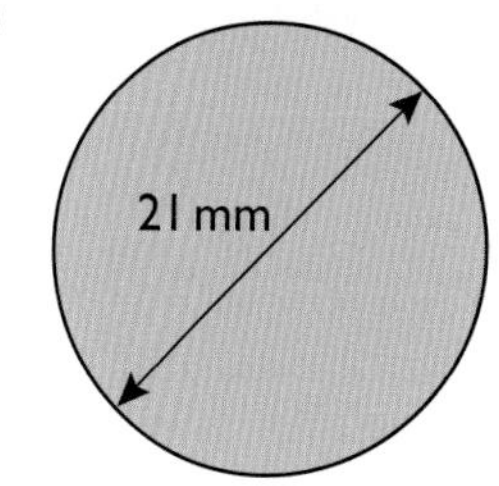

2 Calculate the area of each of these circles. Each circle is drawn accurately.

a

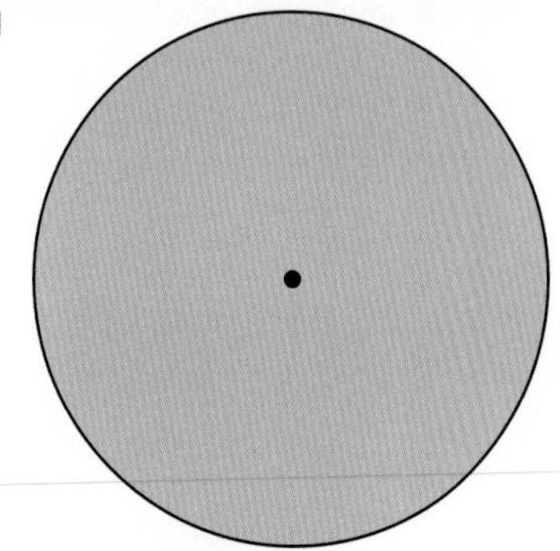

b

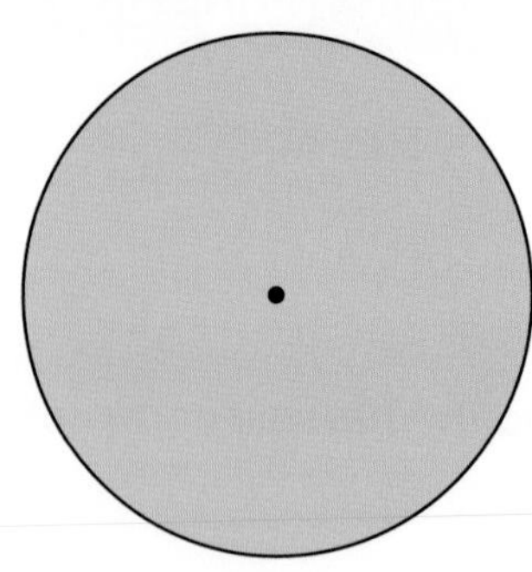

c

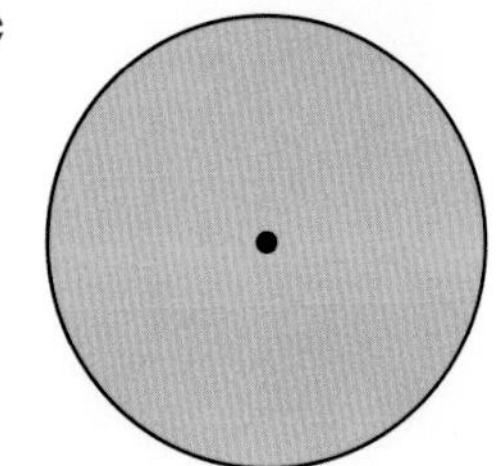

d

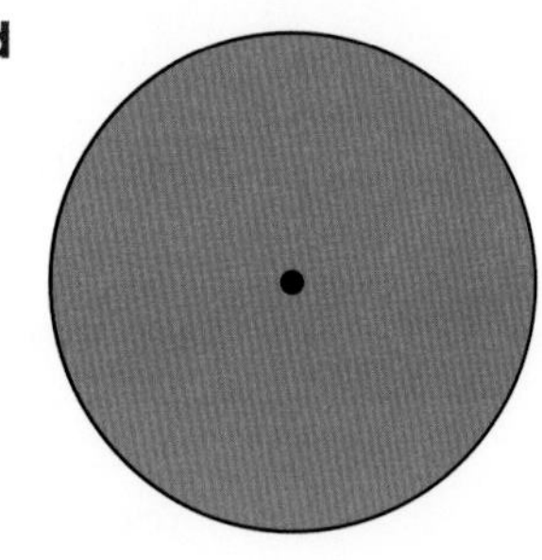

3 A circle has an area of 35 cm^2.

Use the formula $A = \pi r^2$ to calculate its radius to one decimal place.

4 A circle has an area of 60 cm^2.

a Calculate its radius to one decimal place.

b Write down its diameter.

c Calculate the circle's circumference.

5 Copy and complete this table which gives some measurements for circles.

	Radius	Diameter	Area	Circumference
a	7 cm			44.0 cm
b		17 cm		
c			80 cm^2	
d			250 cm^2	56.0 cm

Developing fluency

1. Calculate the area of each shape.

a

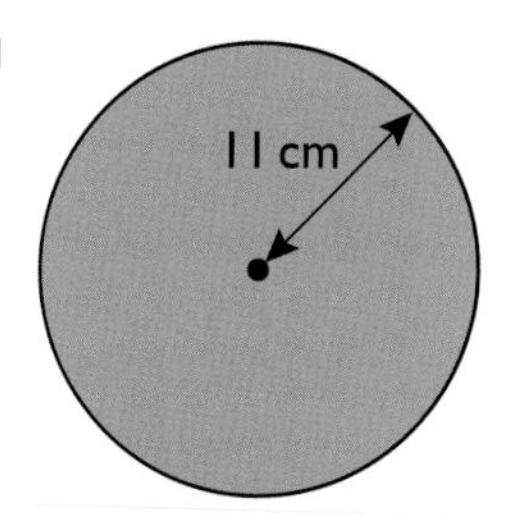

b

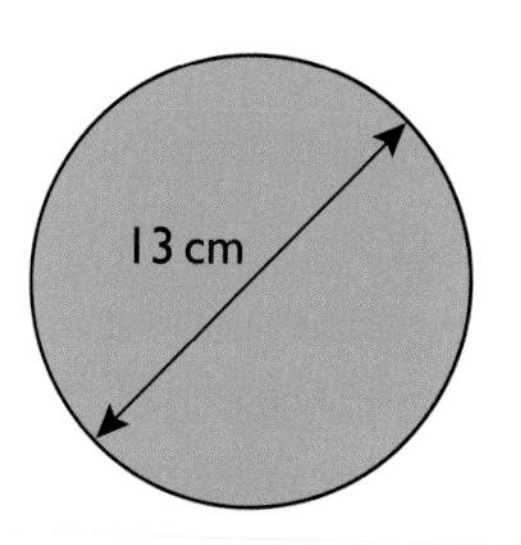

c

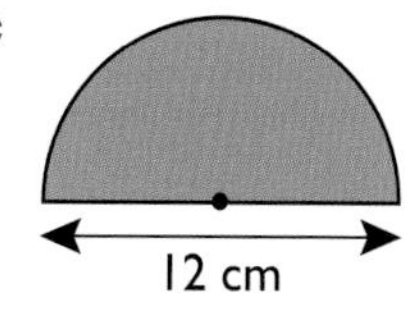

d

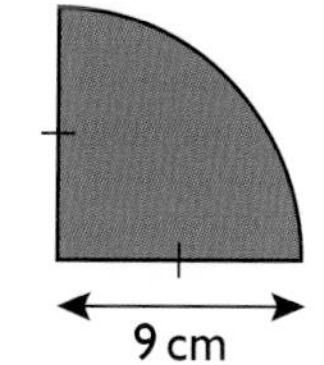

e

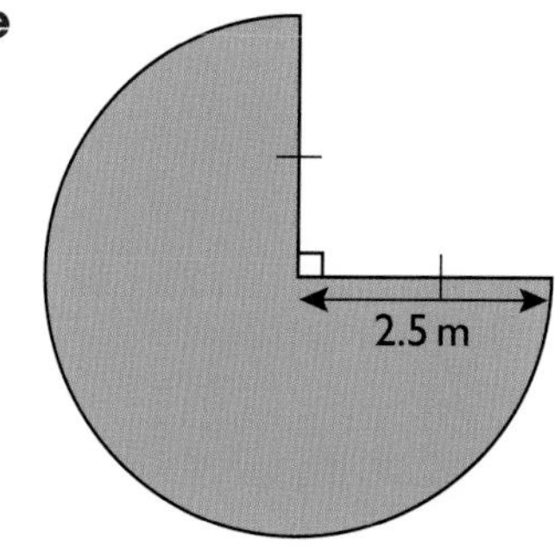

f

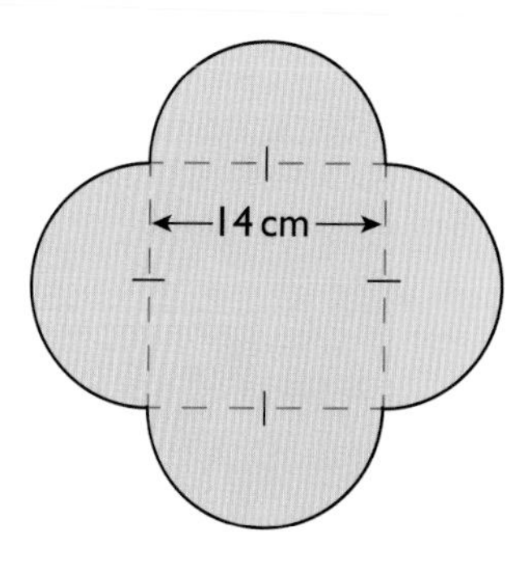

2. A circle has area 200 cm^2. What is its circumference?

3. A circle has circumference 30 cm. What is its area?

Exam-style

4. This semicircle has area 45 cm^2.

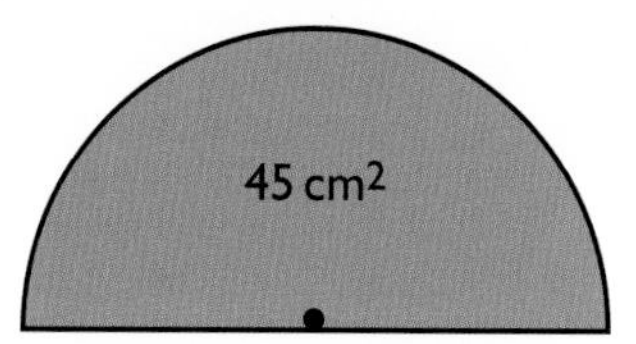

Work out the perimeter of the semicircle.

Extension

Exam-style

5. Work out the difference between the red area and the blue area of this target board.

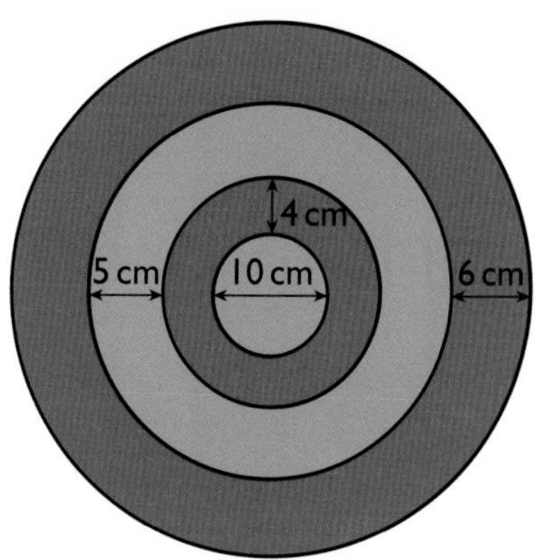

6) Candice is making biscuits. She has a rectangular piece of biscuit mixture measuring 32 cm by 16 cm.

She cuts out eight circular biscuits of diameter 8 cm.

She uses the remainder of the mixture to make one last biscuit.

What is its diameter, assuming all the biscuits have the same thickness?

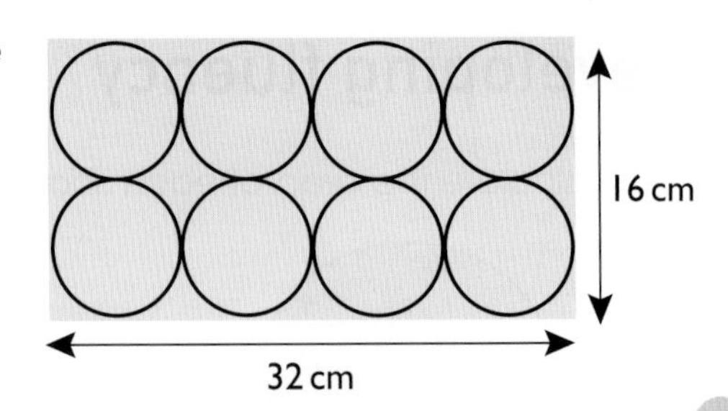

Exam-style

7) Adam ties his goat to the corner of his garden shed.

The shed is 19 m by 8 m. The rope is 12 m long.

Calculate the total area that Adam's goat can graze using this diagram.

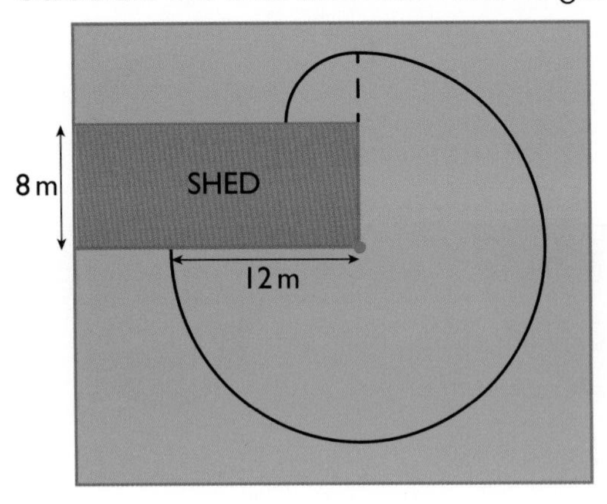

Extension

8) Betsy is designing a display board for the entrance to her company.

The board is in the shape of a rectangle and two semicircles.

The board has a single line of symmetry.

a Work out the area of the display board.

b She uses very expensive paint. It costs her £73.

What is the cost of the paint per square centimetre?

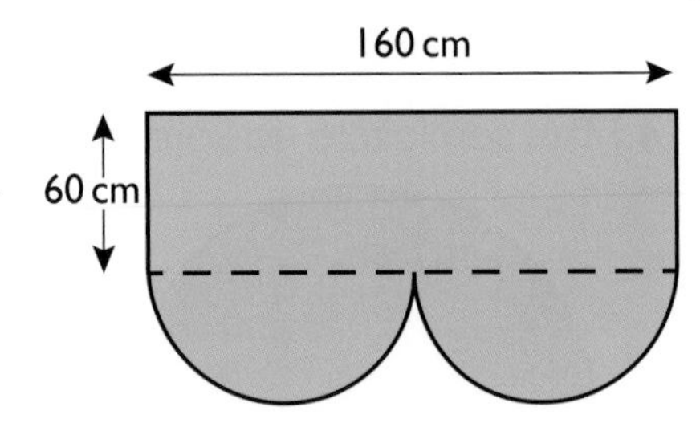

Problem solving

Exam-style

1) Here is a plan of Sophie's garden.

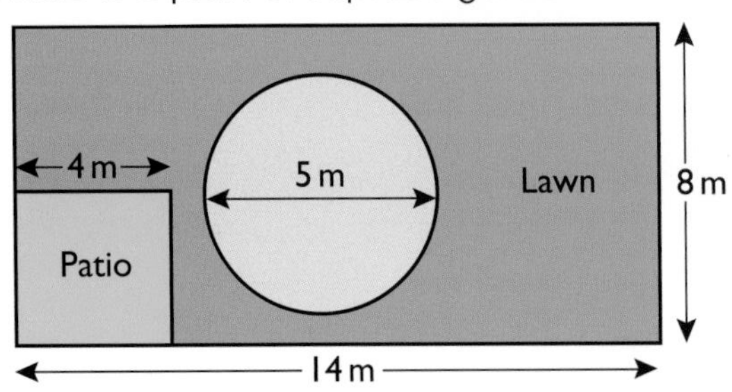

The patio is a square 4 m by 4 m.

The pond is a circle of diameter 5 m.

The rest of the garden is lawn.

Sophie wants to put new turf on the lawn.

Sophie buys 80 m^2 of turf.

Show that she has enough turf for the whole lawn.

What assumptions has Sophie made?

Exam-style

2. Here is a plan of a circular path.

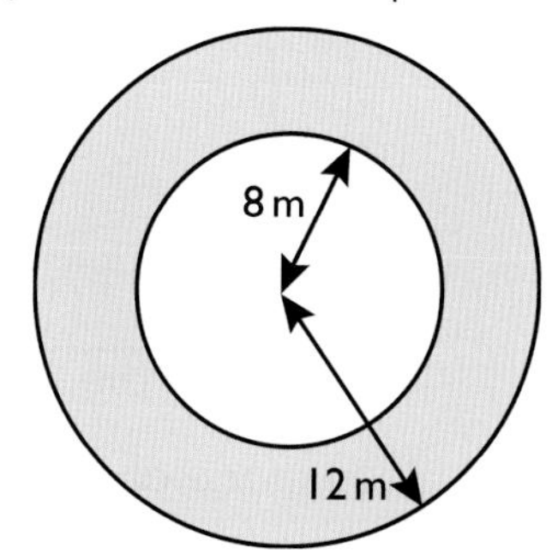

The inner edge of the path is a circle with radius 8 metres.
The outer edge of the path is a circle with radius 12 metres.
Both circles have the same centre.
Helen is going to lay rubber matting on the path, for the children to play on.
The rubber matting costs £30 per square metre.
Work out the total cost of the rubber matting.

Extension

Exam-style

3. The diagram shows a garden plot in the shape of a semicircle.

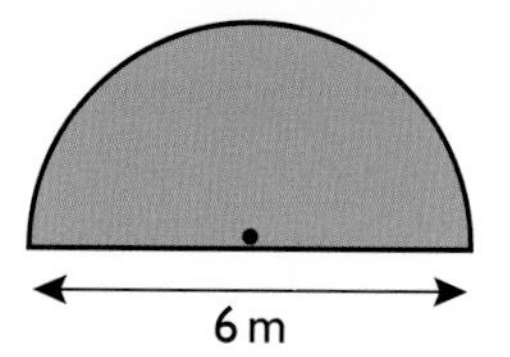

The diameter of the semicircle is 6 m.
Louise wants to cover the garden plot with wood bark chippings.
One bag of chippings covers an area of $0.6\,m^2$ and costs £3.
Louise has £75 for the chippings.
Does she have enough money?
Show all of your working.

Extension

Exam-style

4. Here is a diagram of an athletics field.

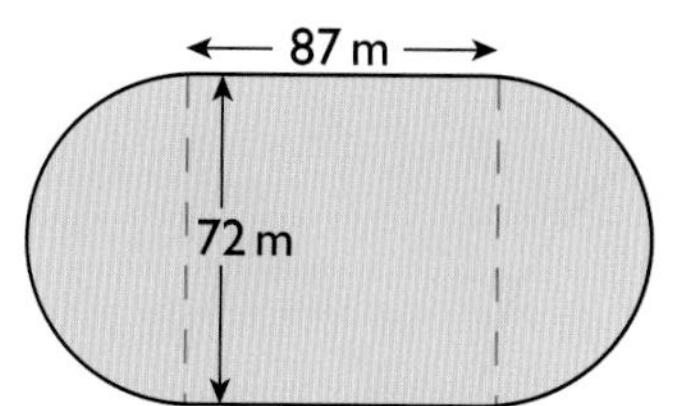

The field has two ends which are semicircles.
Each semicircle has a diameter of 72 m.
The length of each straight part is 87 m.

a What is the length of the field to the nearest metre?

Fatima is going to spray the field with fertiliser.
One litre of fertiliser is enough for $50\,m^2$.

b How many litres of fertiliser does Fatima need?

Extension

5 Here is a diagram of a window made of stained glass.
The window is in the shape of a rectangle and a semicircle.
The width of the window is 60 cm.
The height of the window is 160 cm.
The stained glass in the window needs repairing.
The cost of a repair is £2000 per m^2 of stained glass.
Work out the total cost of the repair.

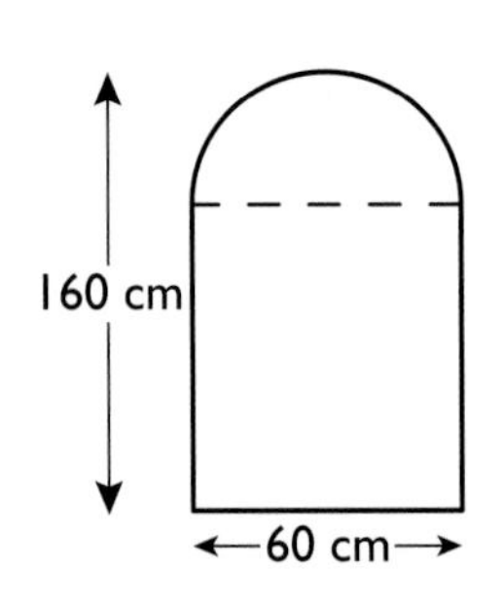

Exam-style

Extension

6 A factory makes metal discs by punching out a circle from a piece of metal.
The piece of metal is a square of sides 25 cm.
Any metal left over when the circle is removed is sent for recycling.
The factory makes 800 discs each hour.
Work out the total area of metal sent for recycling each hour.
Give your answer in square metres.

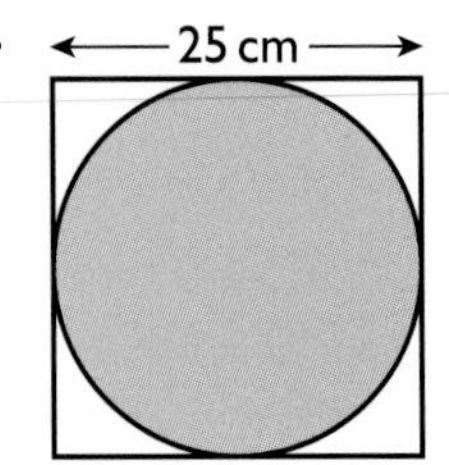

Exam-style

Extension

Reviewing skills

1 Calculate the area of each of these circles. Give your answers to one decimal place.

a

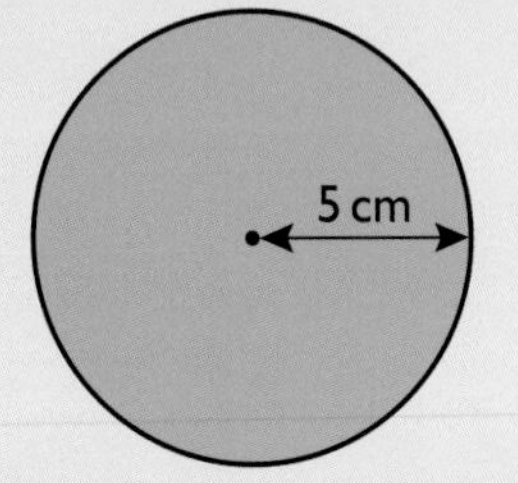

b

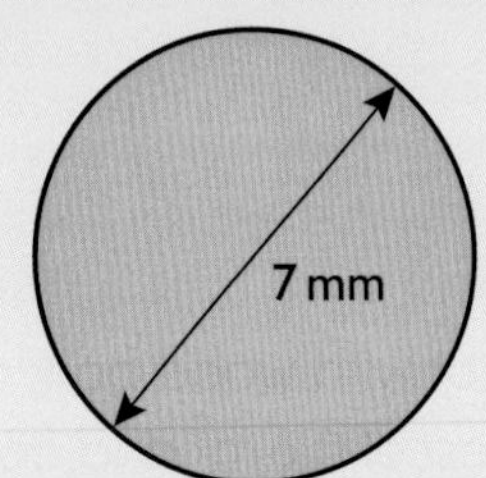

c

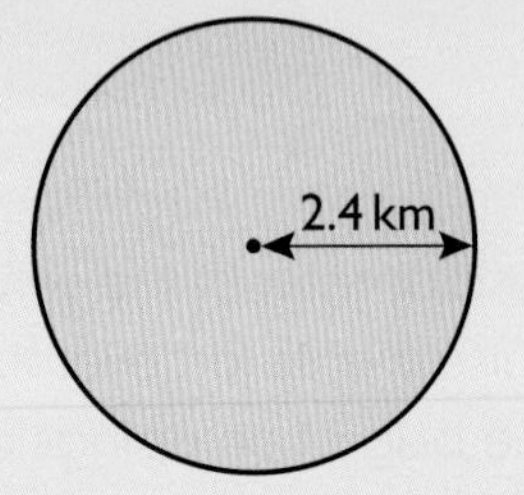

2 A circle has an area of 52 cm^2. Calculate to one decimal place

a the radius of the circle

b the diameter of the circle.

3 Calculate the area of each shape.

a

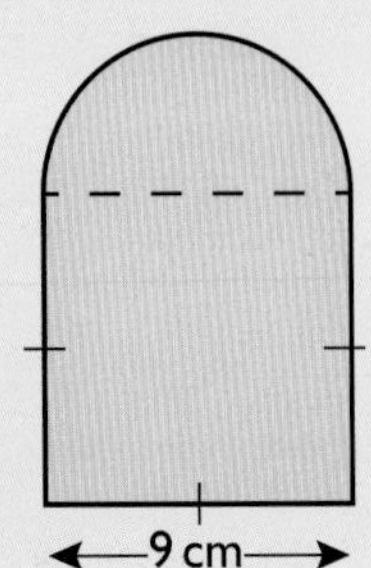

b

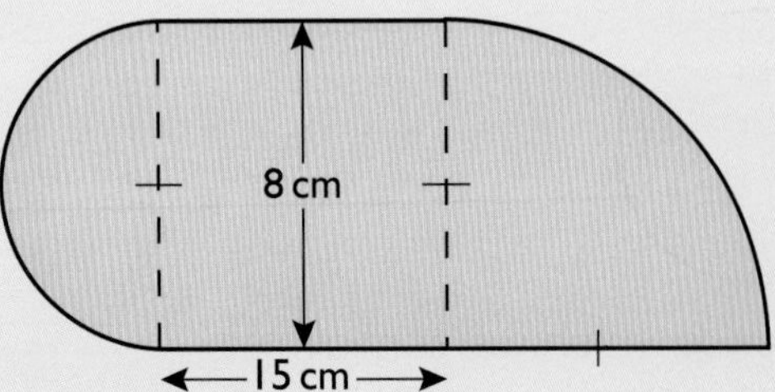

4 The diagram shows part of the cross-section of a building.
The shaded part of the diagram shows a support for the building.
The edges of the support are quarter-circle arcs with the same centre.
The width of the support is 2 metres.
The radius of the inner quarter circle is 3 metres.
Work out the area of the support's cross-section.

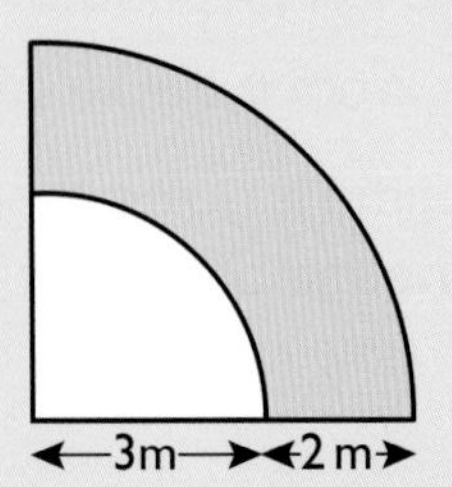

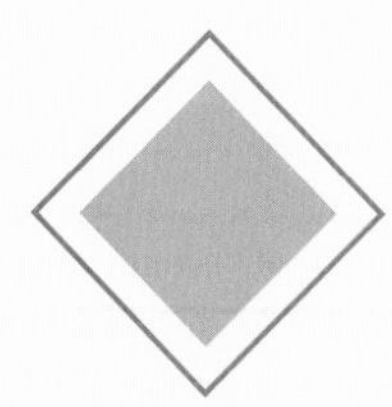

Geometry and Measures Strand 4 Construction

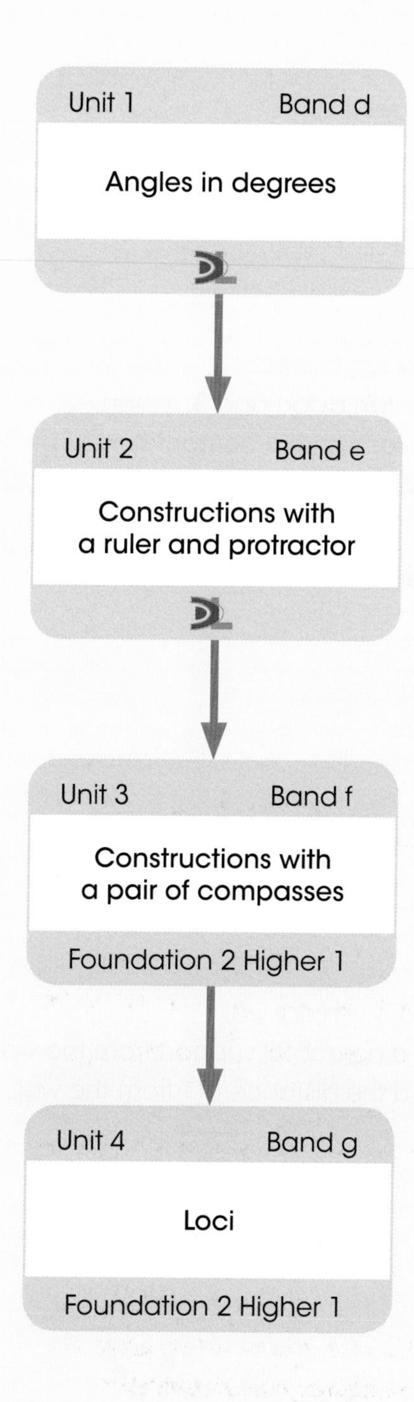

Units 1–2 are assumed knowledge for this book. They are reviewed and extended in the Moving on section on page 262.

Units 1–2 • Moving on

Exam-style

1 Here is a drawing of a water tower.
The angle between the two supporting inside legs is 44°.
The feet of the supporting legs are 10 metres apart.
Use a suitable drawing to find the height h of the supporting legs.

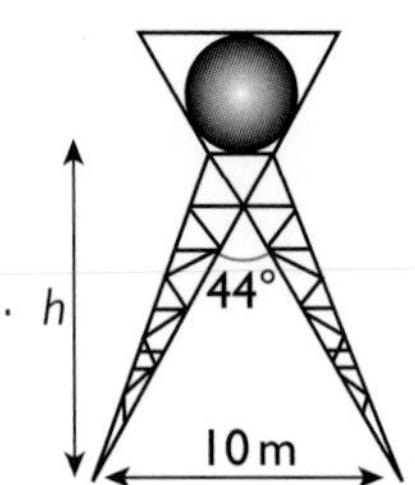

Exam-style

2 Alex wants to put a roof on his log cabin.
The walls of the log cabin are 8 m apart and 3 m high.
The angle which the roof makes with the horizontal must be 30°.
Use a suitable drawing to find the height of the top of the roof.

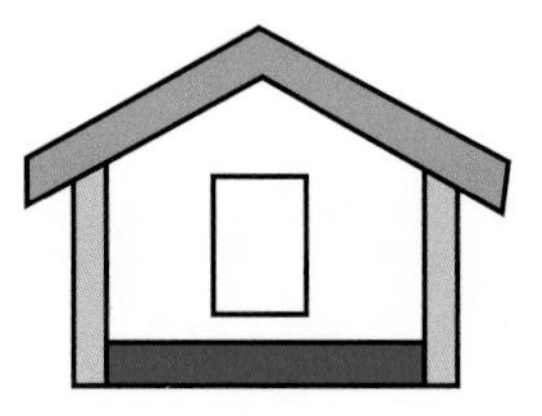

Exam-style

3 The diagram represents the supports for a hoist T.

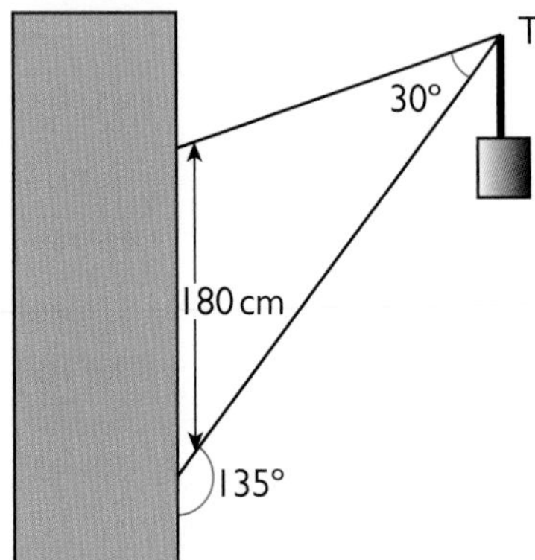

The supports are attached to a vertical wall.
An engineer wants to put in a horizontal support from the wall to T.
Use a suitable drawing to find the distance of T from the wall.

Exam-style

4 The diagram shows a tower.

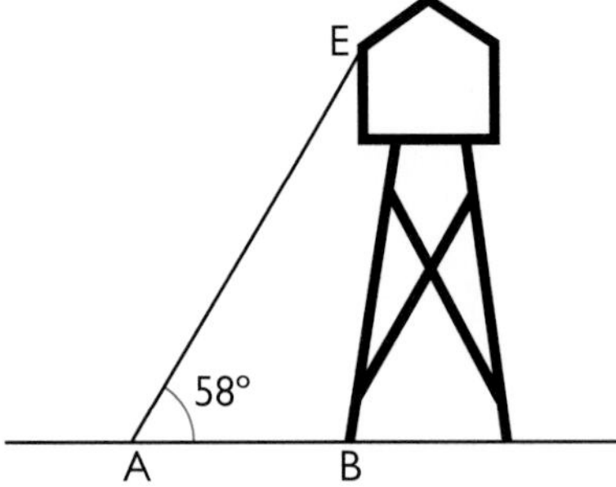

Jim sees the top edge, E, of the tower at an angle of elevation of 58° from the point A.
Jim walks 20 m away from A in the direction BA.
The angle of elevation is now 30°.
Use a suitable drawing to find the height of E.

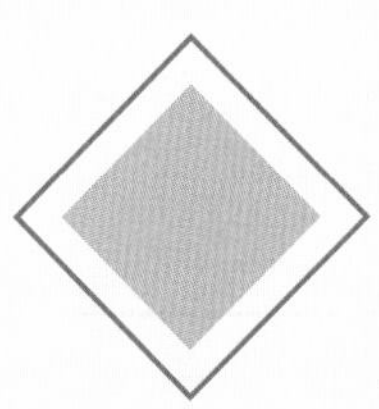

Geometry and Measures Strand 5 Transformations

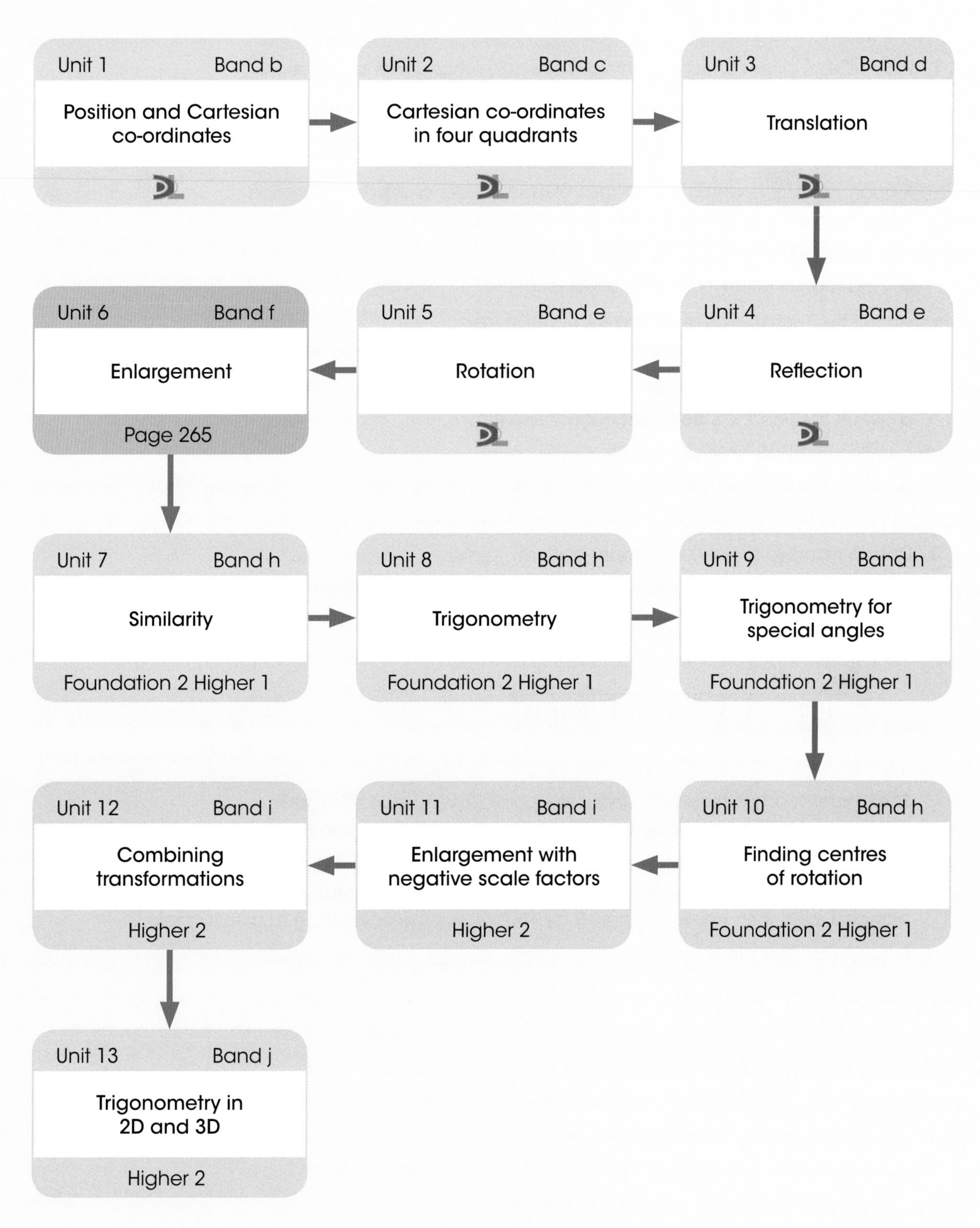

Units 1–5 are assumed knowledge for this book. They are reviewed and extended in the Moving on section on page 264.

Units 1–5 • Moving on

1. P and Q are points which are images of each other under a reflection.
   P has co-ordinates (5, 8).
   Q has co-ordinates (5, 12).

   a Find the equation of the line of reflection.

   S and T are points which are images of each under a different reflection.
   S has co-ordinates (2, 4).
   T has co-ordinates (−2, 4).

   b Find the equation of the line of reflection.

2. PQ is a line of length 6 units parallel to the x axis.
   PQ is reflected in the line $x = 5$ to give the line RS.
   Find the length of the line RS

   a when the line $x = 5$ **does not** cut the line PQ.
   Give a reason for your answer.

   b when the line $x = 5$ **does** cut the line PQ.
   Give a reason for your answer.

3. Donna wants to make a wallpaper pattern.
   She practices first with a simple shape.

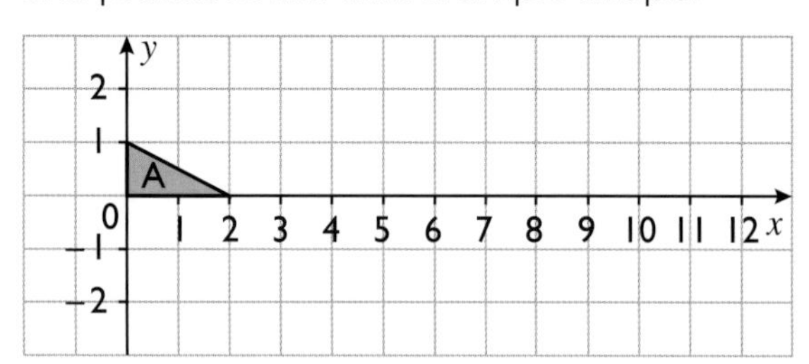

   She rotates triangle A by 180° about the point (2, 0) to give triangle B.
   She then rotates triangle B by 180° about the point (4, 0) to give triangle C.

   a Describe the single transformation that maps triangle A to triangle C.

   For a different design Donna rotates triangle A by 180° about the point (4, 0) to give triangle B and then rotates triangle B by 180° about the point (8, 0) to give triangle C.

   b Describe fully this transformation.

Unit 6 • Enlargement • Band f

Outside the Maths classroom

TV screens

When you watch a film on TV, why do you sometimes get a thin rectangle of black along the top and bottom of the screen?

Toolbox

Enlargement is a transformation that changes the size of an object.

One shape is an enlargement of another if **all the angles in the shape are the same** and the **lengths of the sides** have all been increased by the **same scale factor**.

In this diagram the pentagon ABCDE is enlarged by a scale factor of 3 to A′B′C′D′E′.

The position of the image depends on the centre of enlargement X.

Because the scale factor of enlargement is 3

- side A′B′ is three times as long as AB
- side E′D′ is three times as long as ED, etc.
- the distance XB′ is three times the distance XB
- the distance XD′ is three times the distance XD, etc.

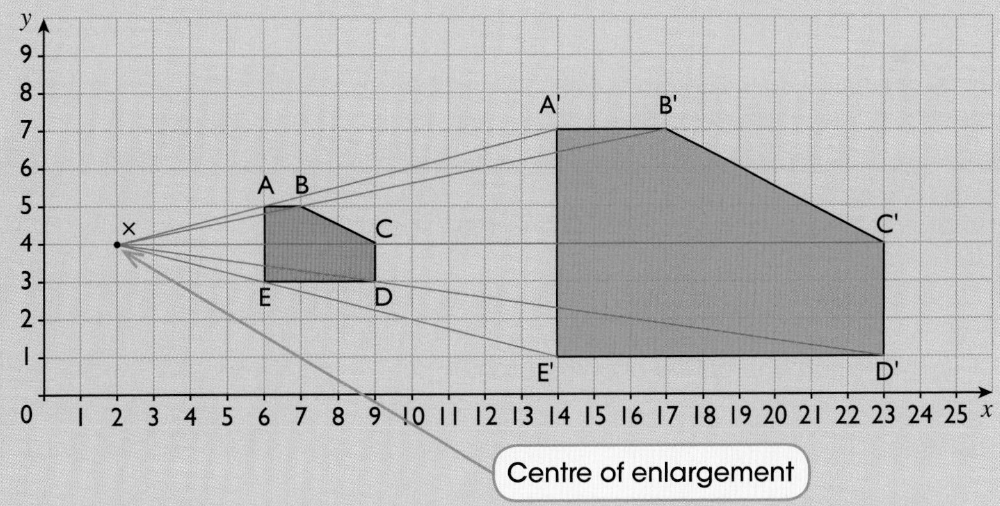

The term enlargement is also used in situations where the shape is made smaller.

In these cases the scale factor is a fraction, like $\frac{1}{2}$ or $\frac{1}{3}$.

In the diagram the scale factor for the enlargement of A′B′C′D′E′ to ABCDE is $\frac{1}{3}$.

Example – Finding the scale factor of an enlargement

The shape on the right shows an enlargement from centre P.

a What is the scale factor of the enlargement?

b What happens if you change the centre of enlargement?

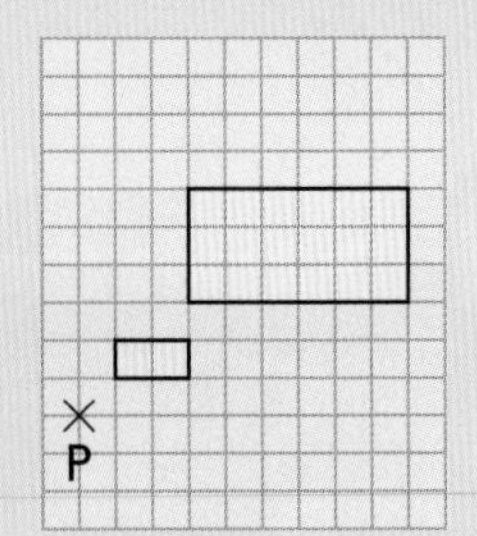

Solution

a The scale factor is 3 because all the sides of the second shape are three times as long as those of the first shape.

b The image would be in a different place.

Example – Enlarging a shape

Plot the points A(2, 4), B(4, 4), C(4, 6) and D(2, 6).

Join the points to form a square.

Enlarge ABCD using (0, 5) as the centre of enlargement and a scale factor of 2. Label the image A′B′C′D′.

a What are the co-ordinates of the points A′, B′, C′ and D′?

b What shape is A′B′C′D′?

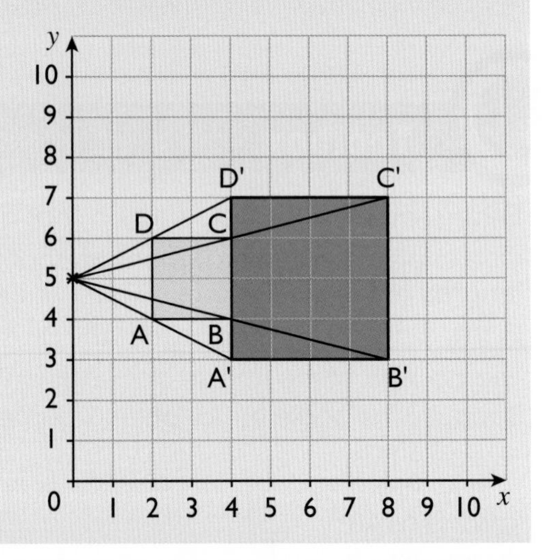

Solution

a A′(4, 3), B′(8, 3), C′(8, 7), D′(4, 7)

b A square

Example – Describing an enlargement

Emma draws the quadrilateral ABCD and then she transforms it as shown to A′B′C′D′.

a Describe fully the transformation from ABCD to A′B′C′D′.

b What can you say about the distance of the object and the image from the centre of enlargement?

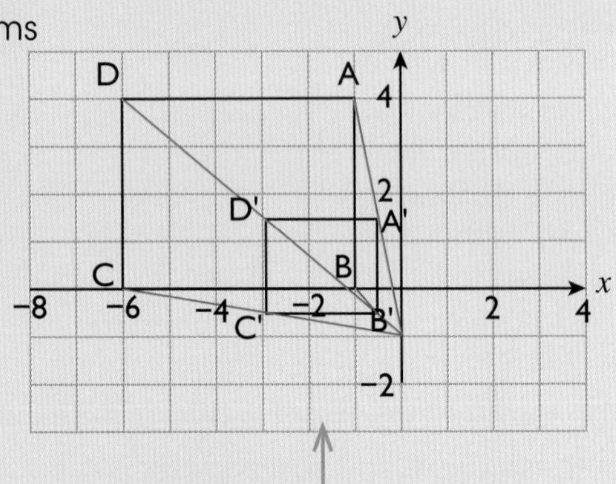

The lines joining the image and the object all meet at the centre of enlargement.

Solution

a The enlargement scale factor is $\frac{1}{2}$.
The centre of enlargement is (0, −1).

b The image is half the distance that the object is from the centre.

Practising skills

1. Identify the pairs of triangles here. In each pair, one triangle is an enlargement of the other.

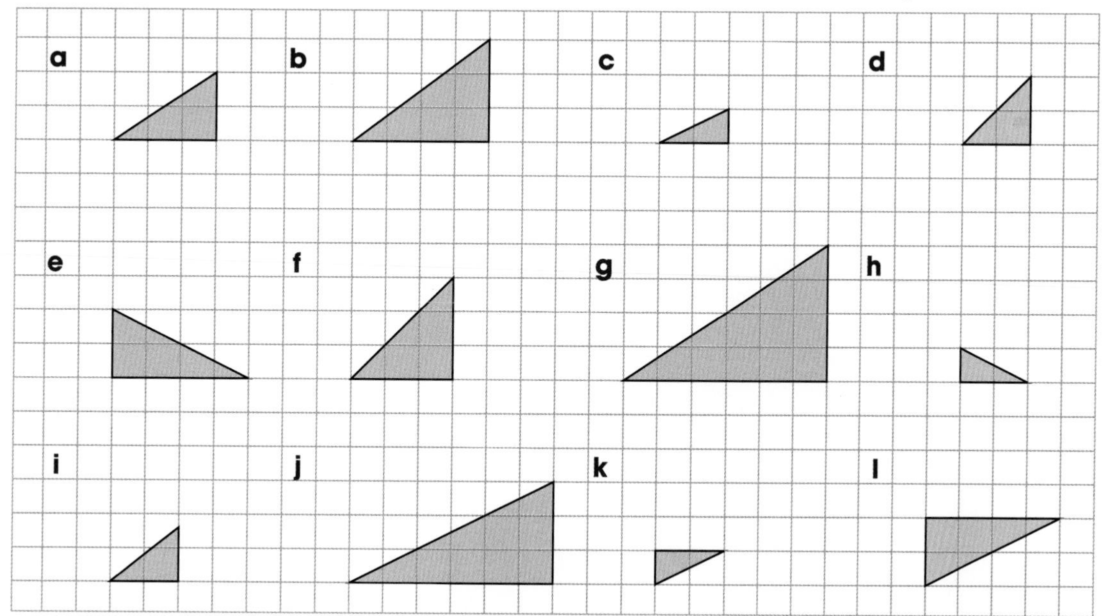

2. Enlarge each shape by a scale factor of 2, using C as the centre of enlargement.

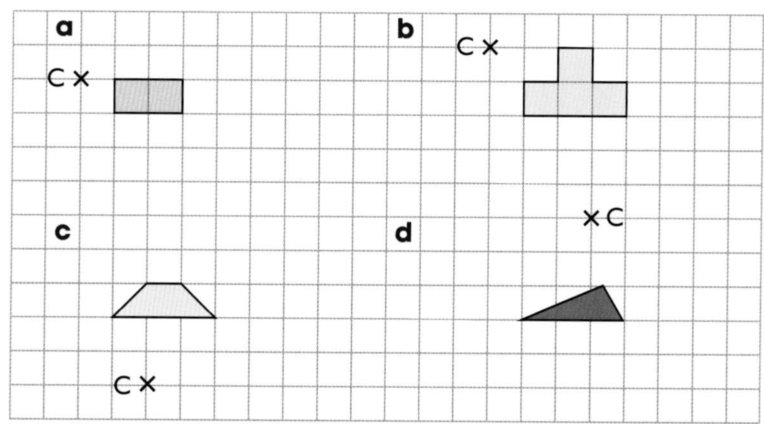

3. Look at this diagram.
   Describe fully the single transformation that maps

   a E to A
   b E to B
   c E to C
   d E to D
   e B to A
   f C to A
   g A to E
   h B to E
   i D to E.

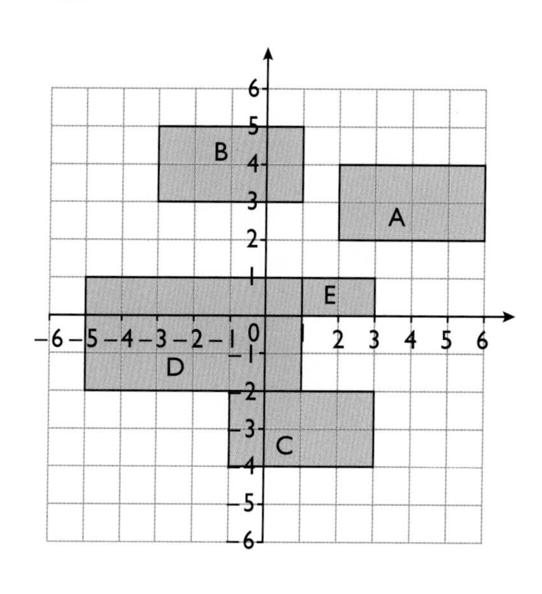

4 **a** An internet company have designed a new logo.

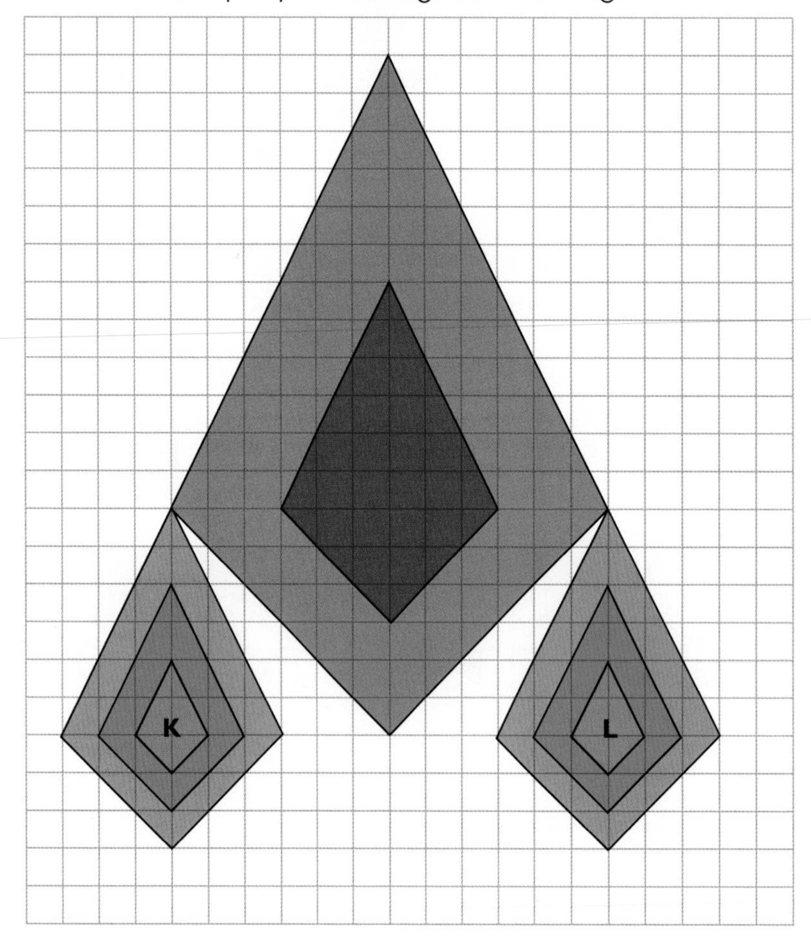

Describe the design fully.

b The green kite K is enlarged to be the same size as the

i blue kite

ii pink kite.

Write down the scale factor of the enlargement for each.

c What transformation will map K to L?

d The blue kite is enlarged to give the purple kite.
Describe the enlargement fully.

e The purple kite is enlarged to give the red kite.
What is the scale factor of enlargement?

Developing fluency

Exam-style

1. Copy this diagram.

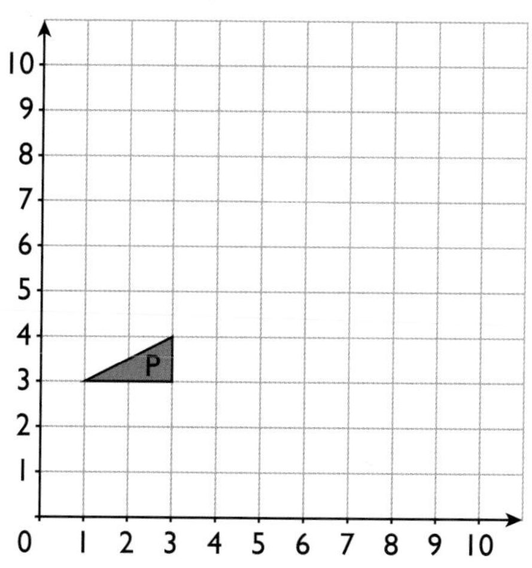

a Enlarge triangle P by a scale factor of 2, using (0, 0) as the centre of enlargement. Label the image A.

b Enlarge triangle P by a scale factor of 2, using (2, 5) as the centre of enlargement. Label the image B.

c Describe the transformation that maps A onto B.

Exam-style

2. Copy this diagram.

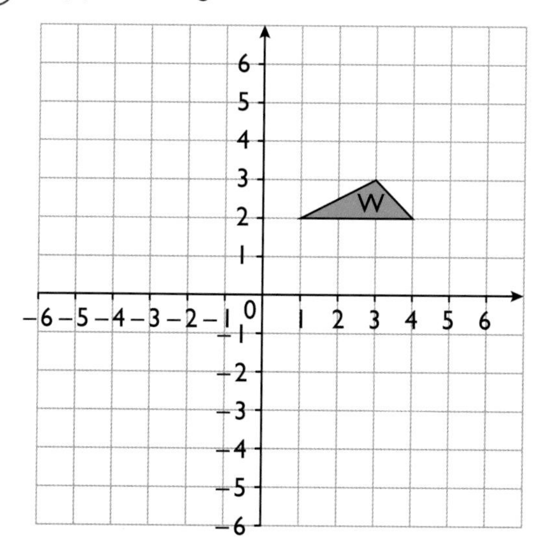

a Enlarge shape W by a scale factor of 2, using (2, 1) as the centre of enlargement. Label the image A.

b Enlarge shape W by a scale factor of 2, using (3, 6) as the centre of enlargement. Label the image B.

c Describe the transformation that maps A onto B.

Exam-style

(3) Copy this diagram.

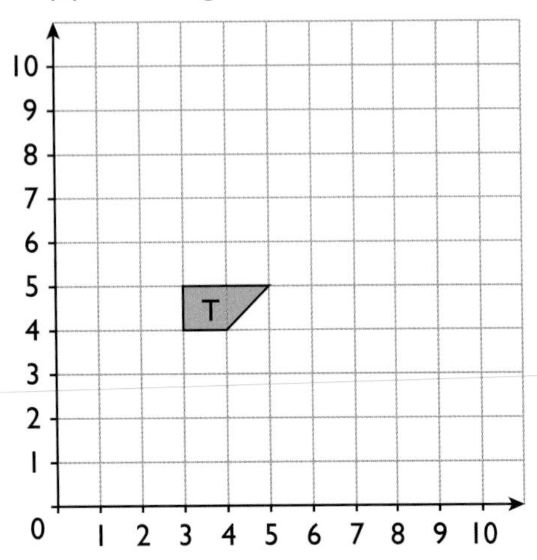

a Enlarge shape T by a scale factor of 2, using (1, 5) as the centre of enlargement. Label the image A.

b Enlarge shape T by a scale factor of 1.5, using (5, 9) as the centre of enlargement. Label the image B.

c Enlarge shape T by a scale factor of 3, using (2, 6) as the centre of enlargement. Label the image C.

d What do you notice about B and C?

(4) In this diagram, triangles B and C are enlargements of triangle A.

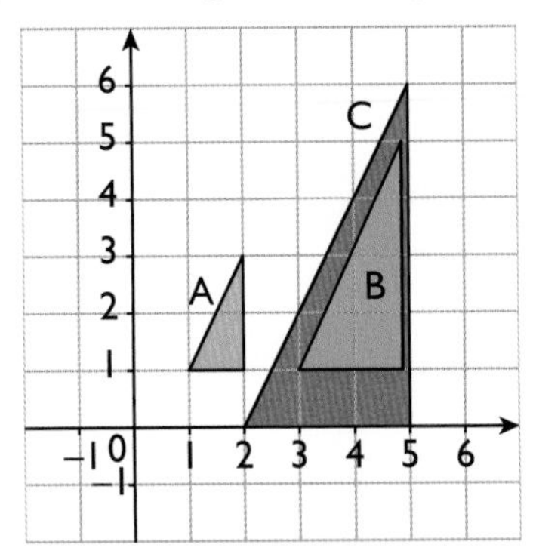

a Find the scale factor, f, and the centre of enlargement for A to B.

b Find the scale factor, g, and the centre of enlargement for B to C.

c The scale factor of the enlargement A to C is h. Use your answers for the scale factors f and g to predict the value of h.

d Now use the diagram to find the centre of enlargement and the scale factor, h, for A to C.

5 Look at this diagram.

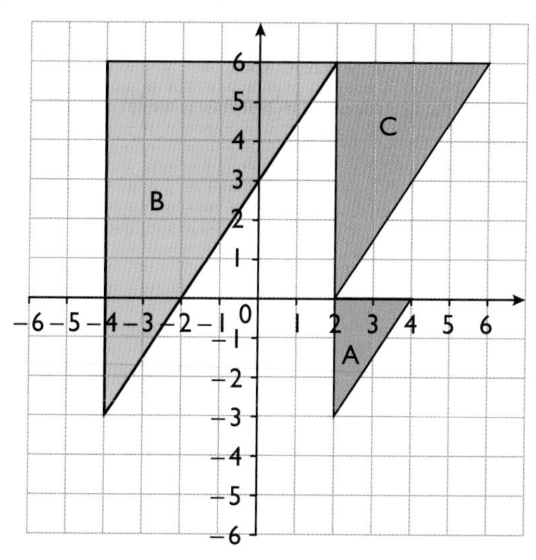

a Describe fully the single transformation that maps A to B.

b Describe fully the single transformation that maps B to A.

c Describe fully the single transformation that maps A to C.

d Describe fully the single transformation that maps C to A.

Exam-style

6 Copy this diagram.

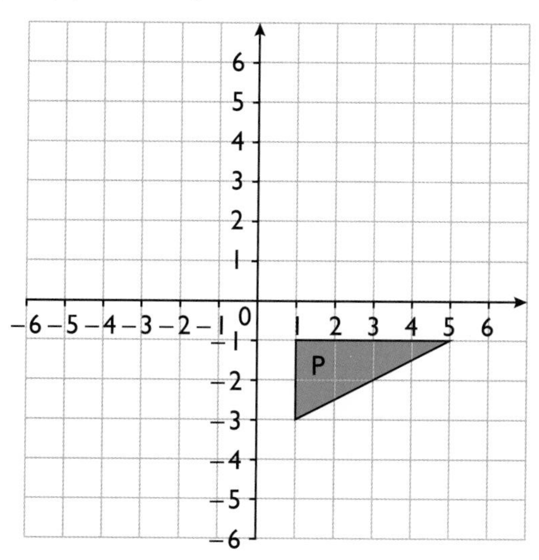

a Enlarge shape P by a scale factor of $\frac{1}{2}$, using (1, 5) as the centre of enlargement. Label the image A.

b Enlarge shape A by a scale factor of 2, using (6, 5) as the centre of enlargement. Label the image B.

c Describe the transformation that maps B to A.

Exam-style

Extension

7) Jim makes this statement about enlargements.

'Start with a shape A, enlarge it with a scale factor k, centre O, to give shape B.

Enlarge shape B with a different scale factor m, centre O, to give shape C.

The single transformation that will map shape A to shape C is an enlargement with a scale factor $k + m$.'

a Give an example to show that Jim is wrong.

b What should Jim have said?

8) **a** Draw an equilateral triangle AXY.

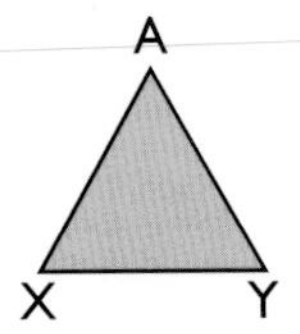

b On the same diagram, draw the enlargement of AXY, centre A, scale factor 2.
Label this triangle ABC.

c Still on the same diagram, draw the enlargement of ABC with scale factor $\frac{1}{2}$ and centres

i B **ii** C.

The new point on BC is Z.

d You have now drawn the net of a solid shape.
What shape is it?

e What happens if your original triangle is not equilateral?

Problem solving

Exam-style

Extension

1) Hannah has this picture of her cat.

She wants to enlarge the picture to fit exactly into the frame.

The dimensions of the picture and the frame are given.

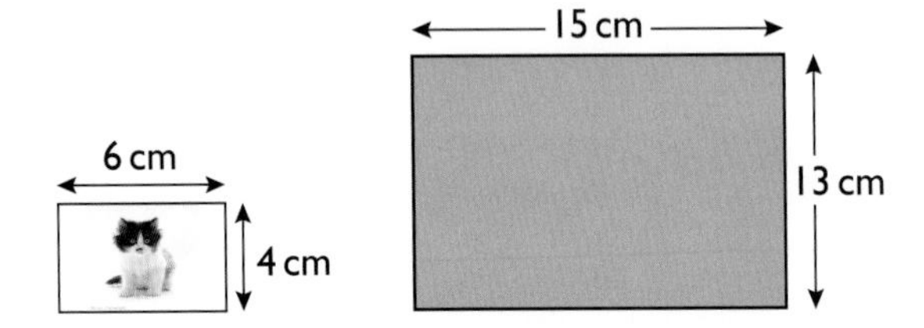

Will she be able to do this? Give a reason for your answer.

Exam-style

2 Here is a scale drawing of a bedroom in a show home.

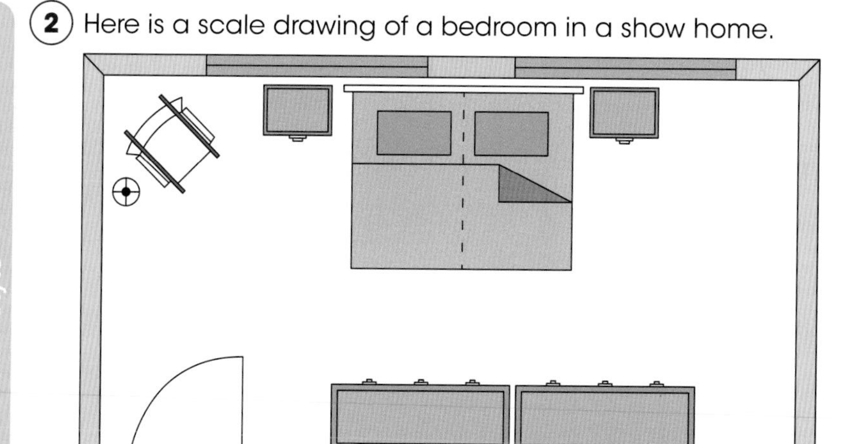

Bedroom

The bedroom has been drawn to a scale of 1 : 125.
Find the area of the bedroom floor.
Give your answer in m^2.

Exam-style E

3 This is a diagram of a film projector from above.

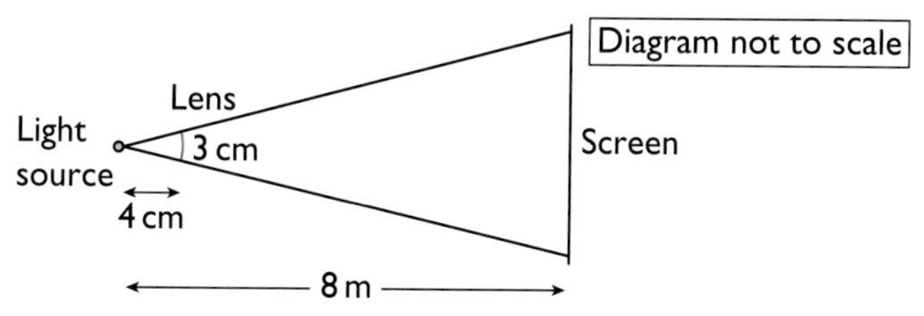

Work out the width of the screen.

Extension

4 The vertices of the square S are (6, 4), (8, 4), (8, 6), and (6, 6).
S is enlarged by scale factor 2, centre (7, 5) to give shape T.
T is enlarged by scale factor 2, centre (7, 5) to give shape U.

a Draw S, T and U on graph paper.

The area of the region between S and T is A.
The area of the region between T and U is B.

b Work out the ratio A : B.

5 The point (3, 2) is a vertex of the square S. The diagonally opposite vertex is at (4, 3).
The square is enlarged by scale factor 3 to give the square T.
The point (3, 2) on S is mapped to the point (7, 4) on T.

a Find the centre of enlargement.

b Find the co-ordinates of the other vertices of T.

6) The diagram shows a rectangle, R, a point A(1, 1) and a point Q(–1, –3).

The point A is one vertex of a rectangle S which you can't see.

The rectangle S is mapped onto R by an enlargement with centre Q.

a Find the scale factor of the enlargement.

b Find the co-ordinates of the other vertices of the rectangle S.

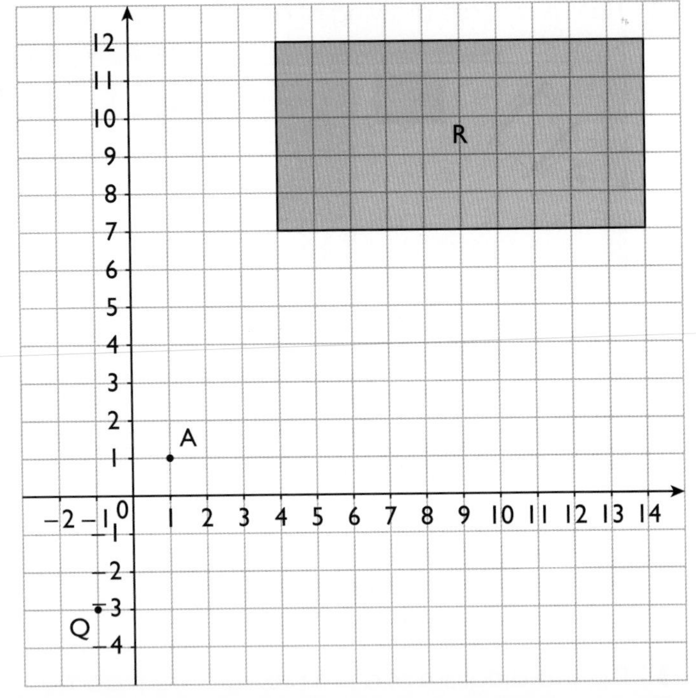

Reviewing skills

1) **a** Enlarge shape **a** by a scale factor of 2, using C as the centre of enlargement.

b Enlarge shape **b** by a scale factor of 2, using D as the centre of enlargement.

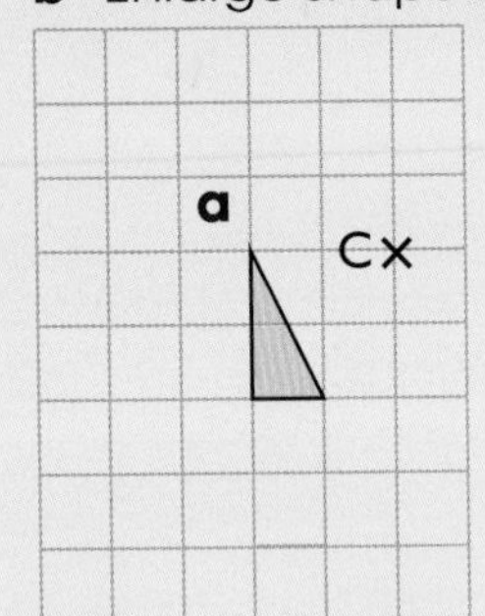

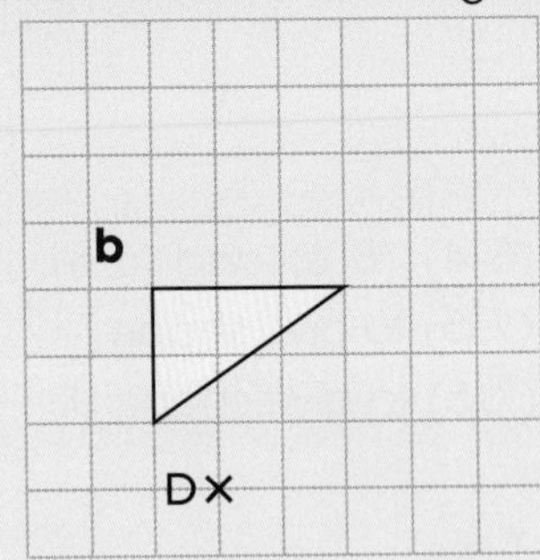

2) Copy this diagram.

a Enlarge shape Q by a scale factor of 2, using (–5, 6) as the centre of enlargement. Label the image A.

b Enlarge shape Q by a scale factor of 4, using (–5, 6) as the centre of enlargement. Label the image B.

c Find the centre of enlargement and the scale factor of the enlargement that maps A to B.

d Find the centre of enlargement and the scale factor of the enlargement that maps B to A.

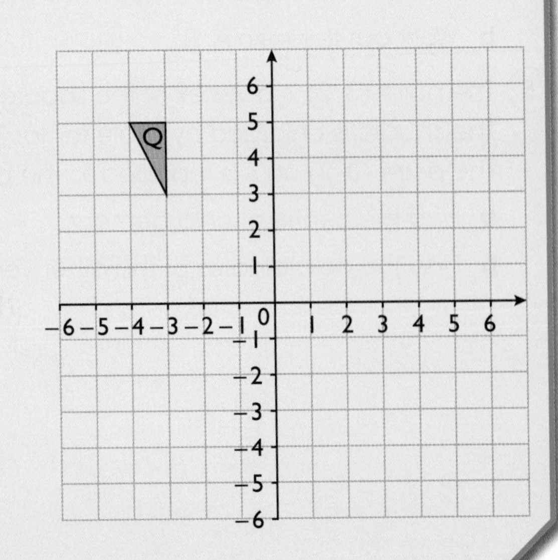

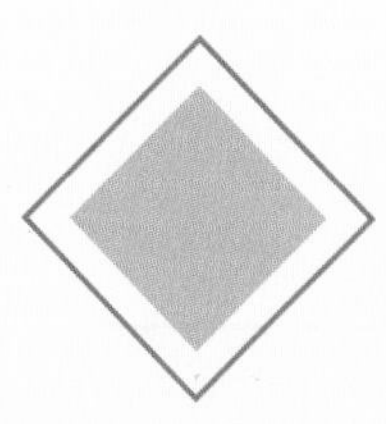

Geometry and Measures Strand 6 Three-dimensional shapes

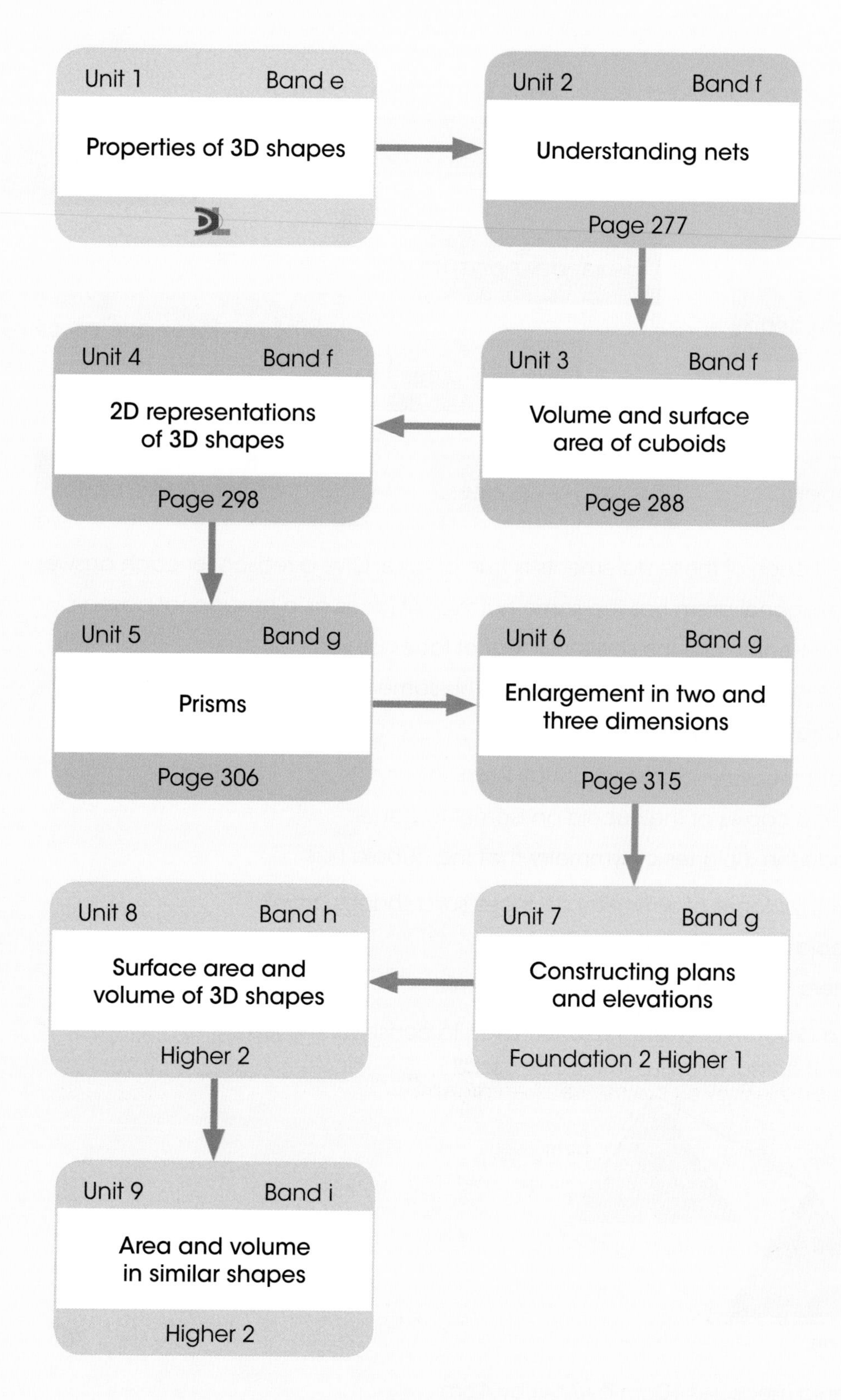

Unit 1 is assumed knowledge for this book. It is reviewed and extended in the Moving on section on page 276.

Unit 1 • Moving on

1. Here are ten cards. Match each shape with its description.

Octagonal prism	I have 4 vertices	I have 3 faces
Tetrahedron	Hexagonal prism	5 of my faces are rectangles
	I have 8 faces	
Cylinder	I have 16 vertices	Pentagonal based prism

2. Decide if each of these statements is true or false. Give a reason for each answer.
   - **a** A hexagonal prism has the same number of faces as a hexagonal-based pyramid.
   - **b** A tetrahedron has the same number of faces as vertices.
   - **c** A tetrahedron's edges are always all the same length.
   - **d** A cone has one plane of symmetry.

3. A cuboid measures 5 cm by 3 cm by 2 cm.
   - **a** Make 3 copies of the cuboid on isometric paper.
   - **b** Shade the 3 planes of symmetry that the cuboid has.

4. How many planes of symmetry do these solid shapes have?
   - **a** cuboid **b** cube
   - **c** sphere **d** square based pyramid

5. This is a drawing of a tent. Poles are used to construct the tent.
   What is the total length of the poles used?

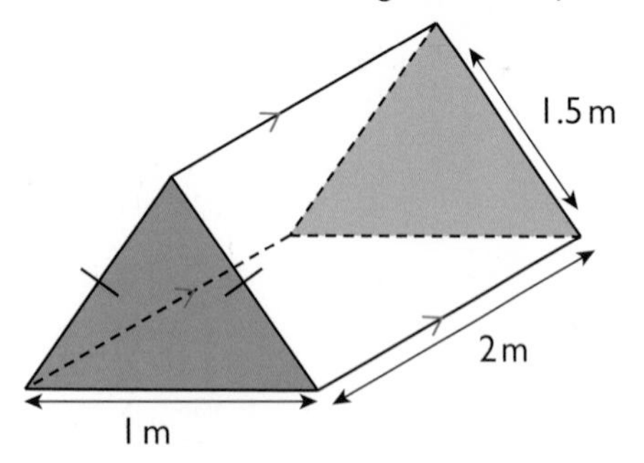

6. A cuboid measures 10 cm by 7 cm by 4 cm.
   Each plane of symmetry forms 2 cuboids.
   What are the dimensions of the cuboids that are formed by each line of symmetry?

Unit 2 • Understanding nets • Band f

Outside the Maths classroom

Packaging

What do manufacturers need to consider when designing packaging?

Toolbox

2D nets are designed to be folded to make 3D objects.

The same object may have different nets.

Whatever their arrangement, the sections of the 2D design must match the faces of the 3D object that is to be made.

Nets need flaps to enable the 3D shape to become rigid.

Clever designers can design the flaps to make the box rigid without glue or staples.

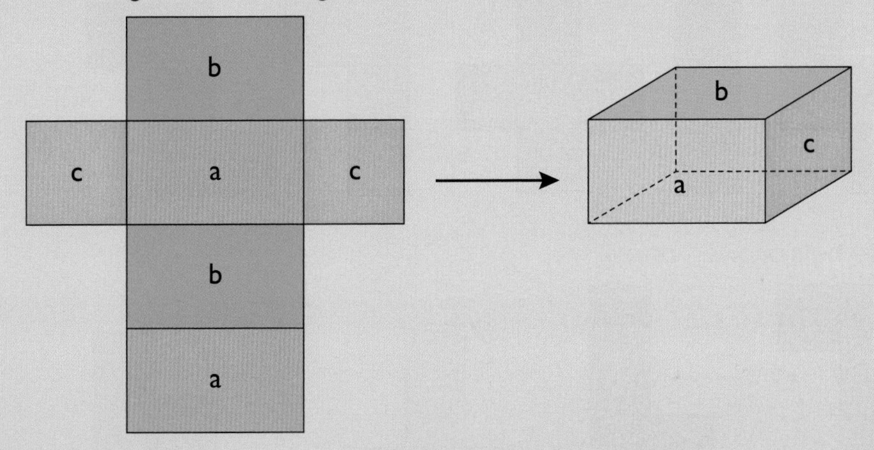

Nets can be used to find the surface area of shapes.

Example – Adding flaps

Is there a relationship between the number of edges of a net and the number of flaps needed to make the solid rigid?

Solution

Yes; the minimum number of flaps needed is half the number of edges that need to be joined because each edge joins with another edge.

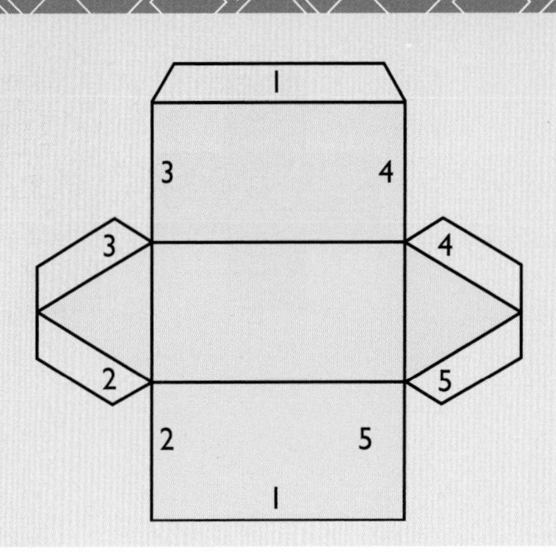

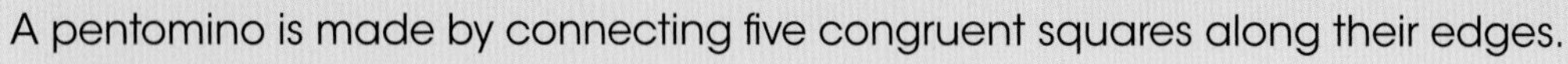

Example – Designing a net

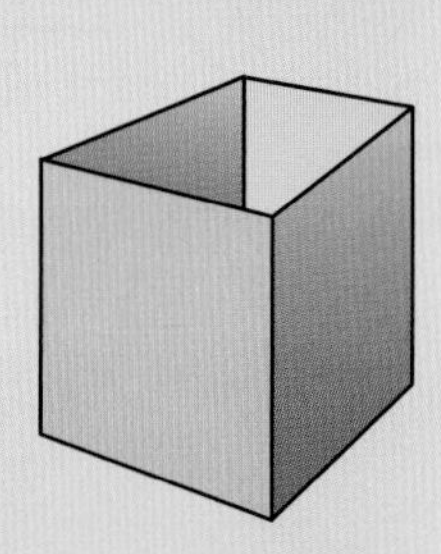

A pentomino is made by connecting five congruent squares along their edges.

a Draw as many different pentominos as you can.

b Which of the pentominos can be a net of an open box?

Solution

a There are twelve different pentominos.

b These pentominos can be nets of open boxes.

Example – Matching the dimensions of a net and its solid

These are the nets of two cuboids.

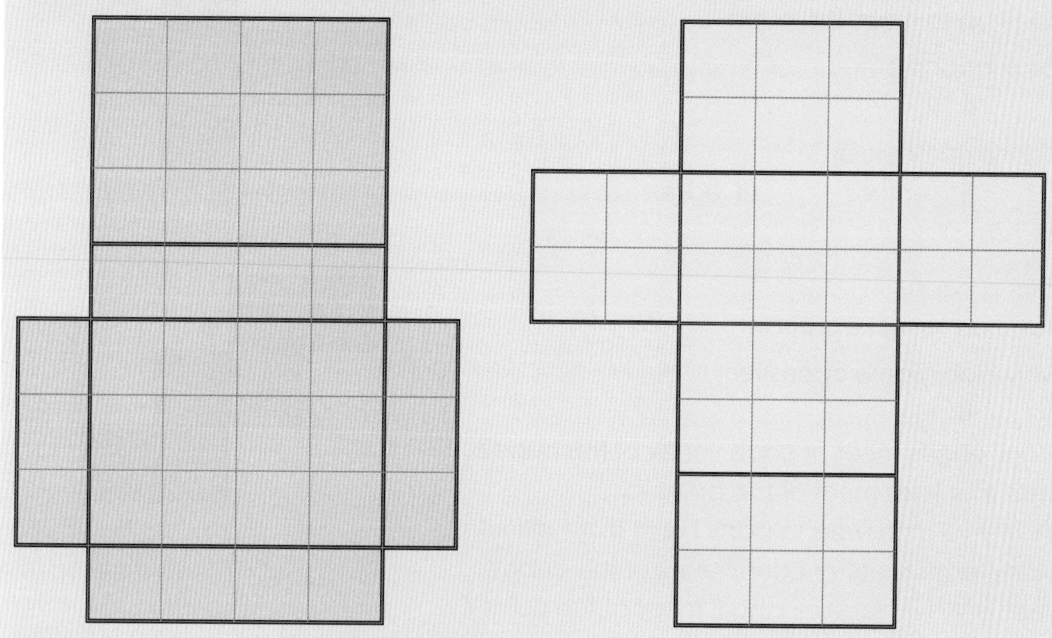

They are made of centimetre squares.
Compare the surface areas of the two cuboids.

Solution

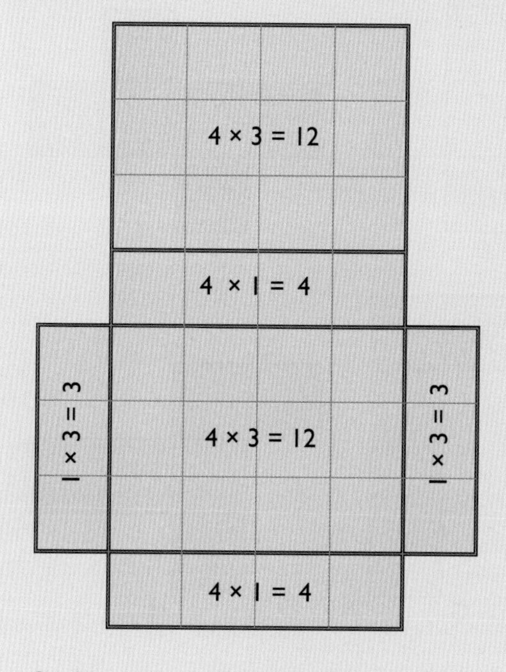

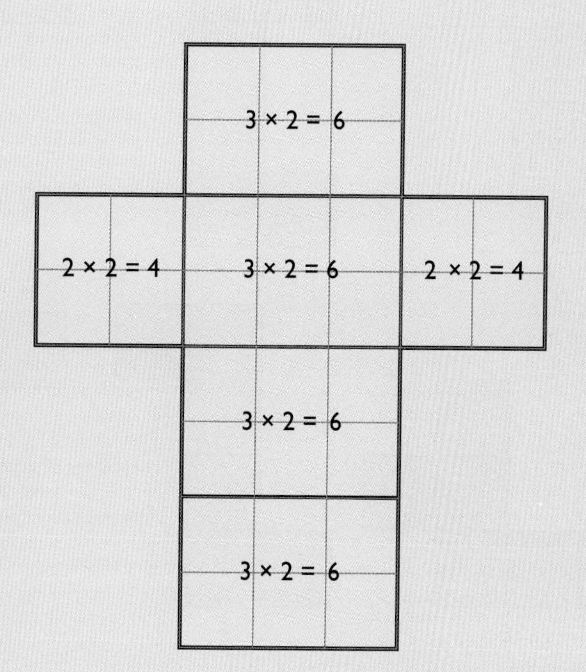

Surface area of pink cuboid $= 12 + 4 + 12 + 4 + 3 + 3$
$= 38\,\text{cm}^2$

Surface area of blue cuboid $= 6 + 6 + 6 + 6 + 4 + 4$
$= 32\,\text{cm}^2$

Practising skills

1 Here is a net for a cube.
The net is folded up to make the cube.

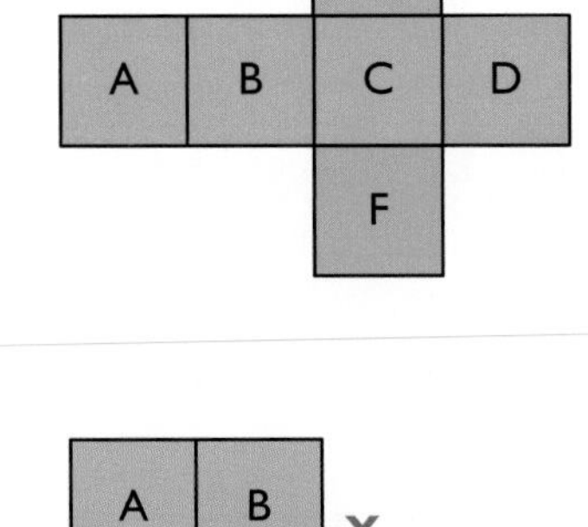

a Which face is opposite
  i B **ii** E.

b Write down the numbers of
  i faces **ii** vertices **iii** edges
that the cube has.

2 Here is a net for a cube.
The net is folded up to make a cube.

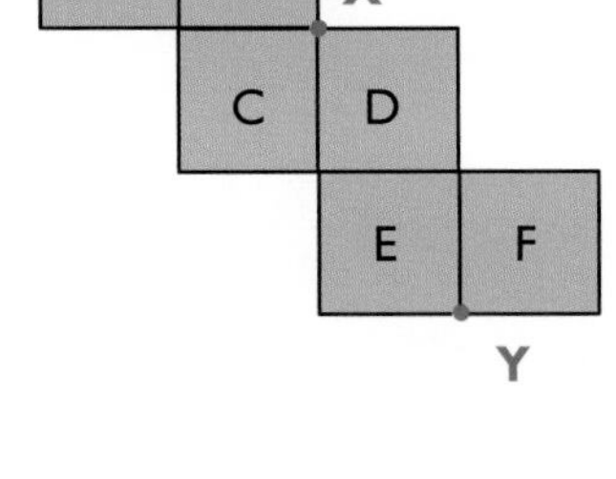

a Write down which face is opposite
  i B **ii** F.

b **i** How many edges meet at each vertex of the cube?
  ii In the net, four lines meet at the point X.
  iii Explain why your answers to parts **i** and **ii** are different.

c **i** How many faces meet at each vertex of the cube?
  ii In the net, only two faces, E and F, meet at the point Y.
  iii Does this alter your answer to part **i**?

3 Which of these are nets for a cube?

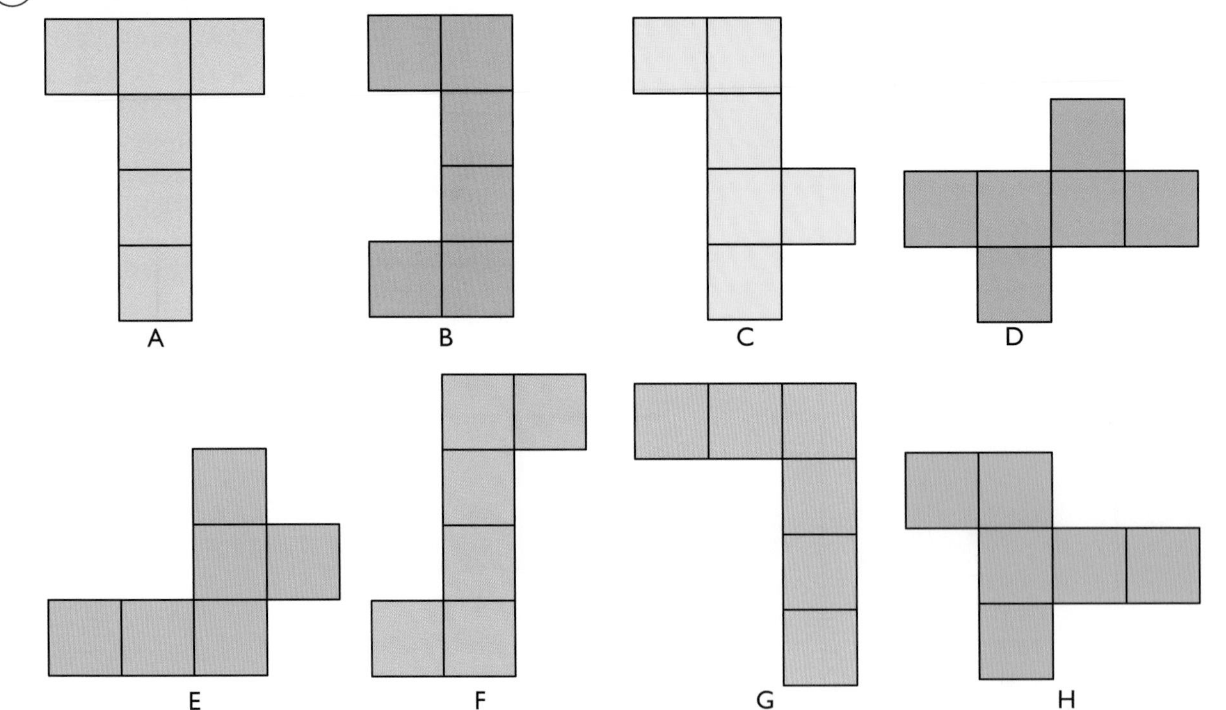

4 Here are two cuboids.

a

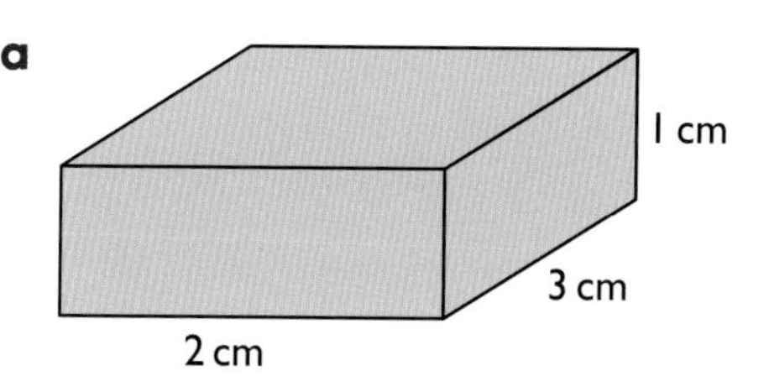

b

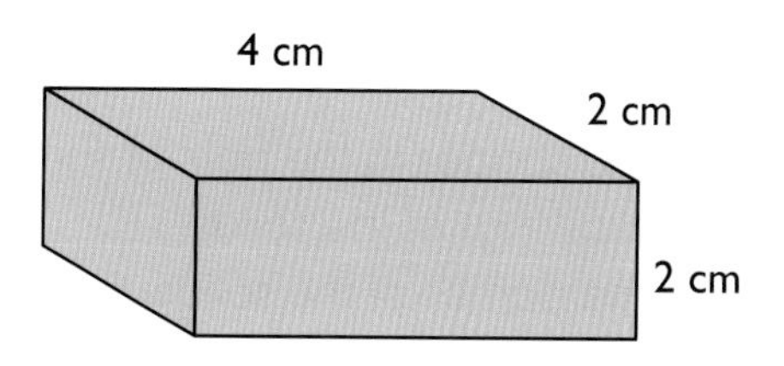

i Use squared paper to draw an accurate net for each of these cuboids.

ii Find the area of each net.

iii What does this tell you about the surface area of each cuboid?

5 Write down the name of the solid that each net will make.
Draw a sketch of each solid.

a

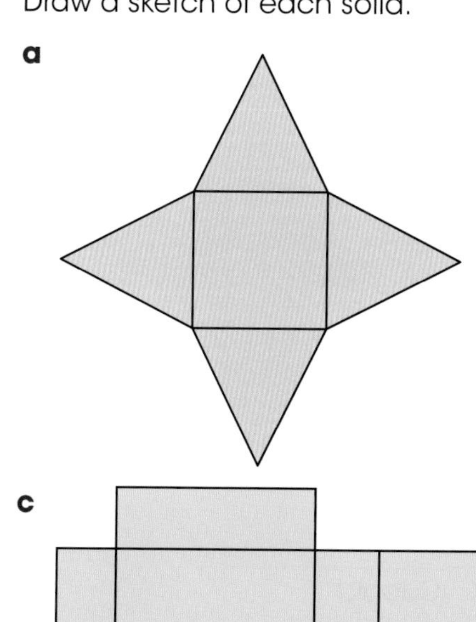

b

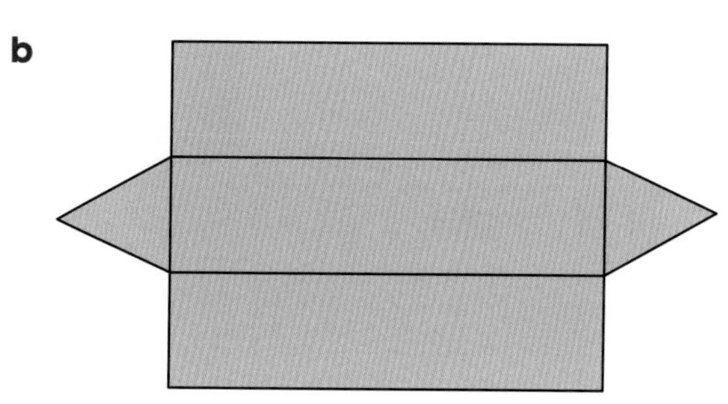

c

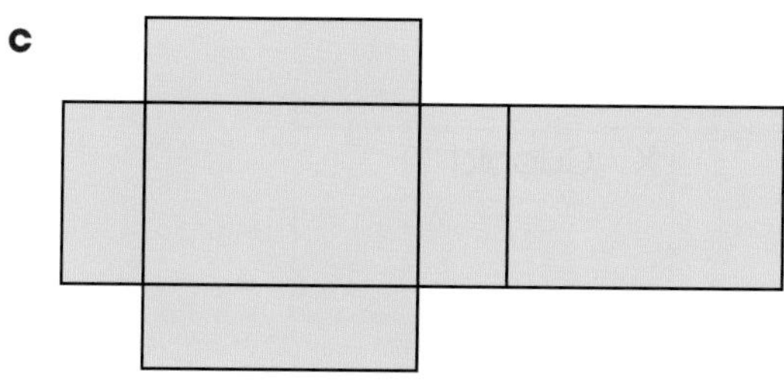

d

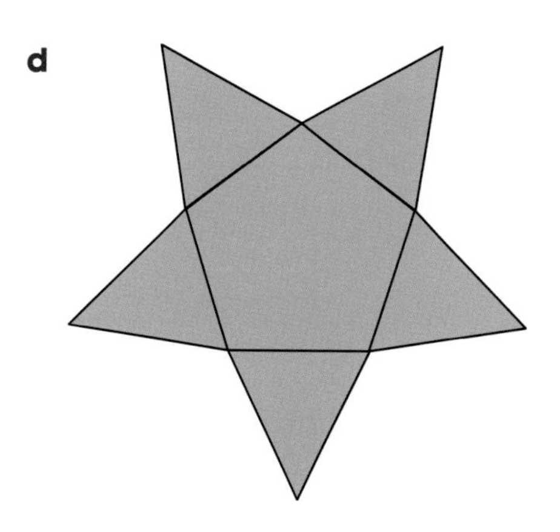

Developing fluency

1. Here are three lists.
   - List 1 gives the shapes in the nets of some solids.
   - List 2 gives drawings of the solids.
   - List 3 gives the names of the solids.

   The lists are not in the same order.

   Match the contents of the three lists.

List 1	List 2	List 3
A 6 rectangles	P	U Hexagonal prism
B 4 triangles	Q	V Cylinder
C 4 triangles and 1 square	R	W Tetrahedron
D 6 rectangles and 2 hexagons	S	X Cuboid
E 2 circles and 1 rectangle	T	Y Square-based pyramid

2 **a** Draw an accurate net for each of these solids. Use a pair of compasses to draw the triangles accurately.

b Then add flaps.

c To check your accuracy, cut the net out, fold it up and glue it together. Does it look right?

i A wedge.

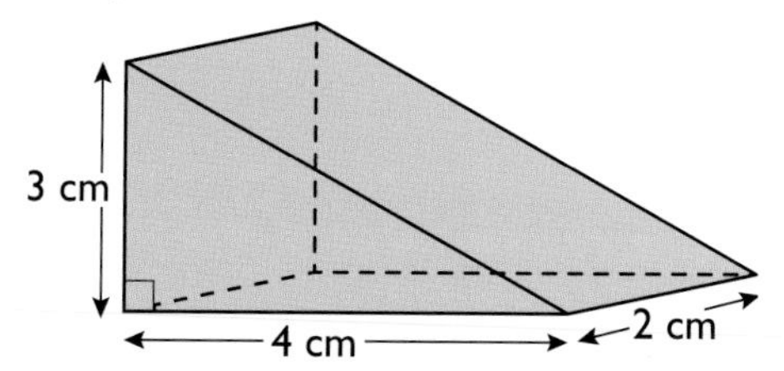

ii A square-based pyramid.

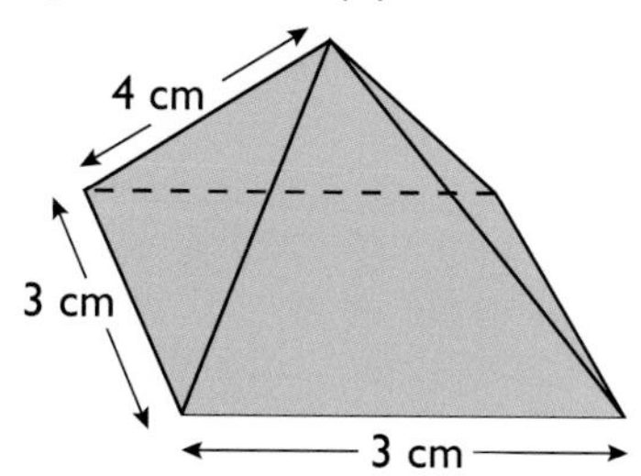

iii A triangular prism.

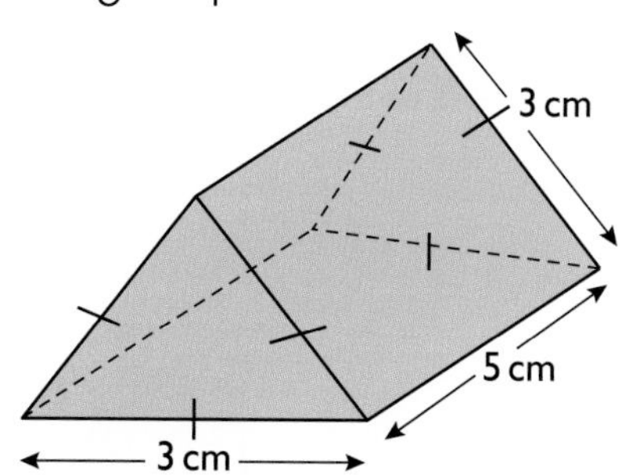

Exam-style

3 The diagram shows the net of a 2cm cube.
It is cut from the piece of cardboard shown.

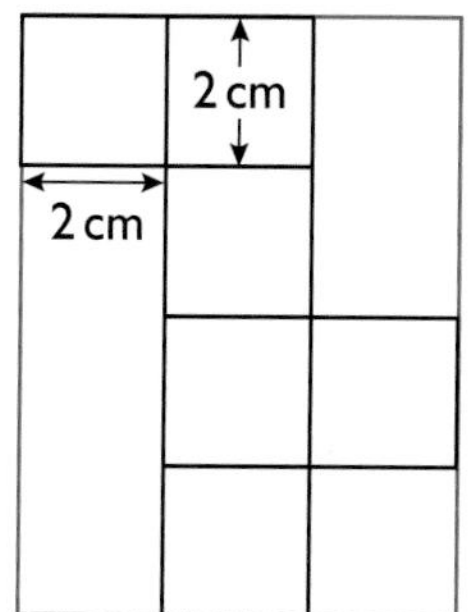

a What is the area of the net?

b What is the area of the piece of cardboard it is cut from?

c What percentage of the piece of cardboard is not used?

d Draw two other nets for the cube. Draw a rectangle round them.
In each case find the percentage of the rectangle that is not used.

4 Here is the net of a regular octahedron.

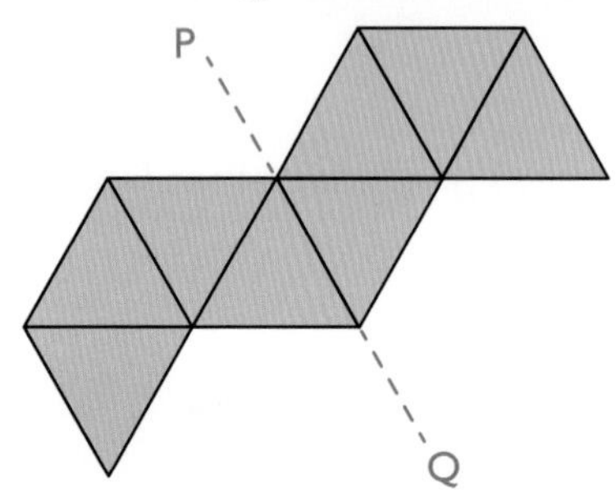

a **i** How many faces does an octahedron have?

ii What shape is each face?

iii How many faces meet at each vertex?

b This solid is a regular tetrahedron.

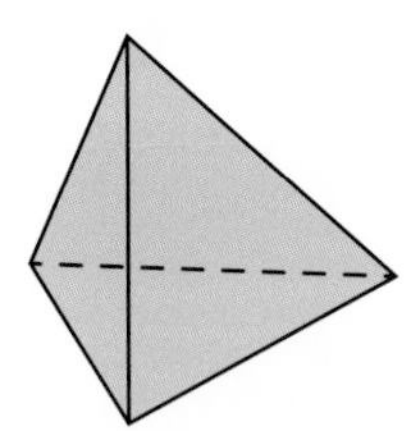

i How many faces does it have?

ii What shape is each face?

c Boris says 'If you cut the net of the octahedron along the line PQ, each piece will be the net for a tetrahedron'.

Is Boris right? Explain your answer.

5 Rosie designs a net of a dodecahedron.

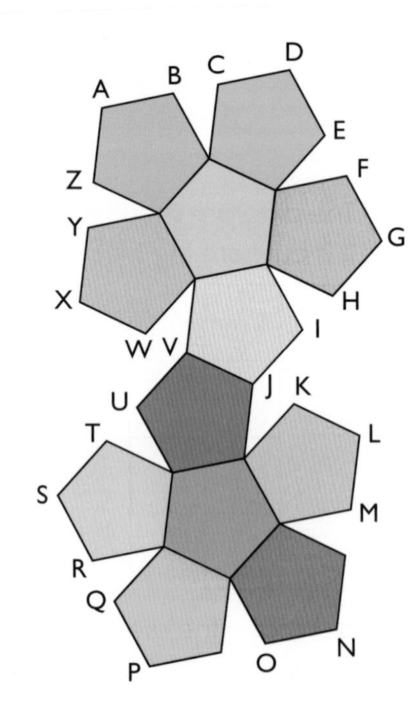

a What shape are the faces?

b Write down the number of

i faces, F

ii edges, E

iii vertices, V, of the dodecahedron.

c Euler's theorem states that $F + V - E = 2$.

Show that the dodecahedron obeys Euler's theorem.

d Rosie adds tabs to her net so that she can stick the dodecahedron together.

What is the least number of tabs that she needs?

Rosie folds up her net to form a dodecahedron.

e Which faces are adjacent to the red face?

f Which edge meets ST?

g Which points meet at point A to form a vertex?

h The area of one face of the dodecahedron is 20cm^2.

What is the surface area of the dodecahedron?

Problem solving

Exam-style

1. Jeff is designing a box for children's toys.

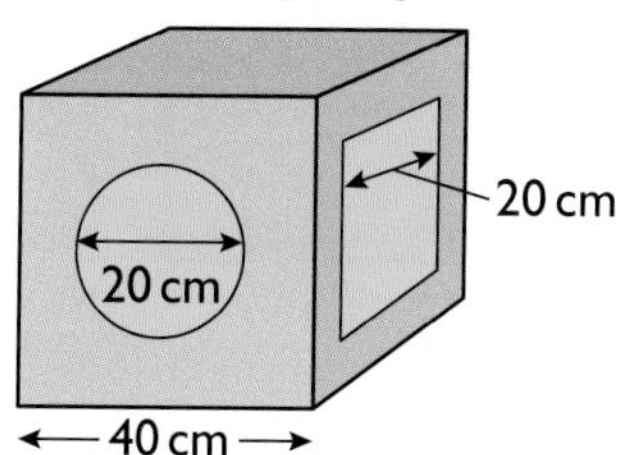

The box is a cube with sides of length 40 cm.
Two opposite faces of the box each have a circle placed centrally on them.
Two opposite faces of the box each have a square placed centrally on them.
The other two faces are left blank.
Draw a scale diagram of the net.
You must include the circles and squares.

Exam-style

2. Kylie makes small boxes for jewellery.

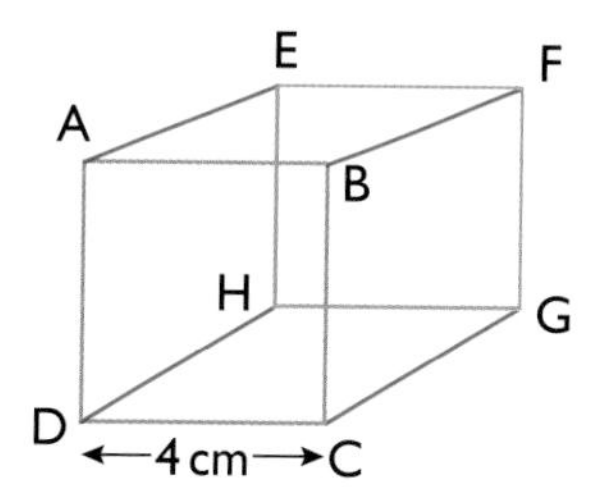

Each box is in the shape of a cube with sides of length 4 cm.
The edges of rectangle ABCD are shaded blue.
The edges of rectangle EFGH are shaded green.
All other edges are shaded red.
Draw a net of the jewellery box.
Label each edge with its shading.

Exam-style

Extension

3. The diagram represents a chocolate in the shape of pyramid.
There is a thread stretched across the sides of the pyramid from A to C, passing through the midpoint of the sloping edge of the pyramid.

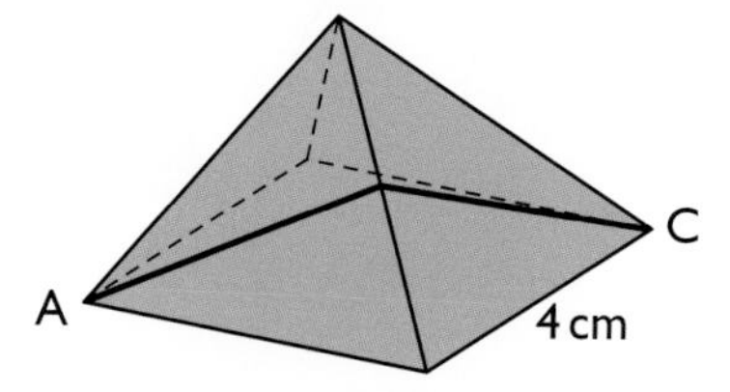

The pyramid has a square base of side 4 cm.
The other faces are equilateral triangles.
Draw a scale diagram to show the net.
Use your diagram to find the length of the string.

Extension

4 The diagram represents a cube used in a seating arrangement.
The edges of the cube have length 40 cm.
A string stretches from A to C.
The string is as short as possible.
Use a scale diagram to find the length of the string.

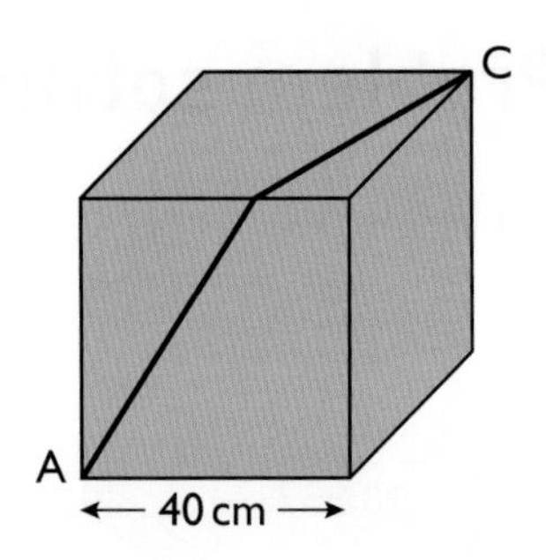

5 Here are 3 views of the same cube. Each face of the cube is a different colour.

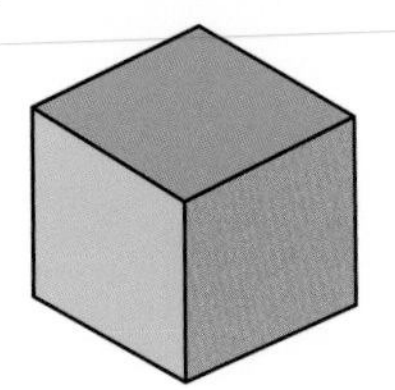 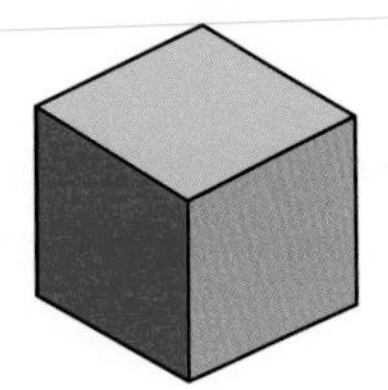 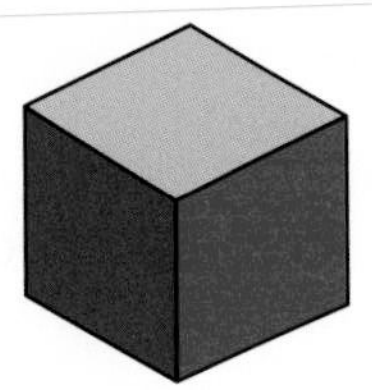

a Draw a net of the cube – label each face with its colour.

b Which faces are opposite each other?

A dot is marked on the centre of each face.
The dots on adjacent sides are joined by straight lines.

c What shape is generated by these lines?

d How many
- **i** faces
- **ii** edges and
- **iii** vertices

does the resulting shape have?

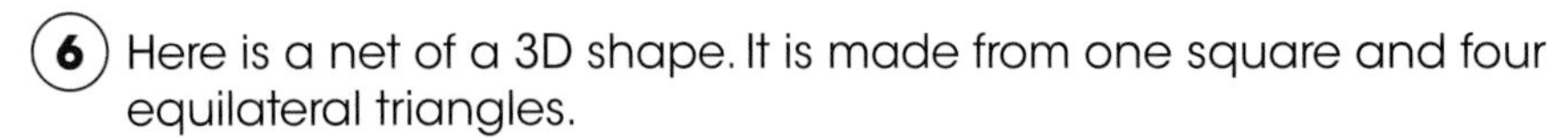

6 Here is a net of a 3D shape. It is made from one square and four equilateral triangles.

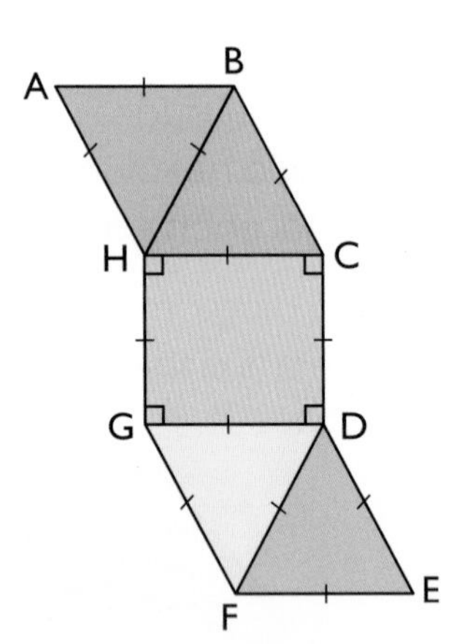

a Describe the symmetry of the net.

b How many
- **i** faces
- **ii** vertices and
- **iii** edges

does the 3D shape have?

c When the net is folded up, which point joins to A?

d Sam says, 'When the net is folded up it will form a prism.'
Is Sam right?
Explain your answer fully.

e AB is 4 cm. Make an accurate construction of the net.

f **i** Measure the perpendicular height (altitude) of one of the triangles.

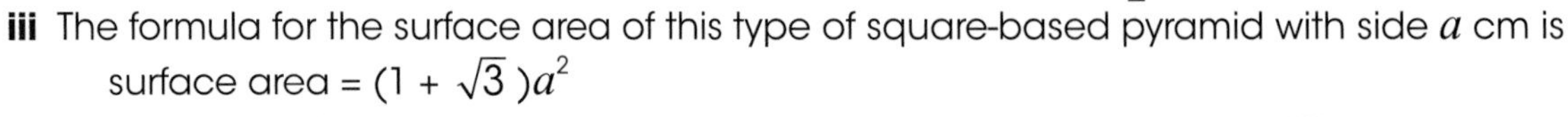

ii What is the area of the net? (Remember: area of a triangle is $\frac{1}{2}$ × *base* × *height*)

iii The formula for the surface area of this type of square-based pyramid with side a cm is

surface area = $(1 + \sqrt{3})a^2$

Show that this formula gives a result consistent with your answer in part **ii**.

Reviewing skills

1. Here are two cuboids

i

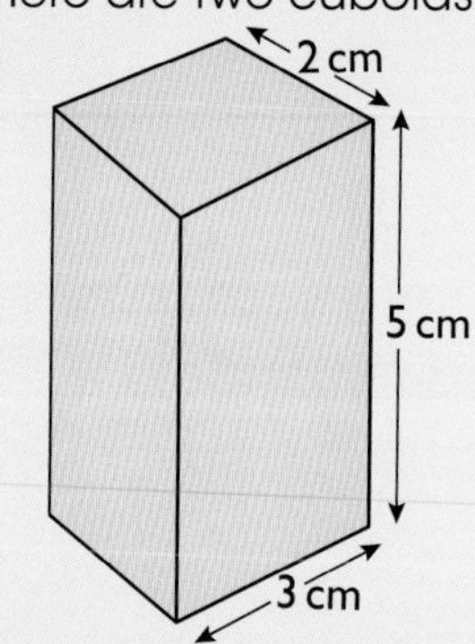

ii

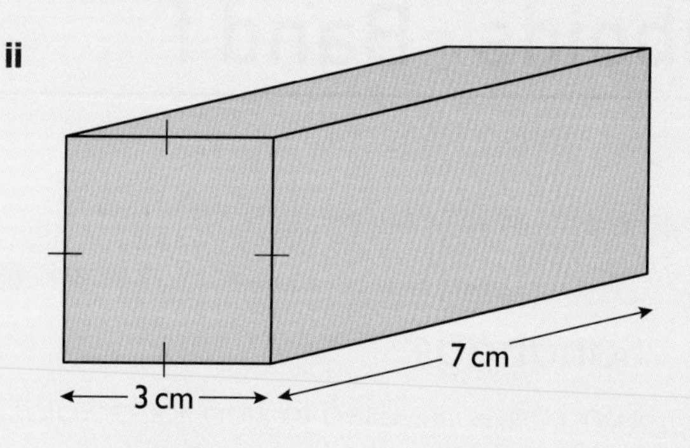

   a Use squared paper to draw an accurate net for each of these cuboids.

   b Find the area of each net.

   What does this tell you about the surface area of each cuboid?

2. **a** Draw an accurate net for this solid.

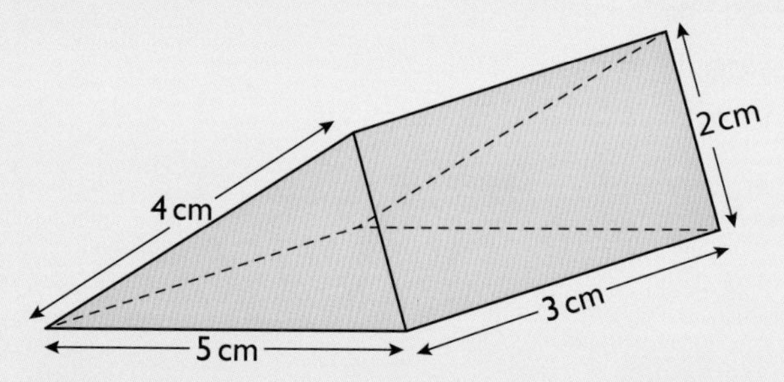

   Use a pair of compasses to draw the triangles.

   b Then add flaps.

3. In this pyramid all the edges are 4 cm.

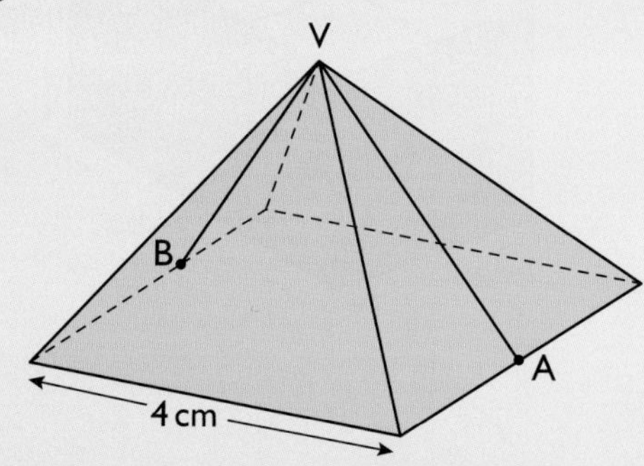

   The base of the pyramid is a square.

   A and B are the midpoints of their sides. Their vertex is V.

   a Use a ruler and a pair of compasses to construct the net of this pyramid.

   b An ant walks on the surface of the pyramid from A to V, and then to point B.

   Use your net to measure how far it walks.

Unit 3 • Volume and surface area of cuboids • Band f

Outside the Maths classroom

Sandbags

Sand bags are used to stop flood water.
How would they be transported?

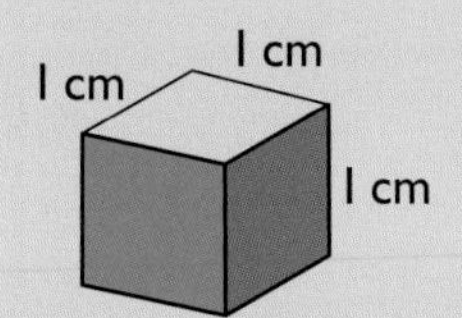

Toolbox

The **volume** of a solid shape is the space inside it.
It is measured in cubic units such as cm^3 and m^3.
This is a centimetre cube.

1 cm 1 cm 1 cm

It has a volume of 1 cubic centimetre (1 cm^3).
You can sometimes find the volume of a cuboid by counting the number of cubes that fit into it.
This cuboid is made of 12 centimetre cubes.
So its volume is 12 cm^3.
Another way to find the volume of a cuboid is to use the formula

volume of a cuboid = length × width × height.

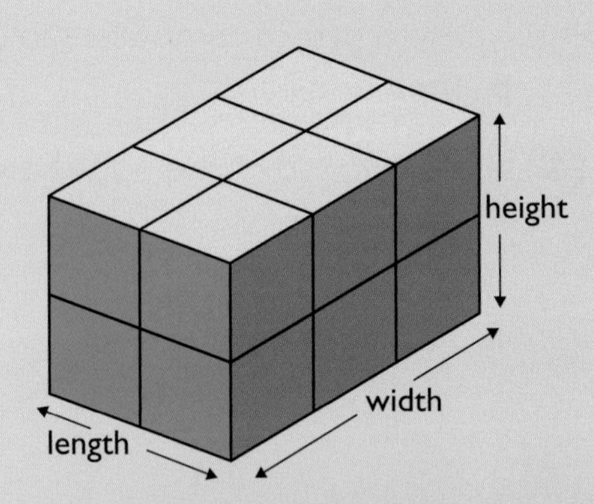

The surface area of a solid is the total area of all its faces.
In a cuboid, the front and back are the same; so are the top and bottom, and the two sides. So

surface area of a cuboid
= 2 × (area of base + area of front + area of one side)

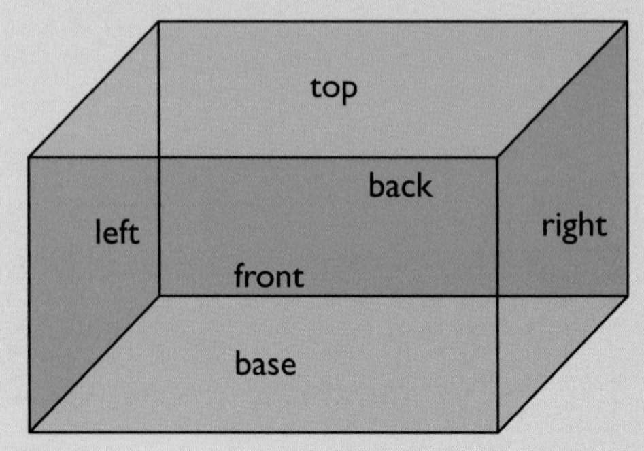

Example – Finding the volume and surface area of a cuboid

Find the volume and surface area of this cuboid.

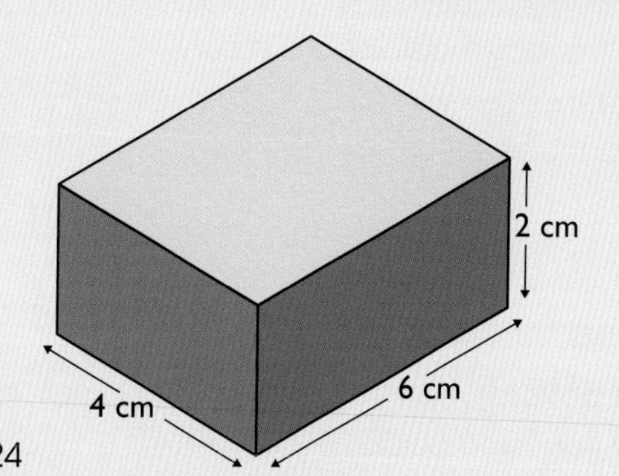

Solution

Volume of a cuboid = length × width × height
= 4 × 6 × 2
= 48

The volume of the cuboid is $48\,\text{cm}^3$.

The surface area of its faces is

Base:	4 × 6 = 24	Top:	4 × 6 = 24
Front:	4 × 2 = 8	Back:	4 × 2 = 8
Right:	6 × 2 = 12	Left:	6 × 2 = 12

So the total surface area = 24 + 24 + 8 + 8 + 12 + 12
= 2 × (24 + 8 + 12)
= 2 × 44
= 88

The surface area of the cuboid is $88\,\text{cm}^2$.

Example – Finding surface area from a net

Here is a net of a cuboid.
Find the surface area of the cuboid.

Each rectangle makes a face of the cuboid.
The green rectangle is the bottom face.

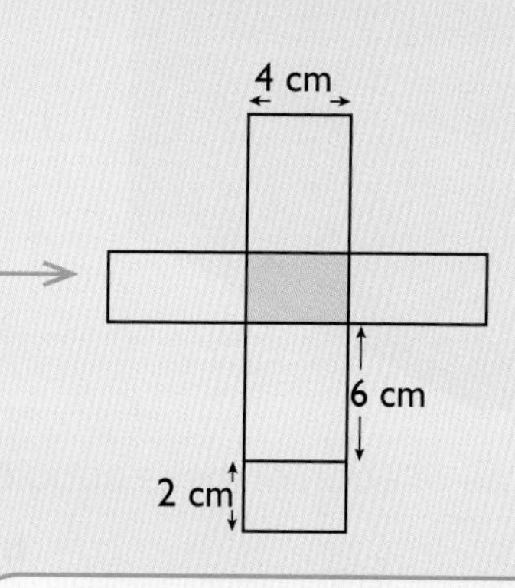

Solution

Work out the area of each face.
Remember that the area of a rectangle is given by the formula area = length × width.

The surface area of the faces is

Base:	4 × 2 = 8	Top:	4 × 2 = 8
Front:	4 × 6 = 24	Back:	4 × 6 = 24
Right:	6 × 2 = 12	Left:	6 × 2 = 12

Total surface area = 2 × (8 + 24 + 12)
= 88

The surface area of the cuboid is $88\,\text{cm}^2$.

Example – Finding the surface area of an open box

Look at this open cardboard box.
It has no top.
Find its surface area.

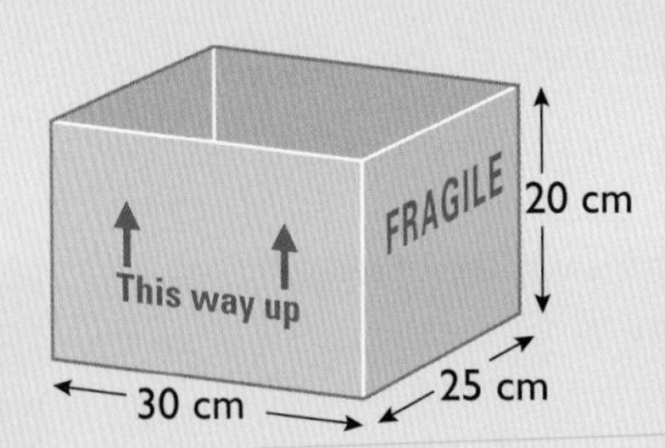

Solution

Area of bottom = 30×25 = $750\,cm^2$

Area of front and back = $2 \times 30 \times 20$ = $1200\,cm^2$ ← The front and back have the same area.

Area of both sides = $2 \times 25 \times 20$ = $1000\,cm^2$ ← Both sides have the same area.

Total surface area = $750 + 1200 + 1000$ = $2950\,cm^2$

Practising skills

1. Each cuboid is made up of $1\,cm^3$ cubes. Work out the volume of each cuboid.

a

b

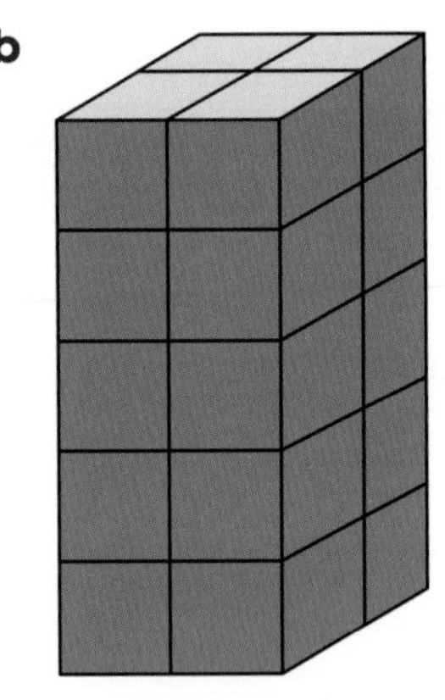

2. Work out the volume of each of these cuboids.

a

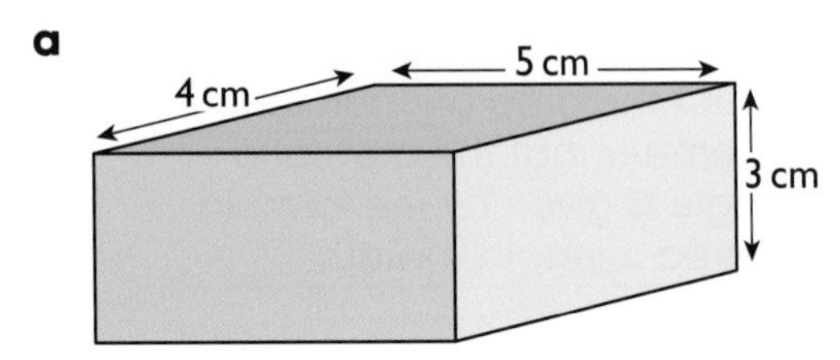

b

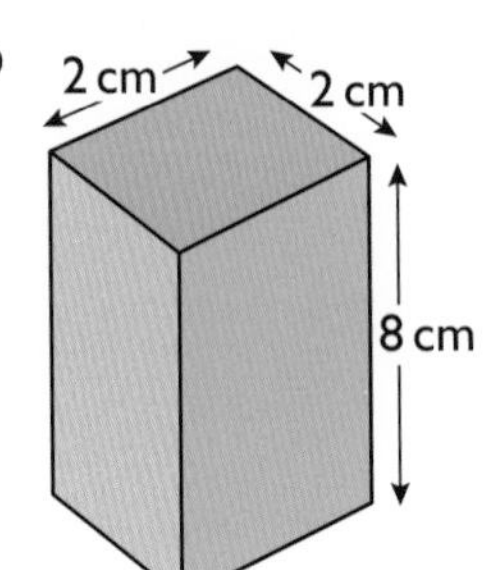

c

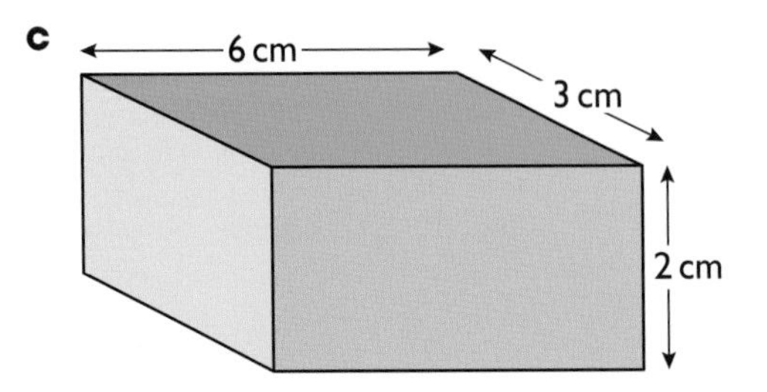

3 Work out the surface area of each of these cuboids.

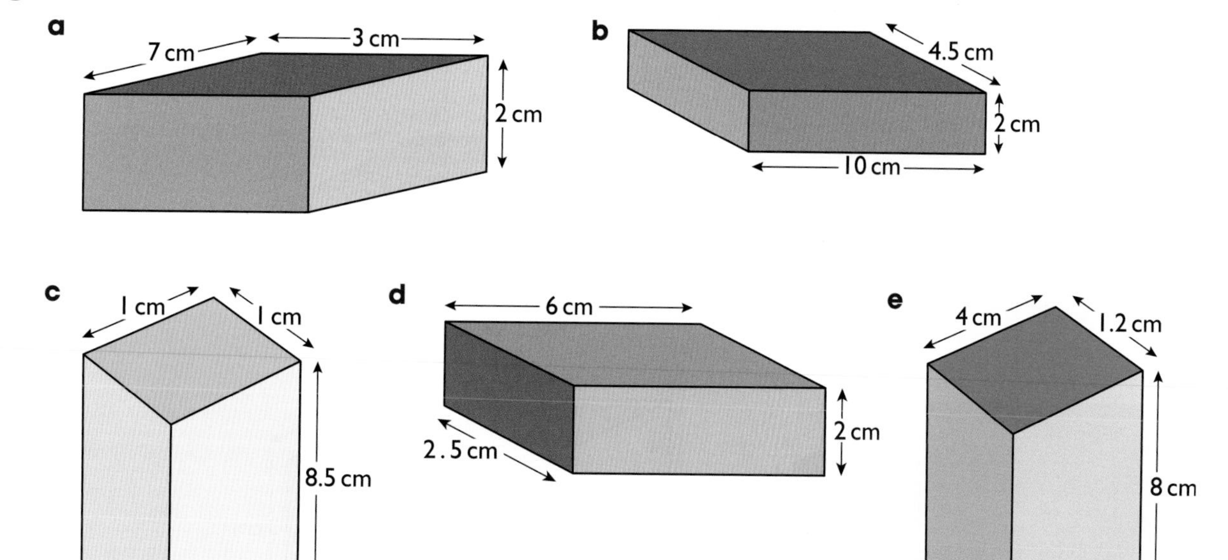

4 A cuboid has volume $72\,cm^3$. Write down the dimensions of five different cuboids which have this volume. Only use whole numbers.

Length	Breadth	Height	Volume
			$72\,cm^3$
			$72\,cm^3$
			$72\,cm^3$
			$72\,cm^3$
			$72\,cm^3$

5 Copy and complete this table for cuboids.

	Length	Breadth	Height	Volume	Surface area
a	5 cm	3 cm	2 cm		
b	6 cm	2 cm	2 cm		
c	5 cm	4 cm		$60\,cm^3$	
d		5 cm	1 cm	$35\,cm^3$	
e	6 cm	4 cm		$36\,cm^3$	

Developing fluency

Exam-style

1. Here are four cuboids.

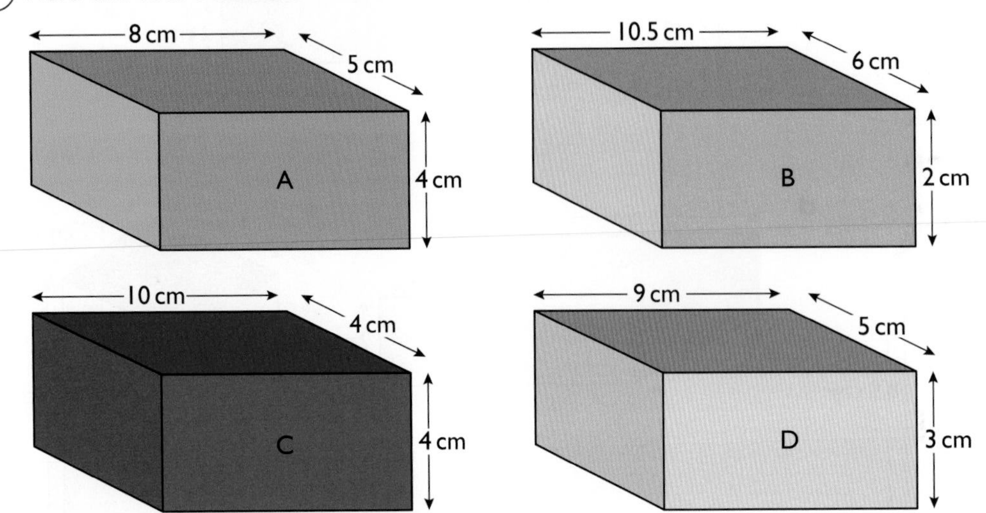

   a Which two cuboids have the same volume?

   b Which two cuboids have the same surface area?

2. Here are four cuboids.

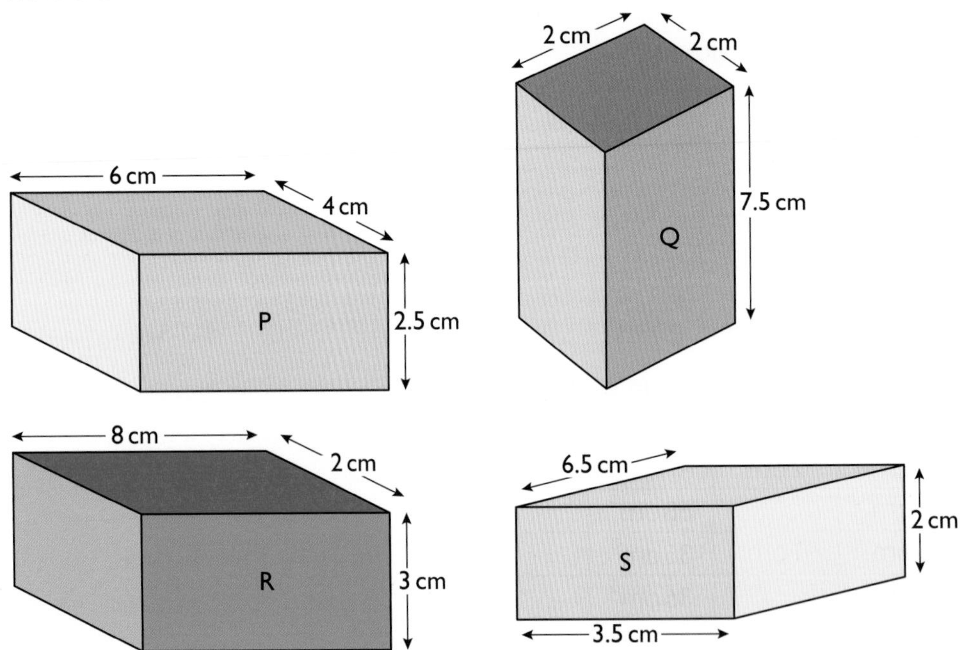

   a Write the cuboids in order of the size of their surface areas starting with the smallest.

   b Name the cuboid with

   i the greatest volume

   ii the smallest volume.

   c Copy and complete these statements.

   The volume of ☐ is double the volume of ☐.

   The volume of ☐ is $\frac{4}{5}$ the volume of ☐.

Reasoning

3 All these statements are false. Explain why they are false.

a A cube of side 1 cm has a volume of 1 cm^2.

b A cube of side 2 cm has surface area 24 cm^3.

c A cube of side 4 cm has a volume of 12 cm^3.

d A cube of side 3 cm has surface area 45 cm^3.

e The volume of a cube of side 2 cm is twice the volume of a cube of side 1 cm.

Exam-style

4 A cube's total surface area is 294 cm^2.

a What is the area of one square face?

b What is the length of each edge?

c What is the cube's volume?

Exam-style

5 Decide if these statements are true or false. If false, explain why.

a The volume of a cuboid 2 m by 1 m by 20 cm is 40 cm^3.

b The volume of a cuboid 9 cm by 3 cm by 1 cm is the same as the volume of a cube of side 3 cm.

c The surface area of a cuboid 6 cm by 4 cm by 2 cm is twice the surface area of a cuboid 3 cm by 2 cm by 1 cm.

6 A carton of juice measures 10 cm by 4 cm by 4 cm.
A box measures 1 m by 40 cm by 20 cm.
How many cartons of juice will fit into the box?

Exam-style

7 A solid wooden table top is in the shape of a cuboid and measures 2 m by 1 m by 4 cm.
The top and the four sides are varnished but not the bottom.
One tin of varnish covers 4 m^2.
How many tins are needed to give the table two coats of varnish?

Exam-style

Extension

8 Two faces of Evi's jewellery box are squares with edges of s cm.

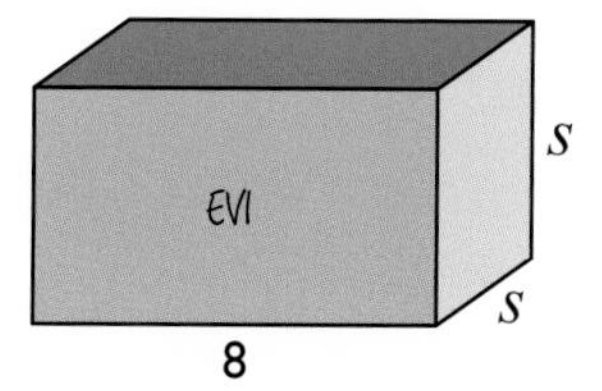

The other faces are rectangles with edges of length 8 cm and s cm.
The volume of the box is 72 cm^3.

a Work out the value of s.

b Find the surface area of the box.

9 The diagram shows the net of a solid shape.

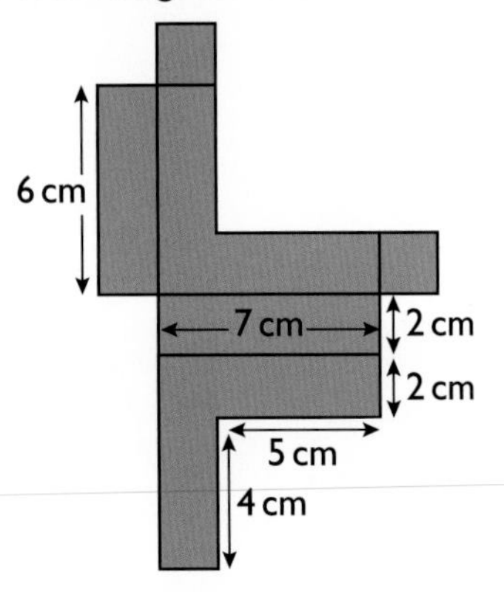

a Draw a sketch of the shape.

b Find the surface area of the shape.

c Find the volume of the shape.

d Write down

i E, the number of edges

ii V, the number of vertices

iii F, the number of faces of the shape.

e Show that the shape obeys Euler's rule $V + F = E + 2$.

Problem solving

Exam-style

Extension

1 Here is a tank for storing water. The tank does not have a top.

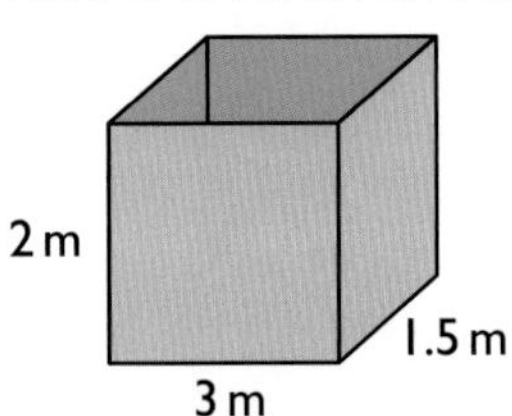

Liz is going to paint the five inside faces and the four outside faces.
One litre of paint covers $20 m^2$. Liz buys the paint in 2.5 litre tins.
Will one tin be enough?

Exam-style

Extension

2 Jo has a swimming pool.

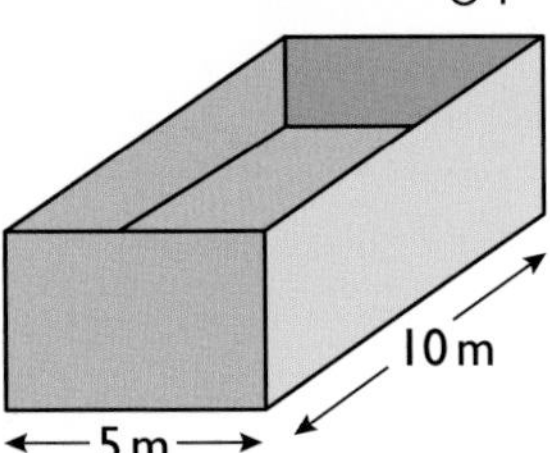

Jo wants to fill the pool to a depth of 180 cm.
The pump will pour water into the pool at a rate of 40 litres per minute.
Jo's friend says it will take more than a day to fill the pool from empty to a depth of 180 cm.
Show that Jo's friend is correct.

Exam-style

Extension

3 Iqbal is studying boxes for keeping shopping cool for a school science project.

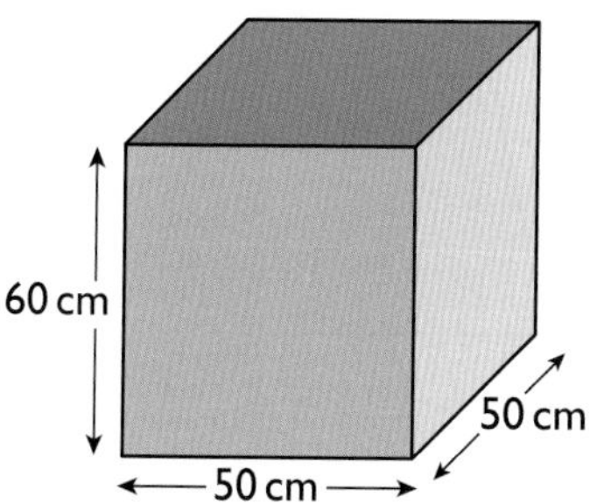

Iqbal measures the edges of the outside of a cool box.
The height is 60 cm.
The width is 50 cm.
The length is 50 cm.

a Use Iqbal's figures to estimate the capacity of the cool box.
Give your answer in litres.
($1\text{ m}^3 = 1000$ litres)

b State an assumption you have made and how it will affect your answer to part **a**.

c The sides, top and bottom of the cool box are all 2 cm thick.
What is the true value of its capacity?

Exam-style

Extension

4 The diagram shows a box on a table.
A and B are opposite faces.

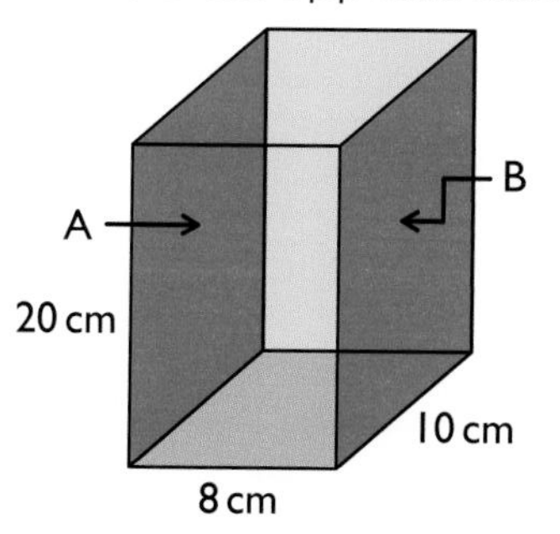

The box has length 10 cm, width 8 cm and height 20 cm.
Lawrence pours sand into the box until the level of the sand is 2 cm below the top.
Then he puts the lid on and turns the box over so that face A is on the table.
Work out how far the level of the sand is below face B.

Exam-style

5 The diagram shows a fish tank partly filled with water.

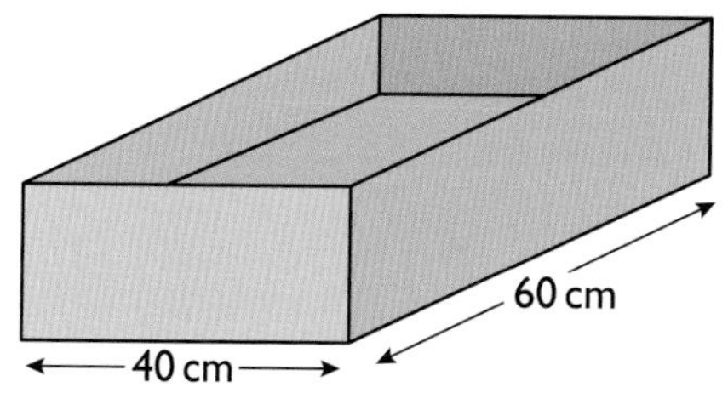

The tank is 40 cm wide and 60 cm long.
The water is 28 cm deep.
A stone is put in the tank and the water completely covers the stone.
The water in the tank is now 30 cm deep.
Work out the volume of the stone.

6 This is the net for a scale model of a pyramid in Central America.

It has a flat top. The perimeter of the real pyramid is 180 m.

Make suitable measurements on the net to answer these questions.

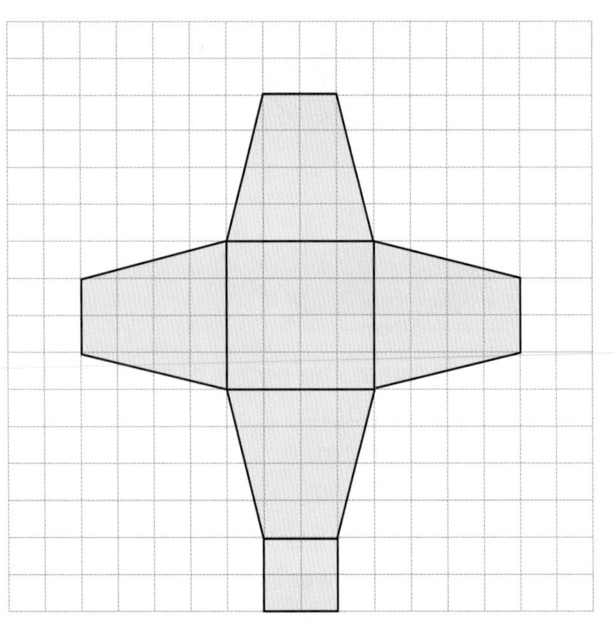

a What is the perimeter of the flat top of the pyramid?

b What is the area of the flat top of the real pyramid?

c How long are the slanting edges of the real pyramid?

d Show that the total area of the four sloping faces is less than 7200 m².

7 Henry designs a net for a gift box on a centimetre grid.

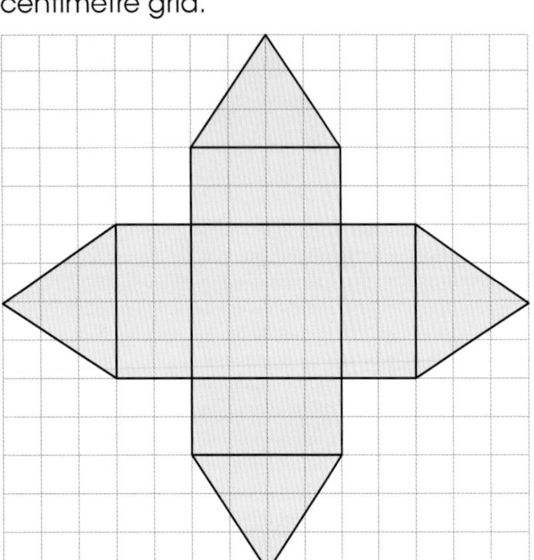

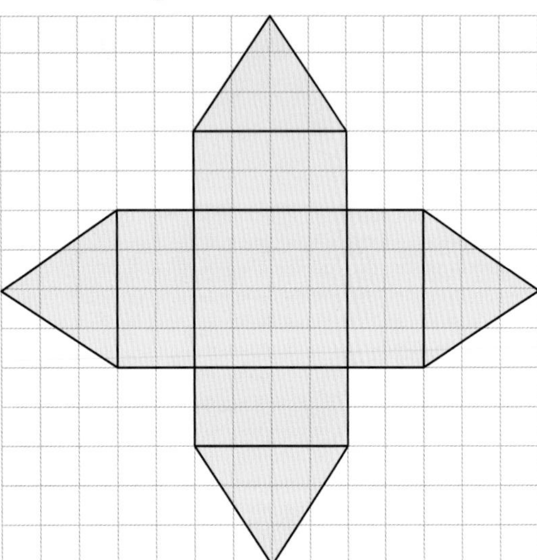

a How many lines of symmetry does the net have?

b What is the order of rotational symmetry of the net?

c Henry cuts the net from a square piece of card measuring 14 cm by 14 cm.

i What area of card does he waste?

Henry folds the net to form the gift box.

ii How many planes of symmetry does the gift box have?

d The gift box has two parts.

Write down the shape of

i the lower part **ii** the top part of the gift box.

e Find the volume of the lower part of the gift box.

f Write down the number of

i faces, F **ii** edges, E **iii** vertices, V of the gift box.

g Euler's theorem states that $F + V - E = 2$.

Show that Henry's gift box obeys Euler's theorem.

Reviewing skills

1. Work out the volume and surface area of each of these cuboids.

a

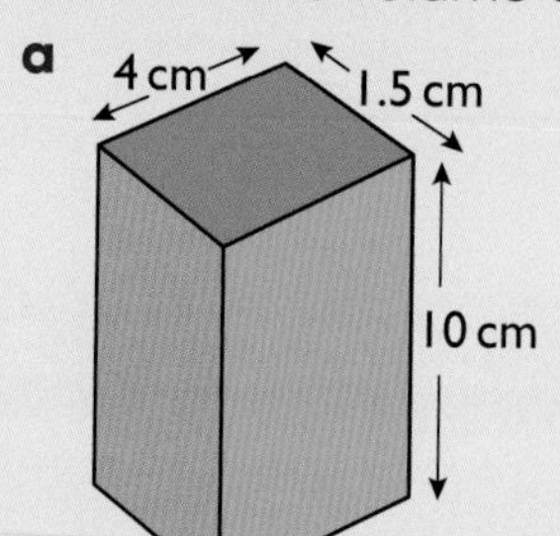

b

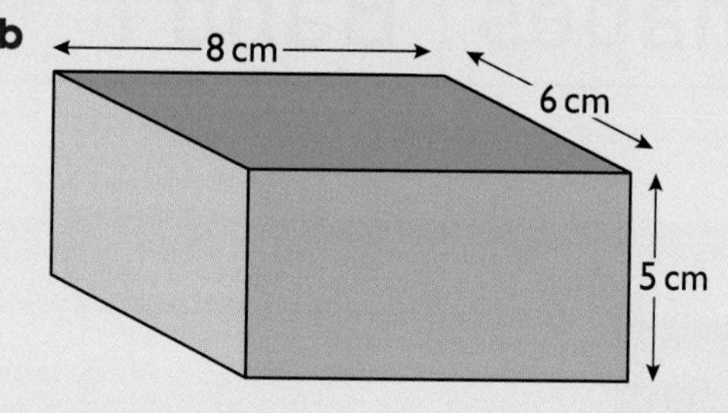

2. A company makes big sugar lumps. Their sides are 3 cm, 2 cm and $\frac{1}{2}$ cm.
The company wants to pack the lumps in a box.
   a Explain why a box 10 cm × 10 cm × 5 cm is not suitable.
   b The company decide the box should be 12 cm × 10 cm × 5 cm.
   How many sugar lumps does it hold?

Unit 4 • 2D representations of 3D shapes • Band f

Outside the Maths classroom

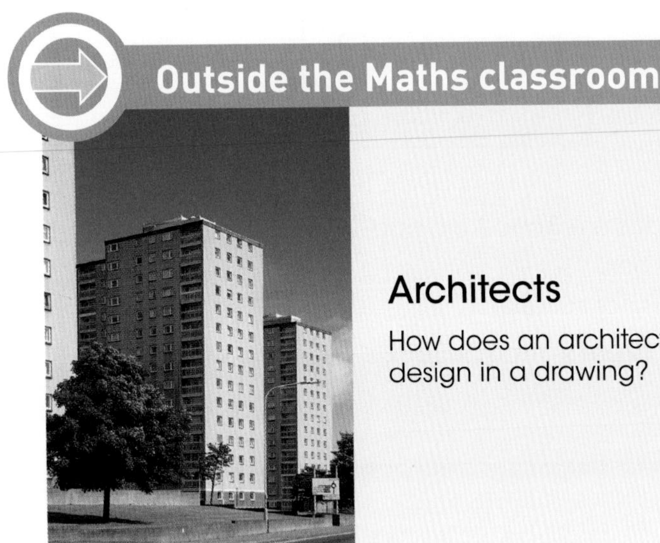

Architects

How does an architect represent all aspects of design in a drawing?

Toolbox

You can represent a three-dimensional (3D) object in two dimensions by drawing its plan and elevations separately.

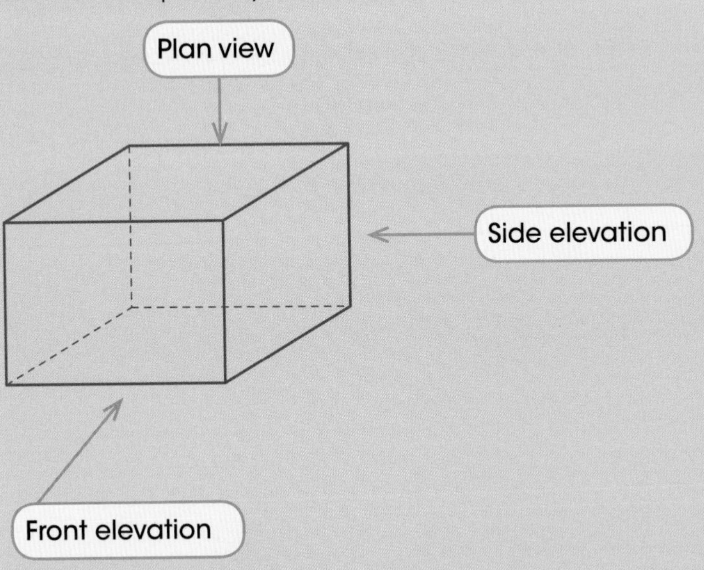

You can also use isometric paper to draw a representation of a 3D object.

Example – Drawing plans and elevations

Draw the plan, front elevation and side elevation of this solid.

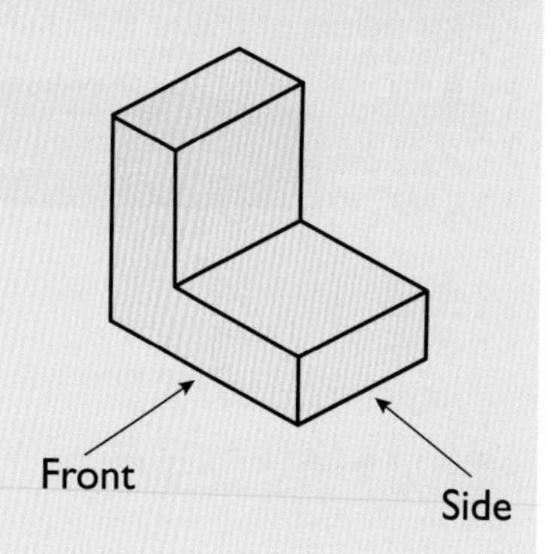

Solution

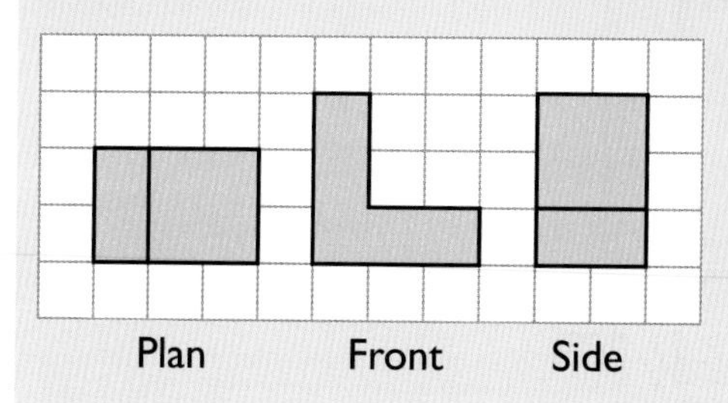

Practising skills

1. These shapes are made up from 1 cm cubes. Using squared paper, draw
   i the front elevation **ii** the side elevation **iii** the plan
   of each shape.

a

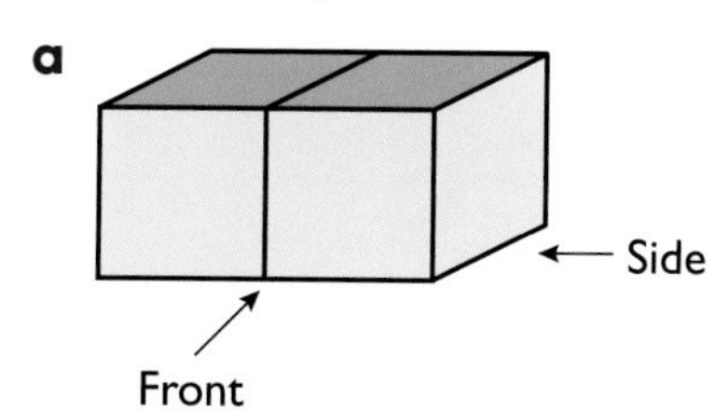

b

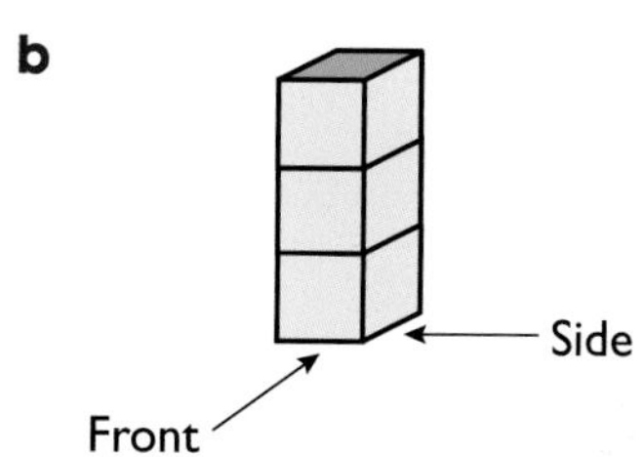

c

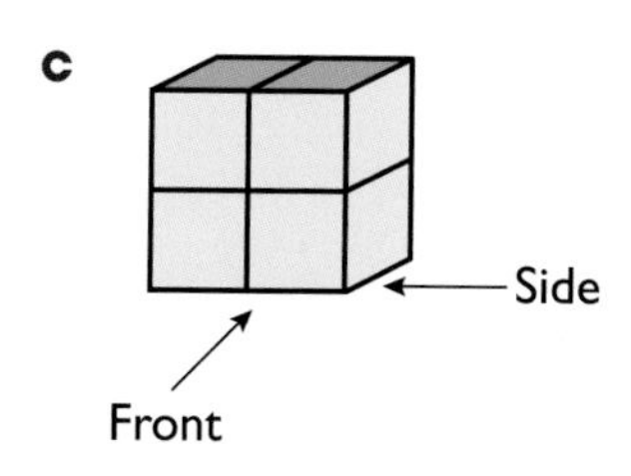

d

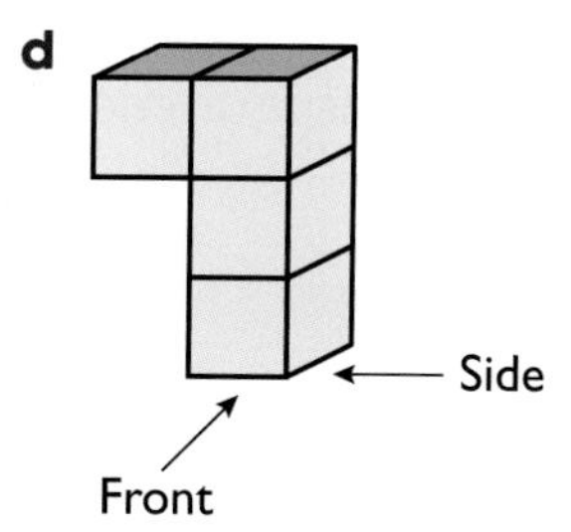

e

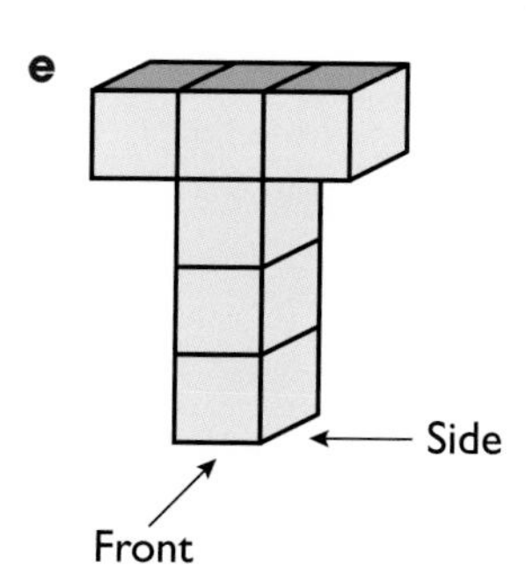

f

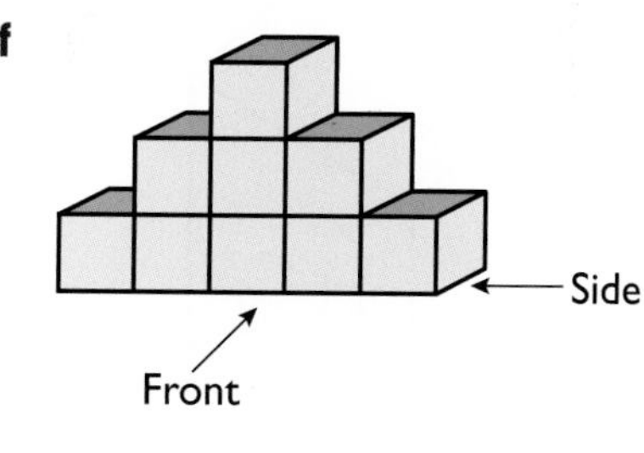

g

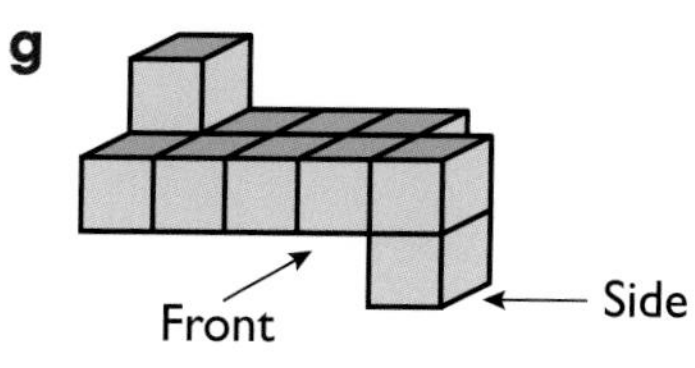

h

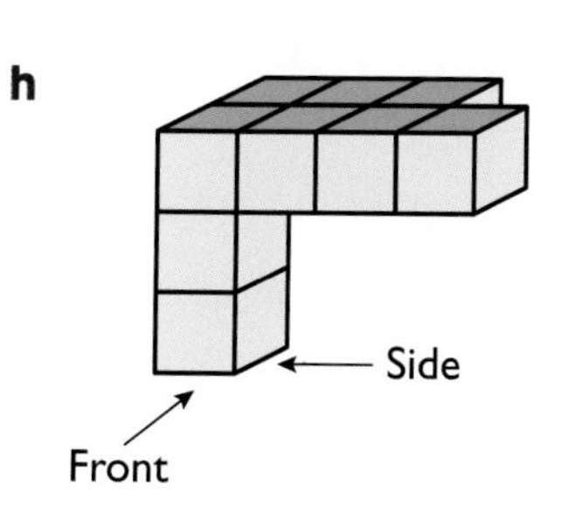

(2) Using squared paper, draw

i the front elevation **ii** the side elevation **iii** the plan

for each of these solid shapes

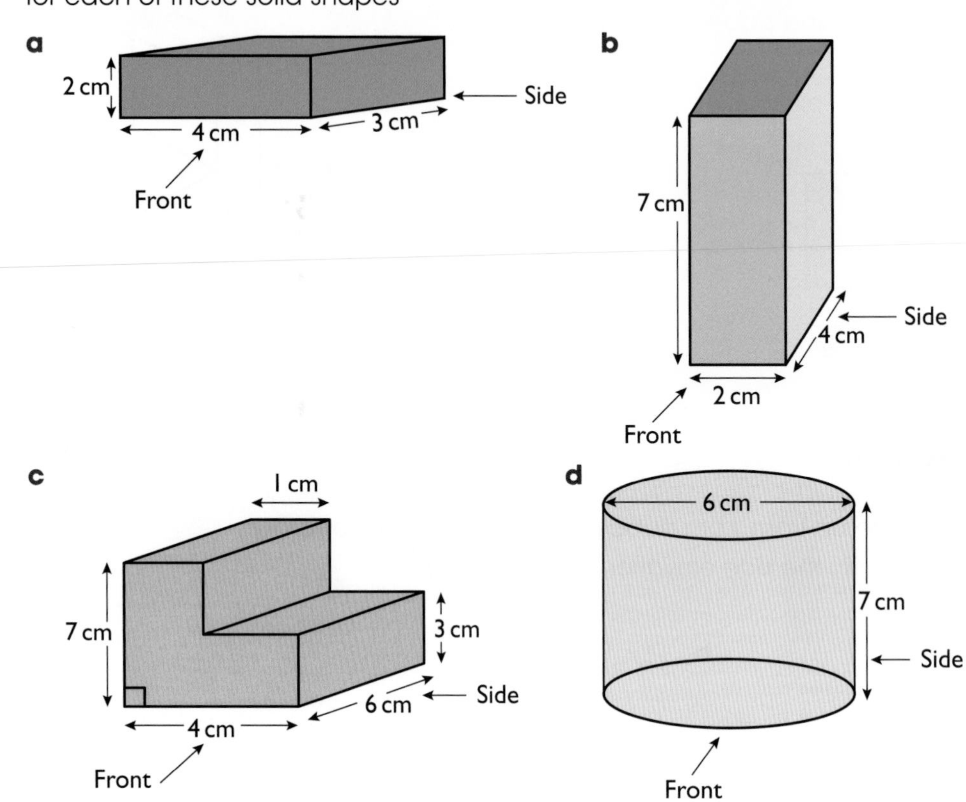

Developing fluency

(1) This is a picture of a house. Some of the measurements are marked on it. Not all the windows are shown.

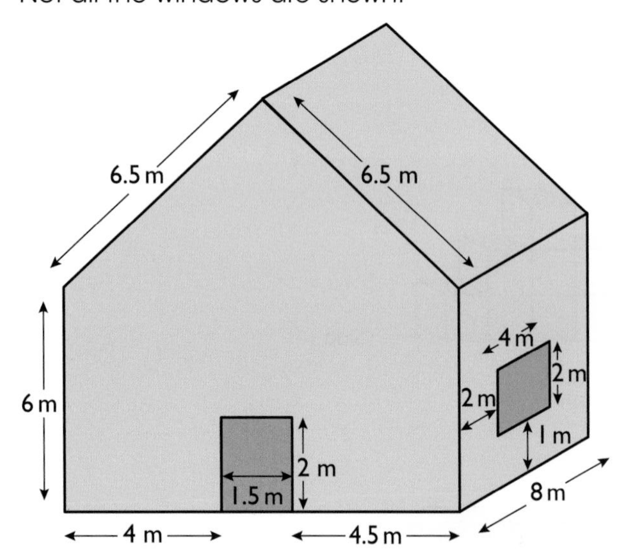

a Make accurate scale drawings of

i the front elevation **ii** the side elevation **iii** the plan.

b Use your drawings to find the height of the house.

c How many bedrooms do you think the house has?

2 The front elevation, side elevation and plan are given for each solid. Draw a sketch of the solid.

	Front	Side	Plan
a			
b			
c			
d			

3 Match together the 3D objects with their plans and elevations.
Some of the plans and elevations are used more than once.

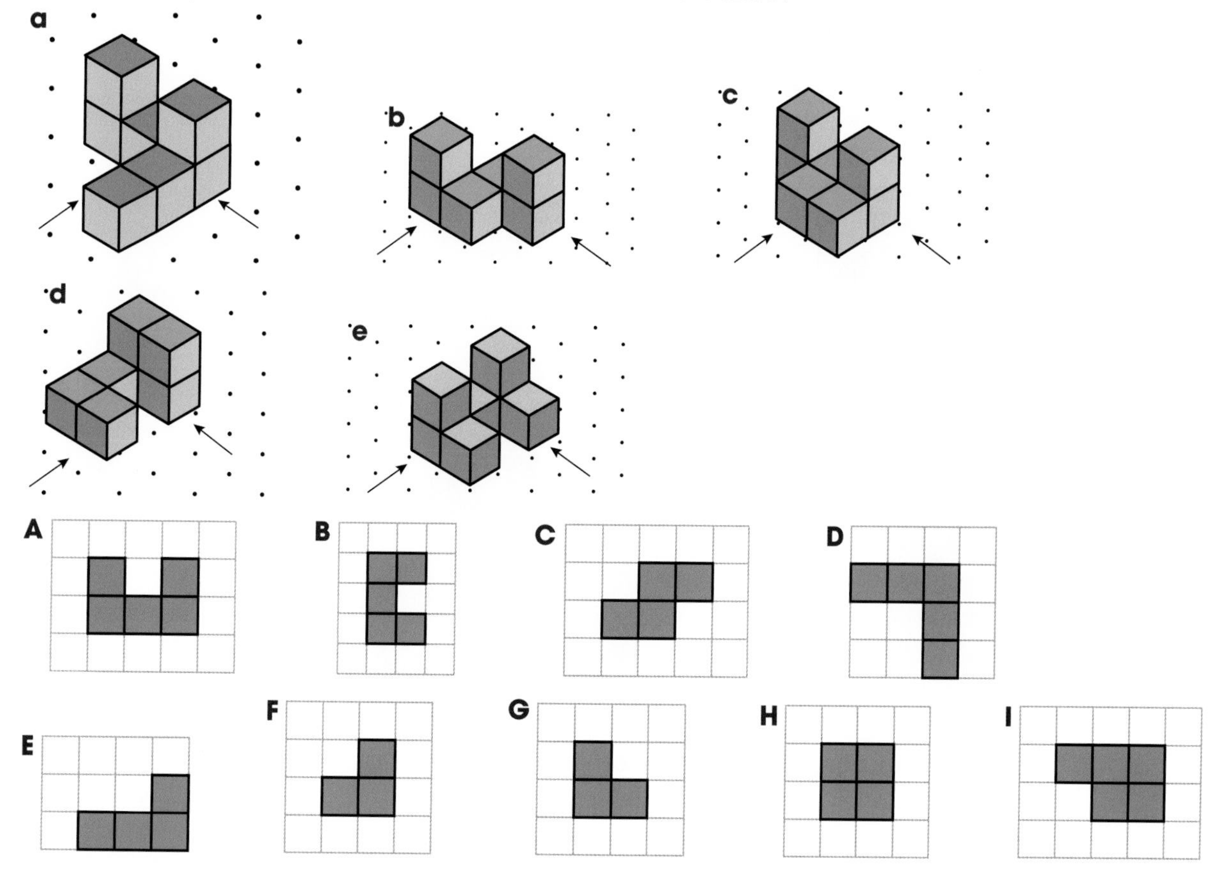

4) Each diagram represents a plan view of a solid made from 1 centimetre cubes in stacks. The number in each square shows how many cubes are in each stack.
For each diagram
   i draw the solid on isometric paper
   ii draw the front and side elevations of the solid.

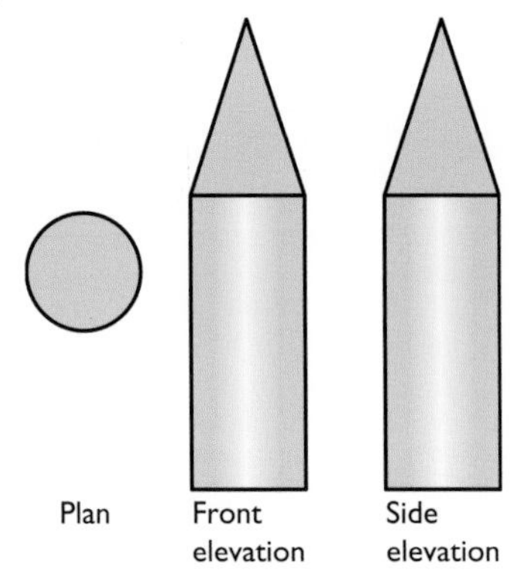

a

3	2
2	2
1	1

b

3	2	1
2	1	
1		

5) Here is the plan, front and side elevations of a child's toy.

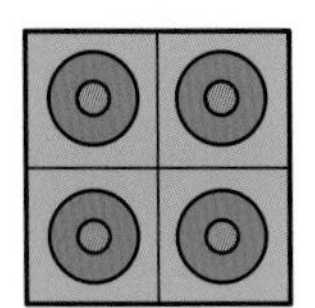
Plan

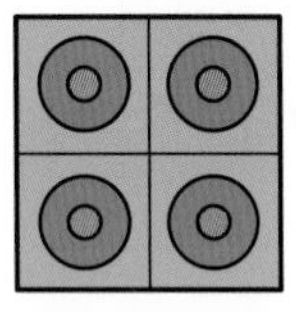
Front elevation

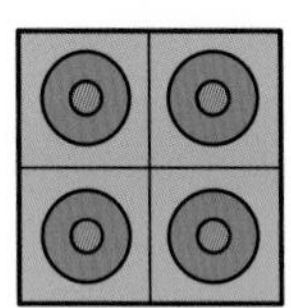
Side elevation

a What two 3D shapes is the toy made from?
b Sketch a net of the top and base of the toy.

Problem solving

1) Here are the three elevations for a 3D shape.

Front elevation   Side elevation   Plan view

The shape could look like this.

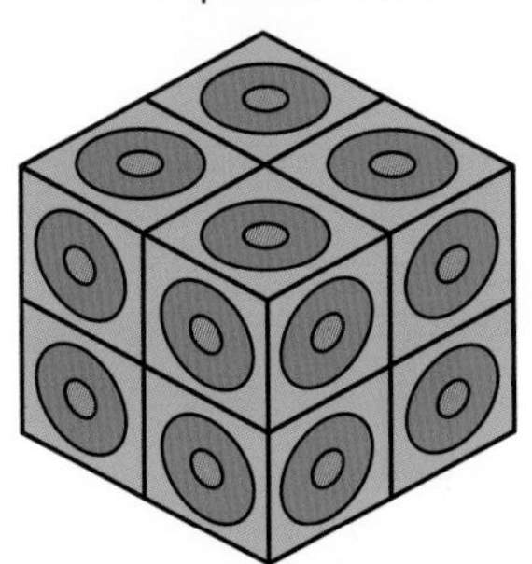

It could also look like this.

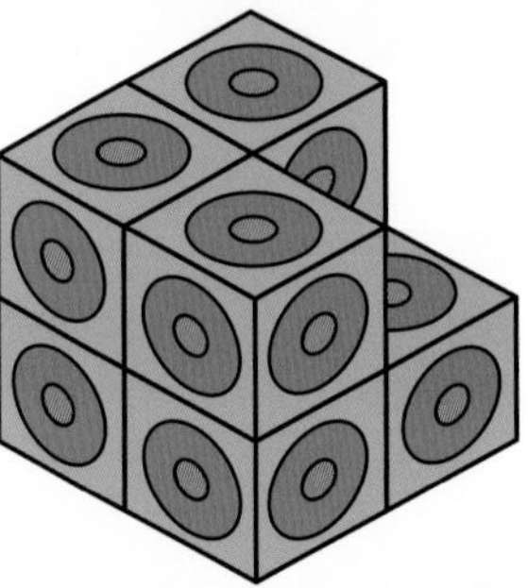

a Are there any other possibilities? Draw some on isometric paper.
b What is the smallest number of cubes that could be used to make the shape?

2 Ben made a shape, drew the three views and then destroyed the shape and made a different one. He did this five times.

Can you reconstruct each of the shapes, 1 to 5, that he made?

If you can, then draw it. If there is more than one possibility, draw them all.

If it is not possible to make it then explain why Ben must have made a mistake.

Shape 1

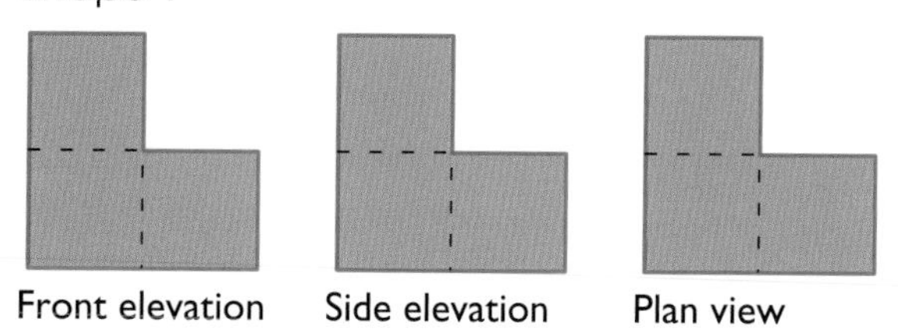

Front elevation Side elevation Plan view

Shape 2

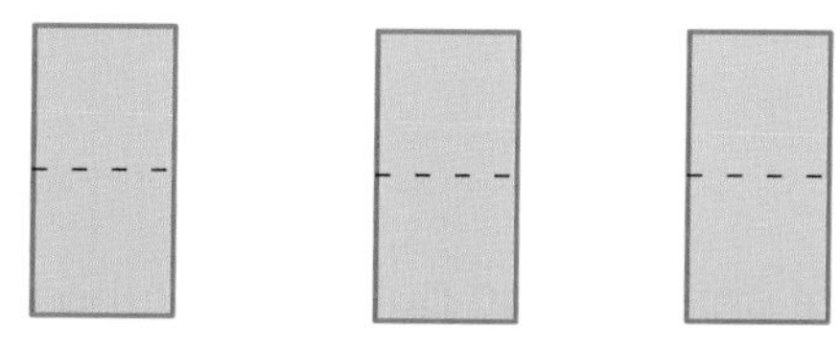

Front elevation Side elevation Plan view

Shape 3

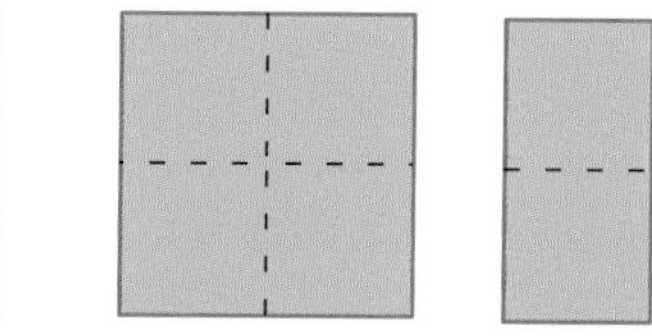

Front elevation Side elevation Plan view

Shape 4

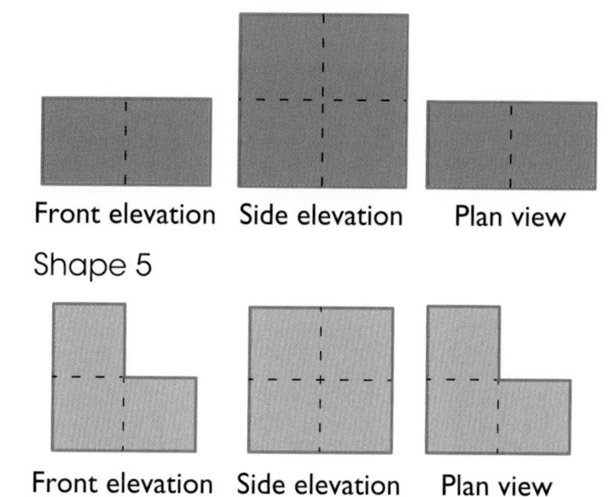

Front elevation Side elevation Plan view

Shape 5

Front elevation Side elevation Plan view

3 The plan and elevations of a 3D shape are drawn on a 1 centimetre grid.

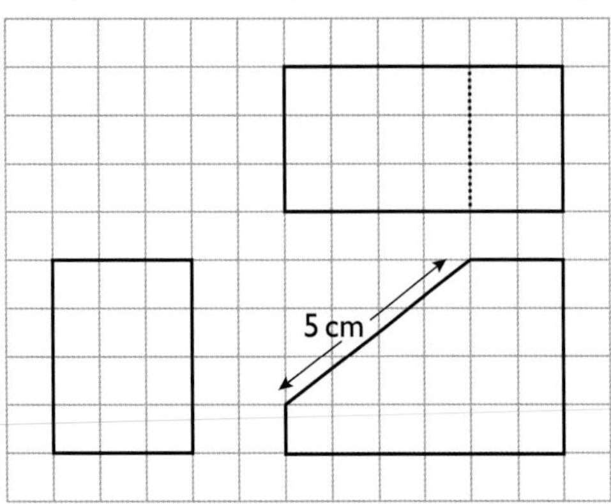

a Draw the shape on isometric paper.

b For the 3D shape, find

  i the surface area

  ii the volume.

4 A toy manufacturer produces large cardboard toys that children can put together themselves. Here is the plan and elevations for a small play house with a garage next to it.

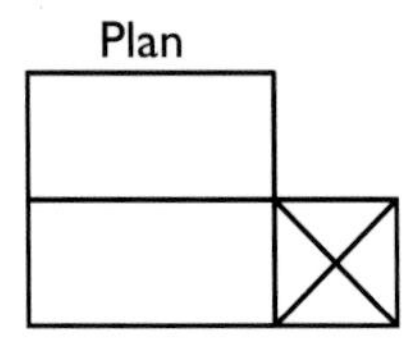

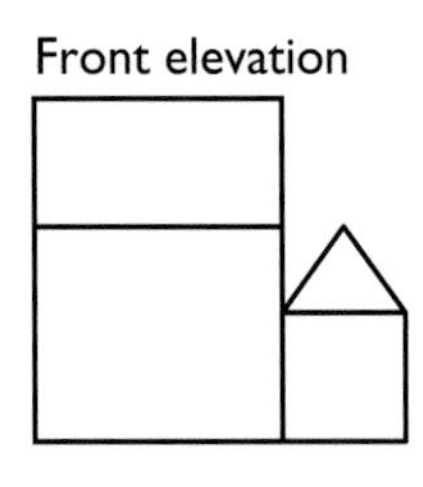

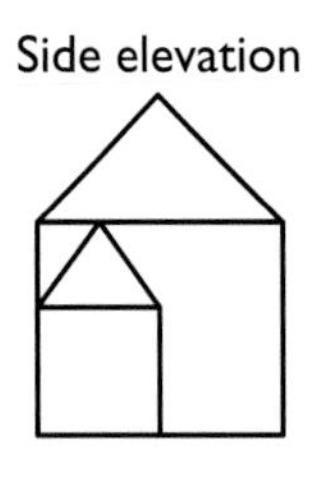

a What shape is the roof of

  i the house

  ii the garage?

b Make a 3D sketch of the house.

c Sketch a net for the garage.

Reviewing skills

1. The shapes are made up of 1 cm cubes.
   Using squared paper, for each shape draw
   i the front elevation **ii** the side elevation **iii** the plan.

a

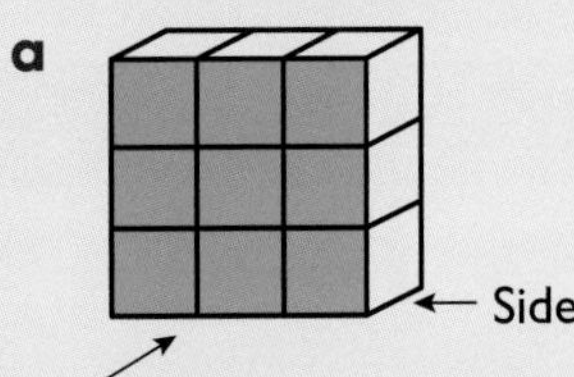

b

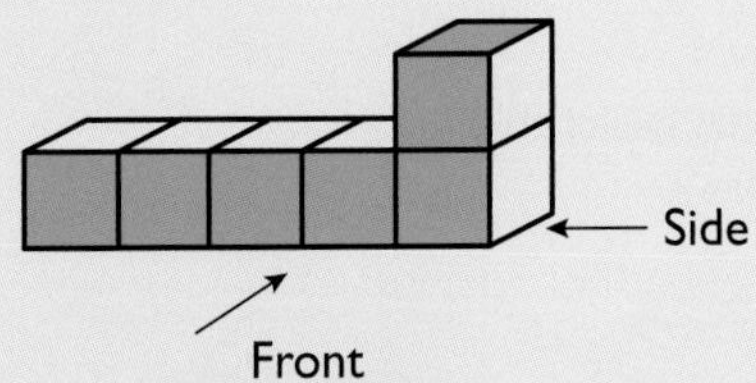

c

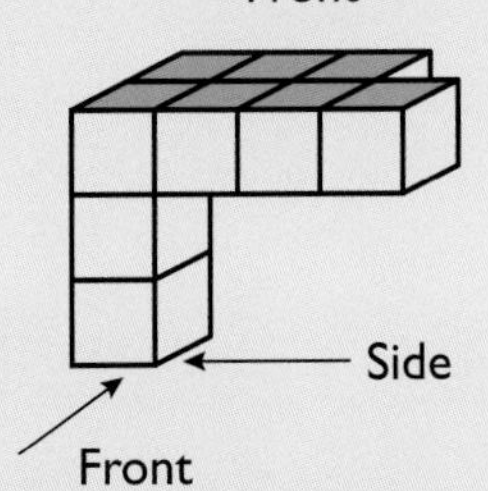

2. Using squared paper, for each solid draw
   i the front elevation **ii** the side elevation **iii** the plan.

a

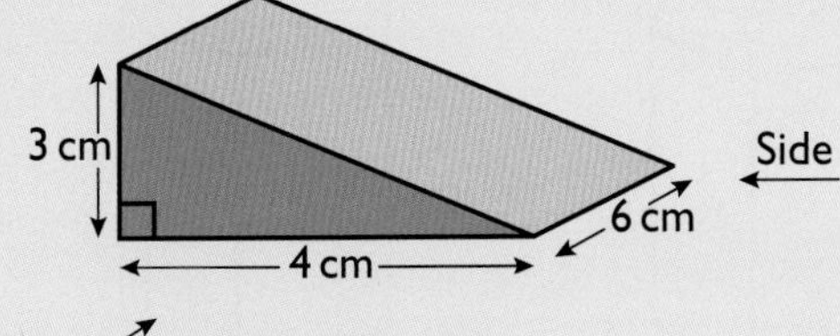

b

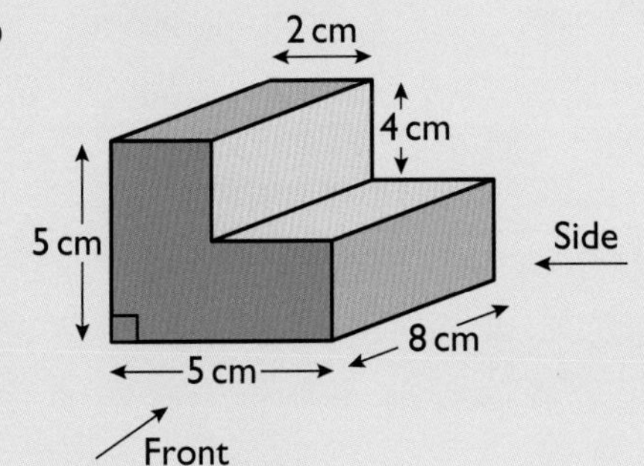

c

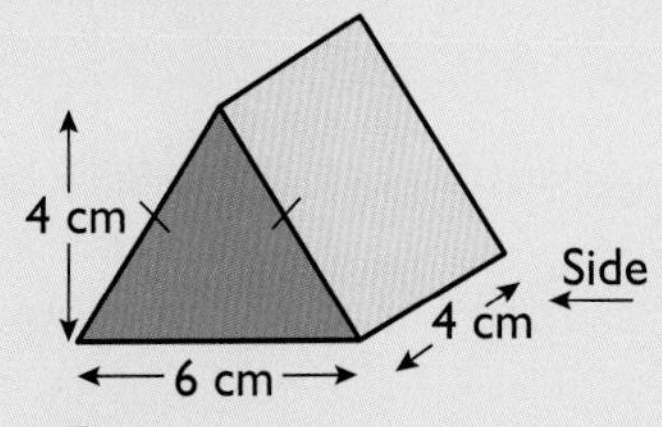

Unit 5 • Prisms • Band g

Outside the Maths classroom

Glass design

How would you design a glass to hold the entire contents of a 250 ml bottle?

Toolbox

A **prism** is a three-dimensional shape with the same cross-section all along its length.

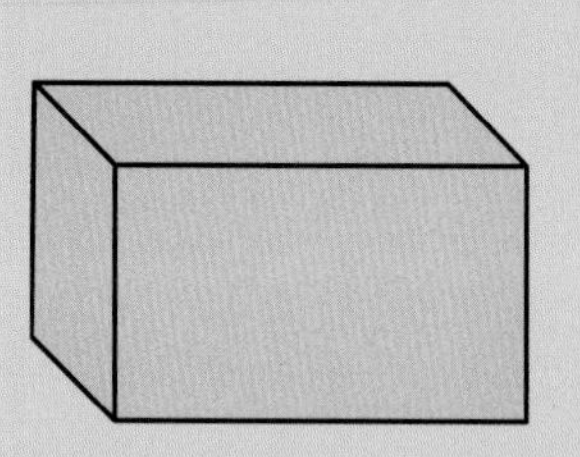
Cuboid

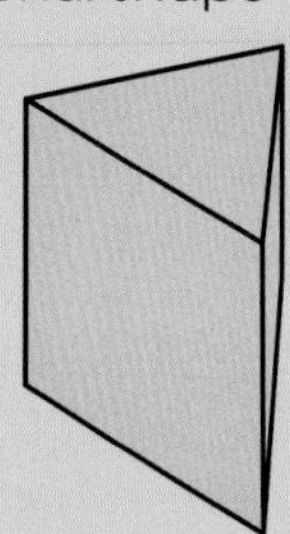
Triangular prism

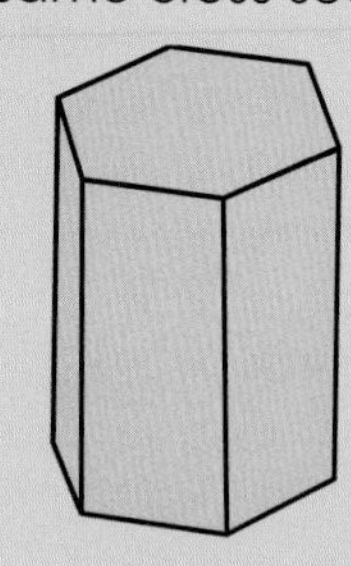
Hexagonal prism

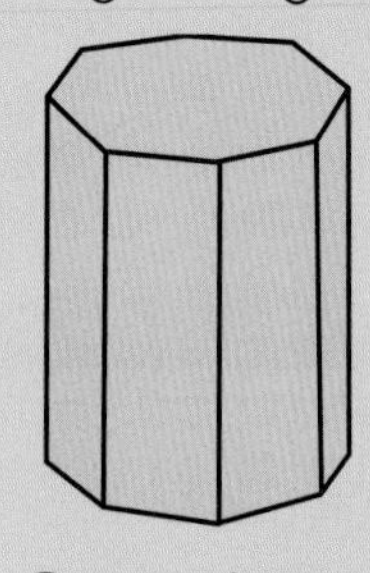
Octagonal prism

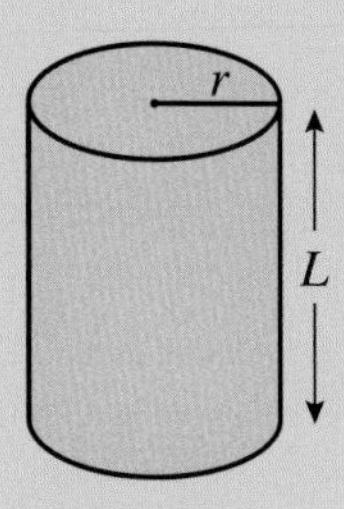

Cylinder

The cylinder is a prism with a circular base.
Its volume is $\pi r^2 L$.
Its surface area is $2\pi r L + \pi r^2 + \pi r^2 = 2\pi r L + 2\pi r^2$

To work out the **volume of a prism**, calculate the area of the cross-section and multiply by the height (or length).
The volume is measured in cubic units such as cubic centimetres (cm^3) or cubic metres (m^3).
The **surface area of a prism** is the total area of all the faces.

Example – Finding the volume of a prism

Work out the volume of this cylinder.

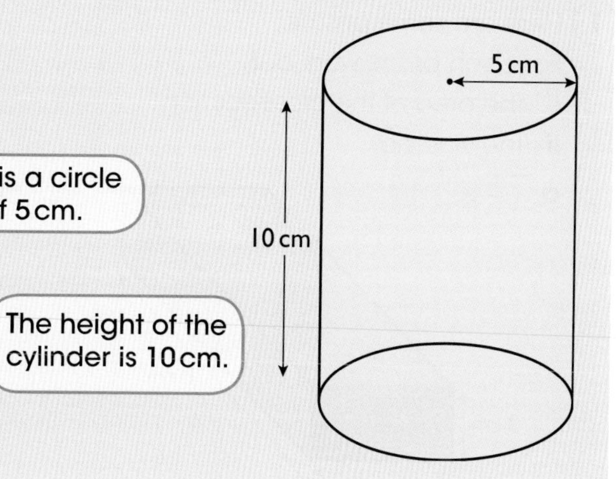

Solution

Area of circle = πr^2

$A = \pi \times 5^2$ ← The cross-section is a circle with a radius of 5 cm.

$= 78.539...$

Volume of cylinder = area of cross-section × height

$= 78.539... \times 10$ ← The height of the cylinder is 10 cm.

$= 785.398...$

The volume of the cylinder is 785.4 cm^3 (to 1 d.p.).

Example – Finding the surface area of a prism

Calculate the surface area of this prism.

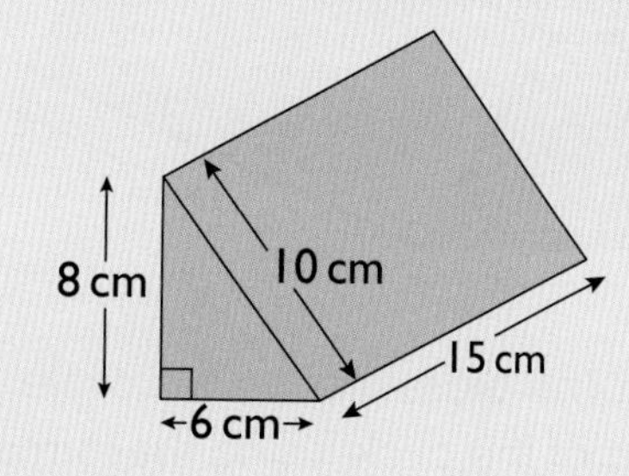

Solution

Work out the area of each face individually.

Area of one triangular face = $\frac{1}{2} \times$ base × height

$= \frac{1}{2} \times 6 \times 8$

$= 24\,cm^2$

Area of other triangular face = $24\,cm^2$ ← The two triangular faces are identical.

Area of rectangular base: $6 \times 15 = 90\,cm^2$ ← The rectangular faces all have a length of 15 cm.

Area of rectangular face: $8 \times 15 = 120\,cm^2$

Area of sloping rectangular face: $10 \times 15 = 150\,cm^2$

Total surface area = $24 + 24 + 90 + 120 + 150$ ← Add the area of all the faces to find the total surface area.

$= 408\,cm^2$

Practising skills

1 Here are some prisms.

For each prism, work out

i the area of the cross-section

ii the volume.

a

3 cm

6 cm

8 cm

b

8 cm

5 cm

12 cm

c

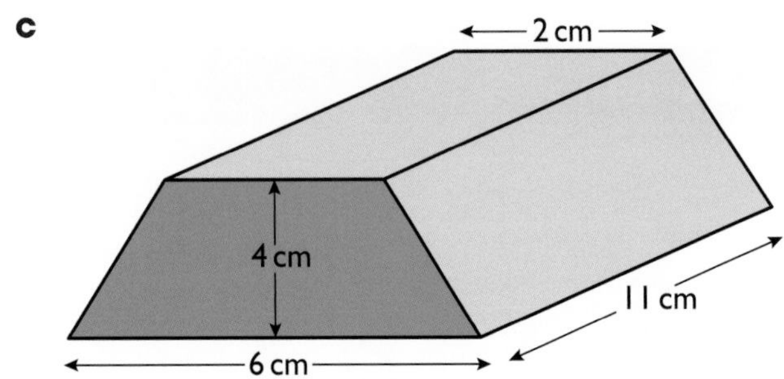

2 A cylinder is a prism. The cross-section is a circle. For each cylinder

i work out the area of the circle using the formula Area = πr^2

ii work out its volume.

a

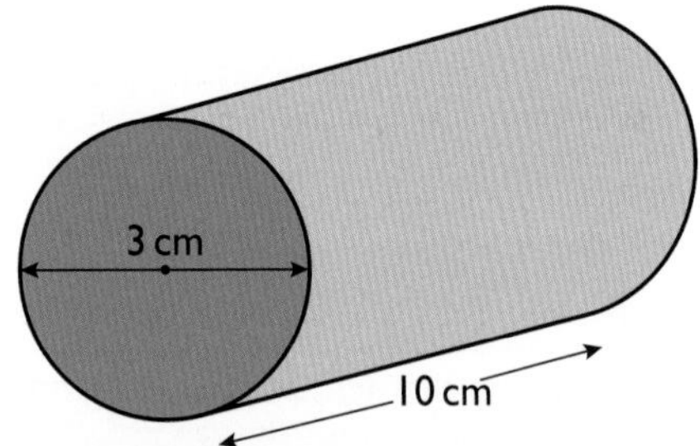

b

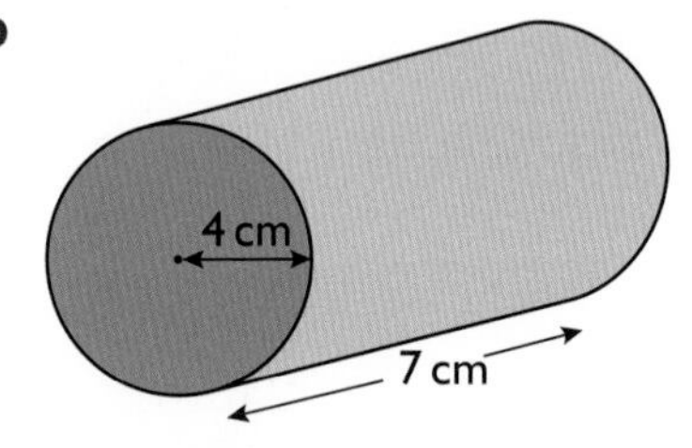

c

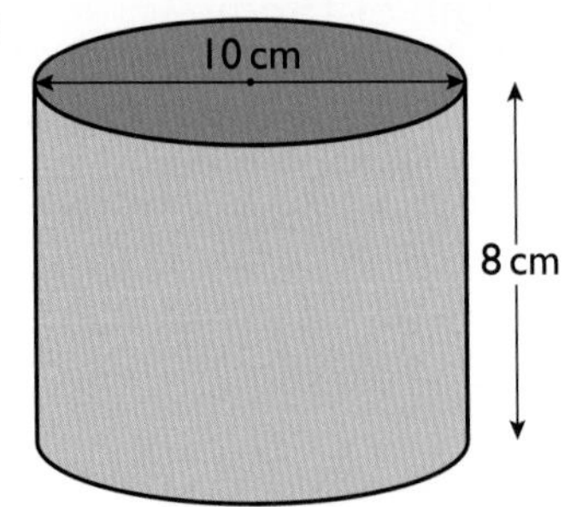

d

14 cm

6 cm

3 A prism has cross-sectional area $18\,cm^2$ and length $3\,cm$.
Work out its volume.

4 A prism has cross-sectional area $15\,cm^2$ and volume $120\,cm^3$.
Work out its length.

5 The volume of this triangular prism is $280\,cm^3$.

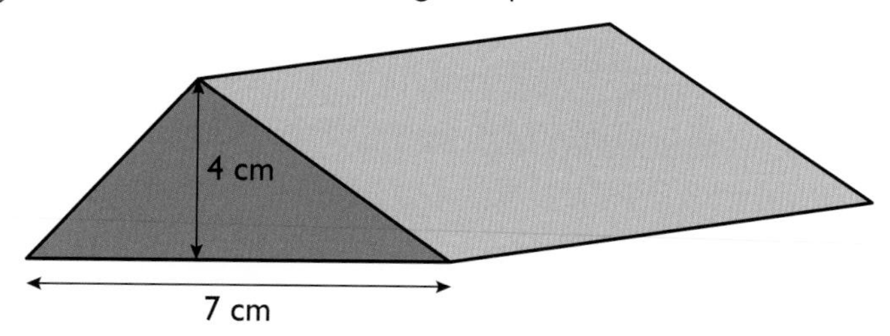

a Work out the area of the triangular face.

b Work out the length of the prism.

6 This diagram shows a triangular prism.

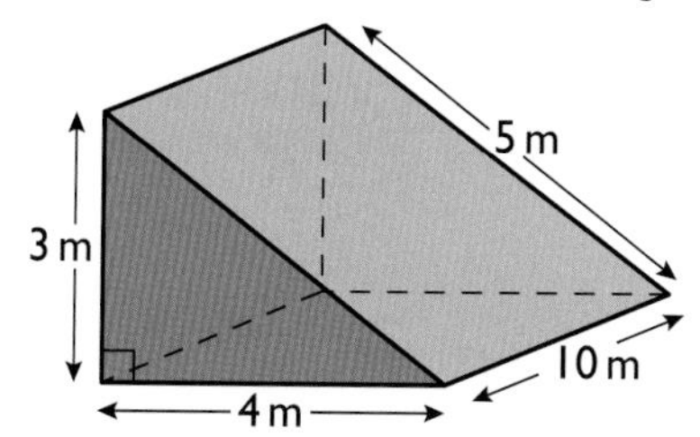

a **i** Find the areas of the three faces that are rectangles.
ii What is the total area of their faces?

b **i** Find the perimeter of the triangular cross-section.
ii Multiply the perimeter by the length of the prism.

c Comment on your answers to **a ii** and **b ii**.

d **i** Find the area of the triangular cross-section.
ii What do you find when you multiply the area by the length of the prism?

7 Work out the volume of each of these prisms in cm^3.

a

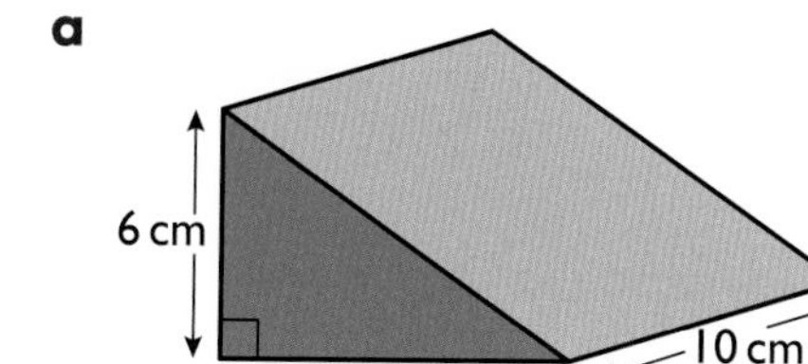

b

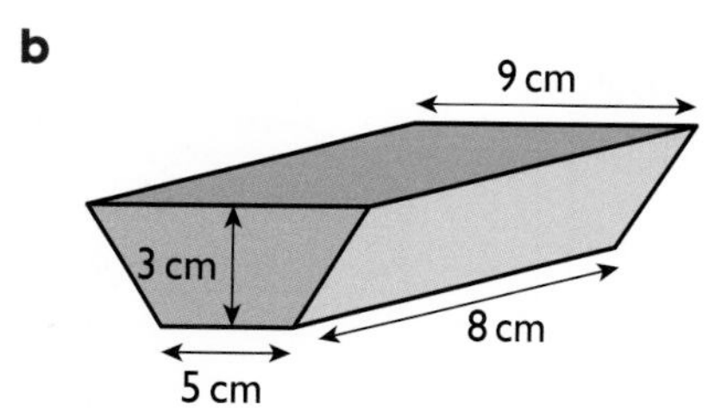

c

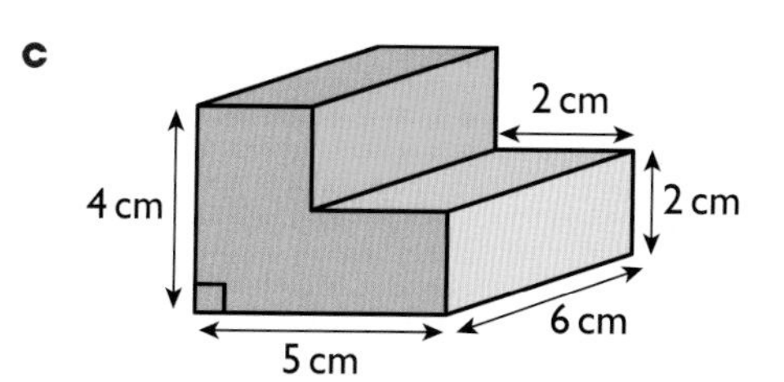

d

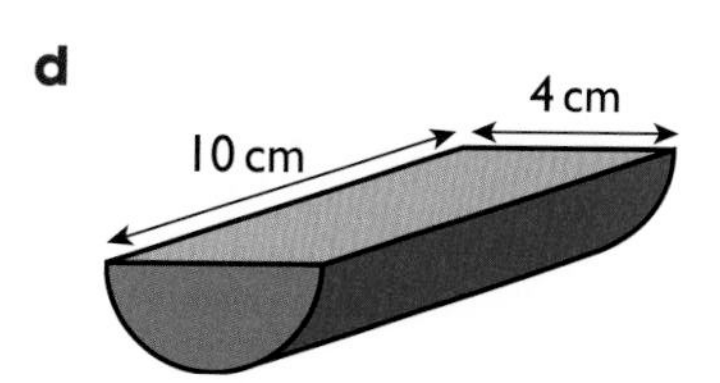

Developing fluency

1. This is a sketch of a child's playhouse.

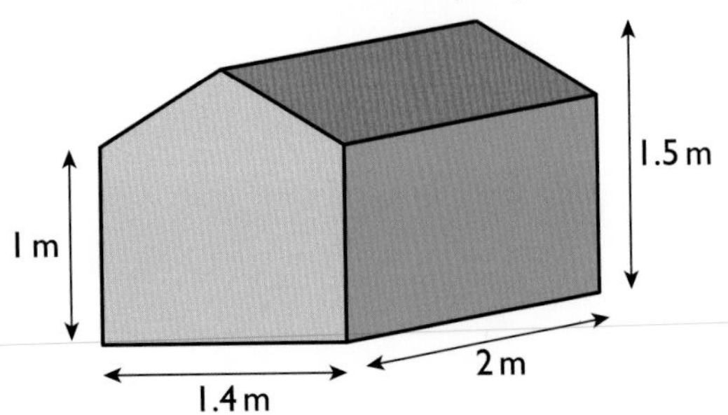

Work out the volume of the playhouse.

2. This cylinder has a radius 10 cm and height 30 cm.

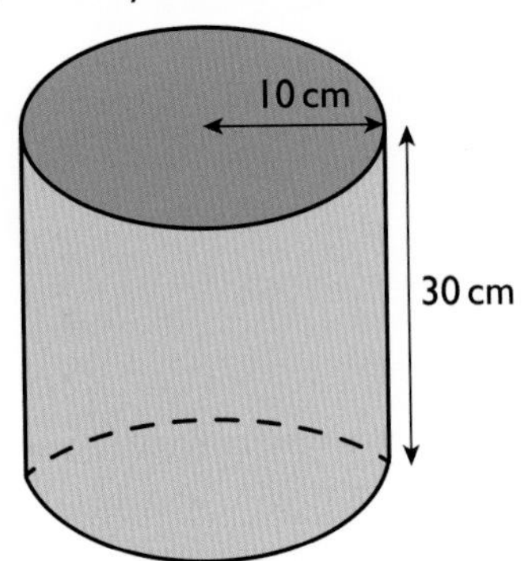

   a Find the circumference of the cross-section.

   b Find the surface area of the walls without the top and bottom.

   c Find the area of the top.

   d Find the total surface area: the walls, top and bottom all together.

3. A tin of beans has diameter 6.8 cm and height 10.2 cm.
   Work out its capacity (volume).

Exam-style

4. Susan has two cylindrical containers, P and Q.

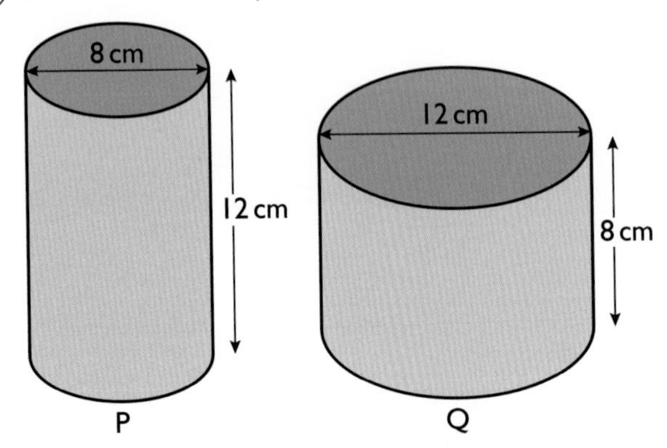

   a Which has the greater volume, P or Q?

   b What is the difference in volume between the two containers?

Exam-style

5. At the park, a cylinder is cut from a wooden cuboid to make a tunnel for children to crawl through.

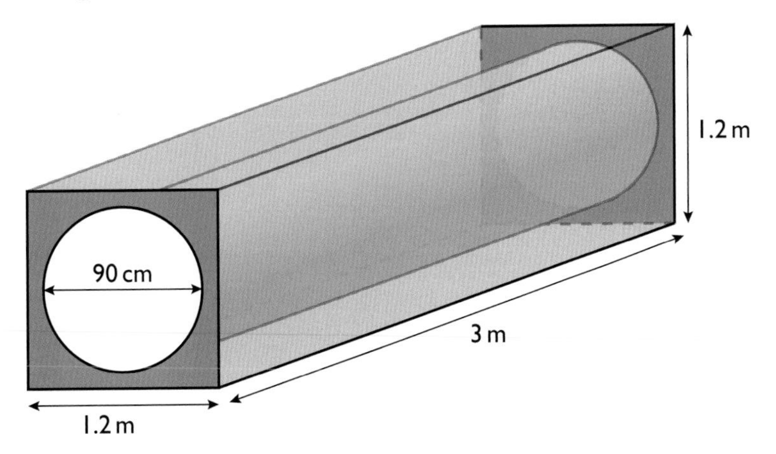

a Work out the volume of wood remaining.

b Work out the surface area of the inside of the tunnel.

Exam-style

6. One summer's day two children, Zara and Anna, use their paddling pools.
Zara's pool is in the shape of a cylinder and Anna's is in the shape of a cuboid.
Zara fills her pool. Anna's pool is three-quarters full.

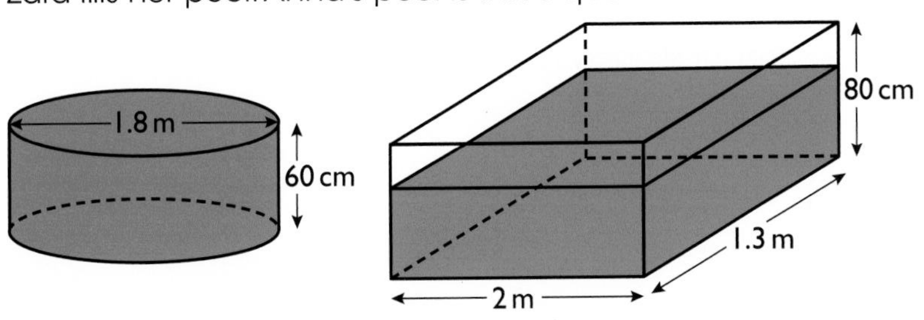

a Which pool contains more water?

b What is the difference in the volume of water in the two pools?

7. For each of these solids, say whether it is a prism.
If it is a prism, write down the shape of its cross-section.

a Cuboid **b** Wedge **c** Cylinder **d** Tetrahedron

e Cone **f** Cube **g** Sphere

Problem solving

1 The diagram shows a design for the roof space of a greenhouse.

The roof space is in the shape of a triangular prism of length 6 m.

The cross-section of the roof space is a right-angled triangle.

The sides of the triangle are 2.4 m, 1.8 m and 3 m.

Find

a the area of the glass used

b the volume of the roof space.

2 A council wants to use bins to store sand for slippery roads.

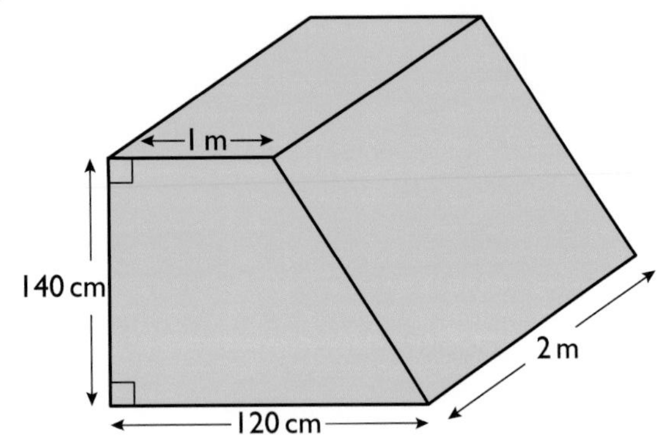

The bins are in the shape of a prism.

The cross-section of the prism is a trapezium; its dimensions are given on the diagram.

The council wants to have enough bins to store 150 m^3 of sand.

Will 50 bins be enough to store this amount of sand?

3 A company digs a tunnel through a mountain.

The tunnel is a cylinder with diameter of 8 metres.

The length of the tunnel is 1 kilometre.

a The material removed is taken away by lorries. Each lorry can carry 30 m^3 of material. How many lorry trips are needed to take away all the material?

b The walls inside the tunnel are covered in concrete.

i What is the area of the walls?

ii The concrete is 10 cm thick. What volume of concrete is needed?

Exam-style

4 The diagram shows a prism-shaped cold frame.
Rectangle BCFE is open. All of the other surfaces are made of glass.
The dimensions of the cold frame are shown on the diagram.

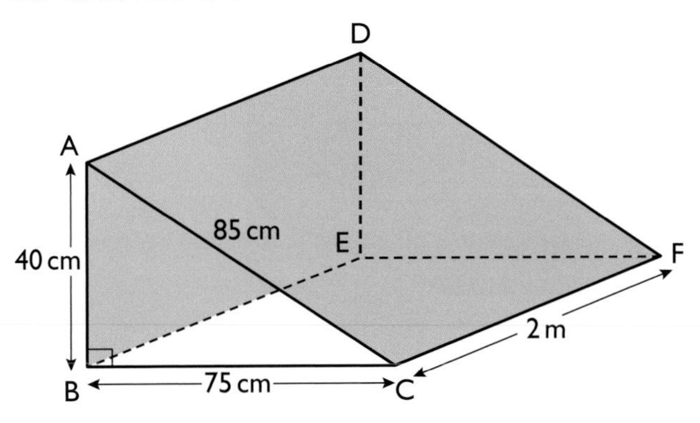

a Find the total area of glass.

b Find the volume of the cold frame.

Extension

Exam-style

5 Here is a container used to mix chemicals. It is a closed cylinder with the dimensions shown.

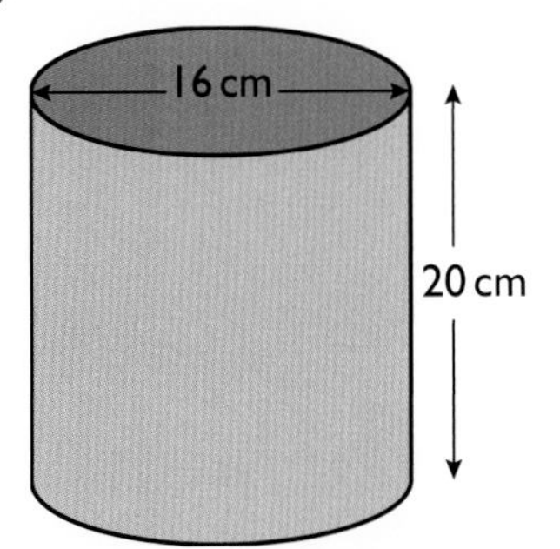

The inside surface of the cylinder has to be covered with a special chemical.
It costs £1.20 to cover each square centimetre with the special chemical.
Work out the total cost of covering the inside surface of the cylinder.

Extension

Exam-style

6 The diagram shows the cross-section of a water channel.
The cross-section is a trapezium.
The channel is 2 km long.

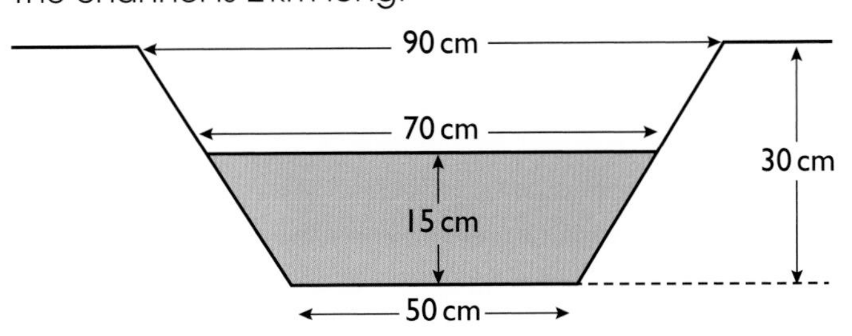

a One day the water is 15 cm deep, as shown in the diagram.

i Find the cross-sectional area of the water.

ii Find the volume of water in the channel.

b On another day, the channel is full. How much more water does it contain?

Reviewing skills

1 Here are some prisms. For each prism, work out

i the area of the cross-section

ii the volume.

a **b**

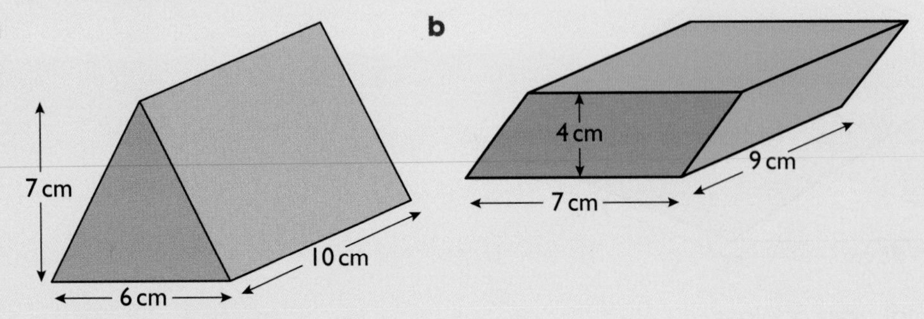

2 A prism has volume $5520\,\text{cm}^3$ and length 1.2 m.
Work out its cross-sectional area.

3 A cylinder is a prism. The cross-section is a circle.
For each cylinder

i work out the area of the circle using the formula Area = πr^2.

ii work out its volume.

a **b** **c**

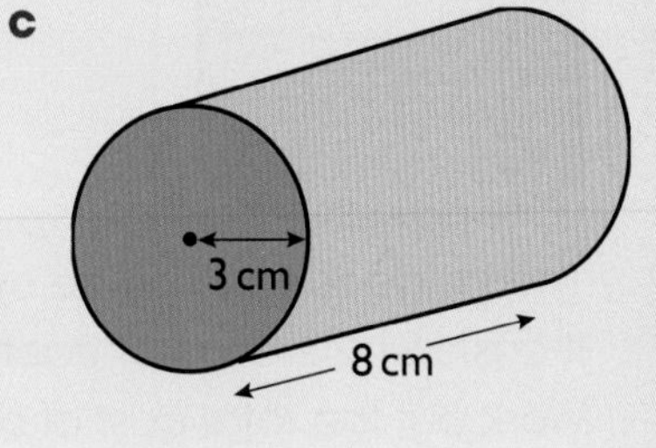

4 Work out the volume of this prism.

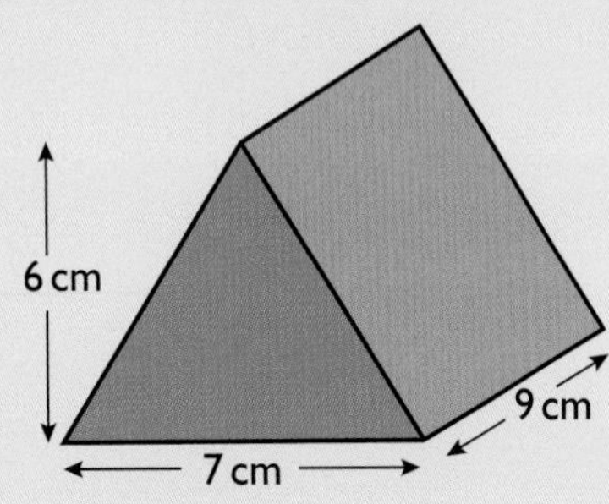

Unit 6 • Enlargement in two and three dimensions • Band g

Outside the Maths classroom

Cost of enlargement

What will the actual cost of a print depend on?

Toolbox

When one shape is an **enlargement** of another, all the lengths are increased by the same scale factor.

The blue cuboid is an enlargement of the red cuboid because all dimensions of the blue cuboid are two times the dimensions of the red cuboid.

Length: 6 cm × 2 = 12 cm

Width: 5 cm × 2 = 10 cm

Height: 2 cm × 2 = 4 cm

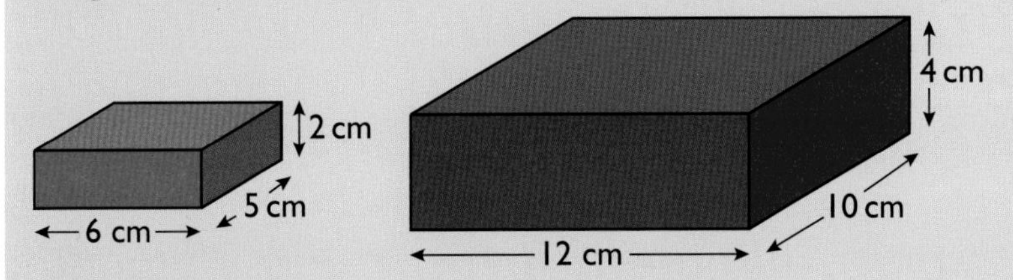

The **surface area** is the area of all 6 surfaces.

Red surface area

$= 2 \times (6 \times 2 + 5 \times 2 + 6 \times 5) = 104\,\text{cm}^2$

Blue surface area

$= 2 \times (12 \times 4 + 10 \times 4 + 12 \times 10) = 416\,\text{cm}^2$

Red volume $= 6 \times 5 \times 2 = 60\,\text{cm}^3$

Blue volume $= 12 \times 10 \times 4 = 480\,\text{cm}^3$

The lengths, areas and volumes can be compared using ratios.

Lengths 6 : 12 or 1 : 2

Areas 104 : 416 or 1 : 4

Volumes 60 : 480 or 1 : 8

Example – Finding area

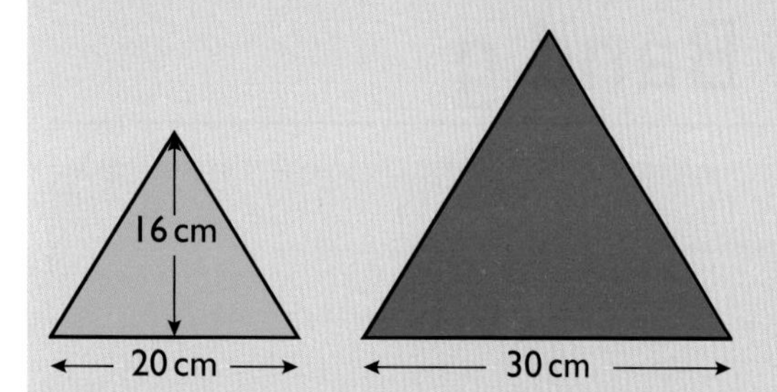

The red triangle is an enlargement of the yellow triangle.
The yellow triangle has a base of 20 cm and a height of 16 cm.
The red triangle has a base of 30 cm.

a Calculate the height of the red triangle.
b Calculate the ratio of the areas of the triangles in its simplest form.

Solution

a First find the ratio of the lengths of the two triangles.

Yellow : Red
Base 20 : 30
2 : 3
or 1 : 1.5

So the height of the red triangle is 1.5 × 16 = 24 cm (this is the height of the yellow triangle)

b Area = $\frac{1}{2}bh$

Yellow triangle area = $\frac{1}{2} \times 20 \times 16 = 160\,\text{cm}^2$

Red triangle area = $\frac{1}{2} \times 30 \times 24 = 360\,\text{cm}^2$

Ratio = 160 : 360 (divide by 40)
= 4 : 9
or 1 : 2.25

Example – Problem solving in 3D

A cuboid has a square base of side 4 cm and a height of h cm. Its volume is 192 cm^3.

a Calculate the height of the cuboid.

b A new cuboid is an enlargement of the green one. Its base has side 12 cm. Find the ratio of their volumes.

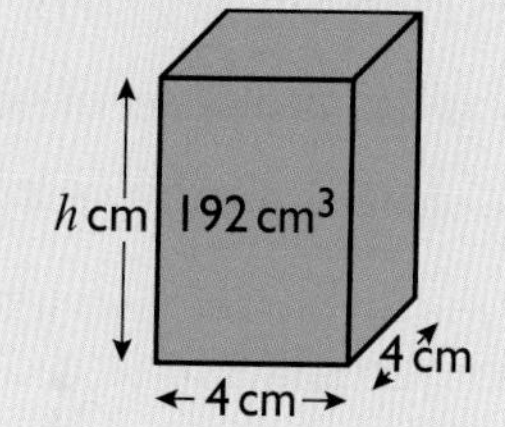

Solution

a Length × width × height = volume

$4 \times 4 \times h = 192$

$16h = 192$

$h = 192 \div 16 = 12$

The height is 12 cm.

b The ratio of the bases is 4:12 ← divide both sides by 3

= 1:3 ← in the simplest form

So the scale factor of enlargement is 3.

Therefore, the height of the new cuboid is 12 × 3 = 36 cm ← multiplying the original height by the scale factor

So, the volume of the new cuboid = length × width × height

= 12 × 12 × 36

= 5184 cm^3

Ratio of the volumes is 192 : 5184

= 48 : 1296 ← dividing by 4

= 12 : 324 ← dividing by 4

= 3 : 81 ← dividing by 4

= 1 : 27 ← dividing by 3

The ratio of the volumes is 1 : 27 in its simplest form.

Practising skills

1 Look at these two rectangles.

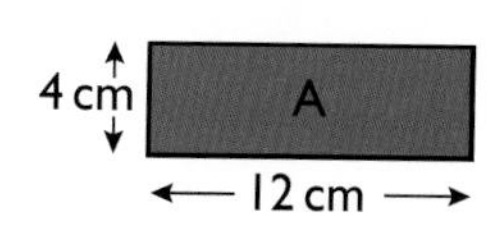

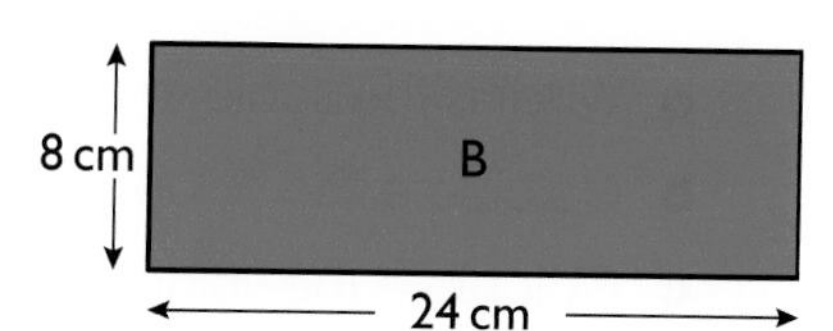

a Write down the ratio of the lengths of the sides of A to the length of the sides of B, in its lowest terms.

b i Calculate the perimeter of each rectangle.

ii Find the ratio of the perimeter of A to the perimeter of B, in its lowest terms.

c i Calculate the area of each rectangle.

ii Find the ratio of the area of A to the area of B, in its lowest terms.

2. The smaller of these two squares has side length of 15 cm.
The ratio of the sides of the two squares is 1 : 2.

15 cm
15 cm

a Calculate the length of the side of the larger square.

b Calculate the area of each square.

c Calculate the ratio of the area of the squares, in its lowest terms.

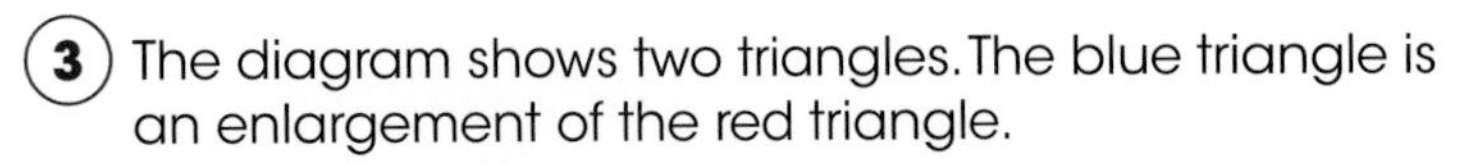

3. The diagram shows two triangles. The blue triangle is an enlargement of the red triangle.

a Calculate the ratio of their heights in its simplest form.

b Calculate the ratio of their areas in its simplest form.

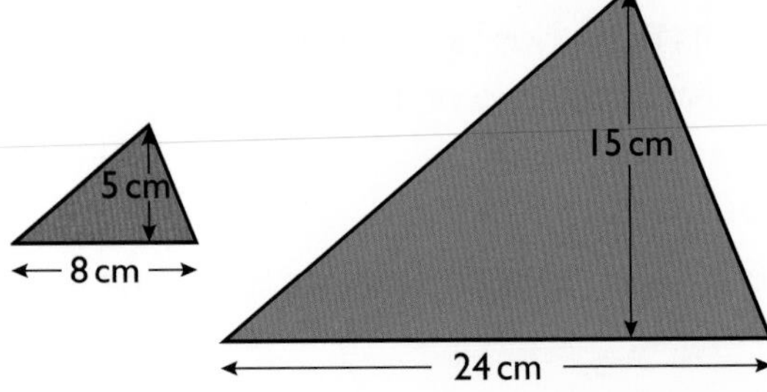

4. A box is in the shape of a cube with sides of 20 cm.
It is filled with small cubes with sides of 4 cm.

a What is the volume of a small cube?

b What is the volume of the box?

c Write down the ratio of the lengths of a side of a small cube to the side of the box, in its lowest terms.

d Write down the ratio of the volume of a small cube to the volume of the box, in its lowest terms.

e How many small cubes fit in the box?

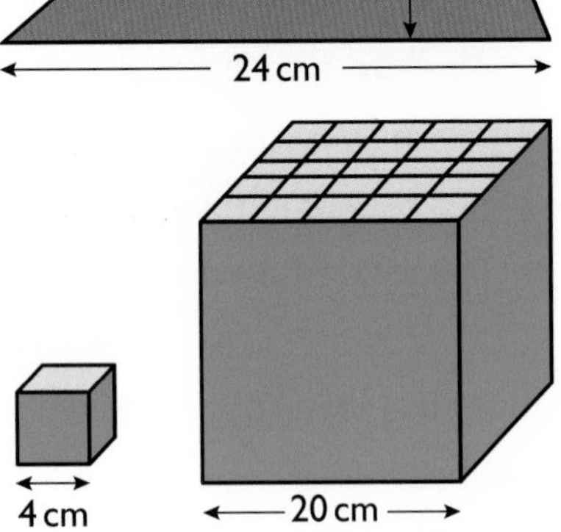

5. A cuboid is enlarged by a scale factor of 2.

a Explain why the surface area is increased by a factor of 4.

b By what factor is the volume increased?

Developing fluency

1. The green shape is an enlargement of the blue shape.

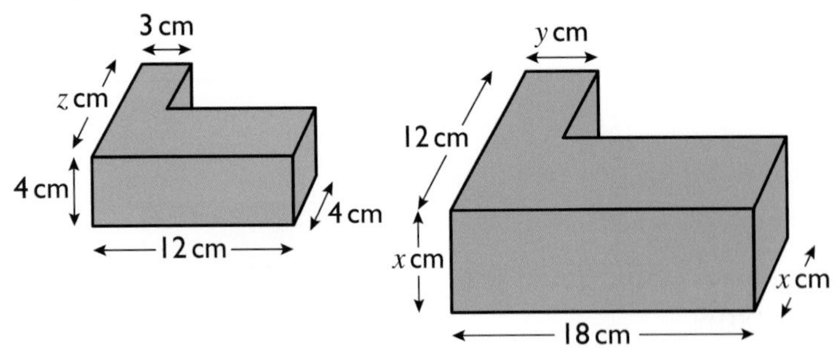

a What is the scale factor?

b Calculate the values of x, y and z.

c Calculate the volume of each shape.

d Calculate the ratio of the volumes in its simplest form.

Exam-style

2 Two squares have areas of $36\,cm^2$ and $100\,cm^2$.

a Find the lengths of their sides.

b Find the ratio of the lengths of their sides in its lowest terms.

3 A map is drawn to a scale of 1 : 500.
On the map, a rectangular building measures 4.8 cm × 3.2 cm.
Calculate

a the dimensions of the actual building

b the area of the building on the map

c the actual floor area of the building

d the ratio of the two areas in its simplest form.

Exam-style

4 This cuboid has a square base of side 6 cm. Its surface area is $162\,cm^2$.

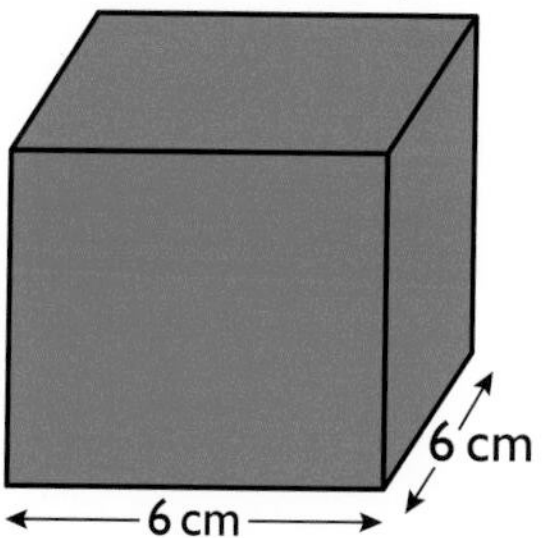

a Calculate

i the height of the cuboid **ii** the volume of the cuboid.

An enlargement of the cuboid is made with a scale factor of 3.

b Calculate

i the surface area **ii** the volume
of the enlargement.

Exam-style

5 Here are two cylinders.

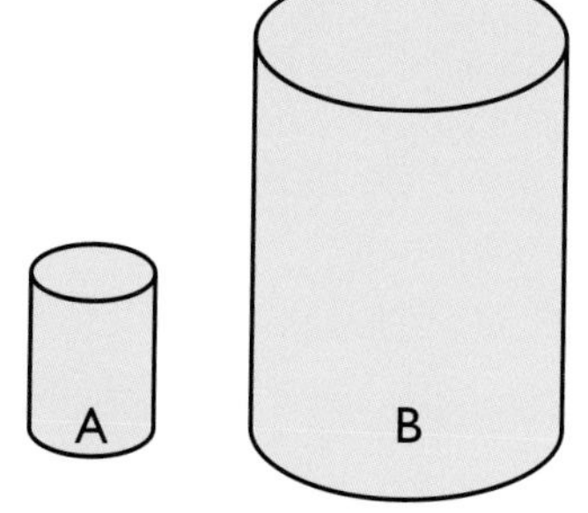

Cylinder A has radius 6 cm and height 15 cm.
Cylinder B is an enlargement of A.
The area of the circular top of B is 4 times the area of the top of A.
Find

a the radius of cylinder B

b the height of cylinder B

c the scale factor of the enlargement

d the ratio of the volumes of the two cylinders, in its simplest form.

Problem solving

Exam-style

1. The diagram represents the two sails of a boat.

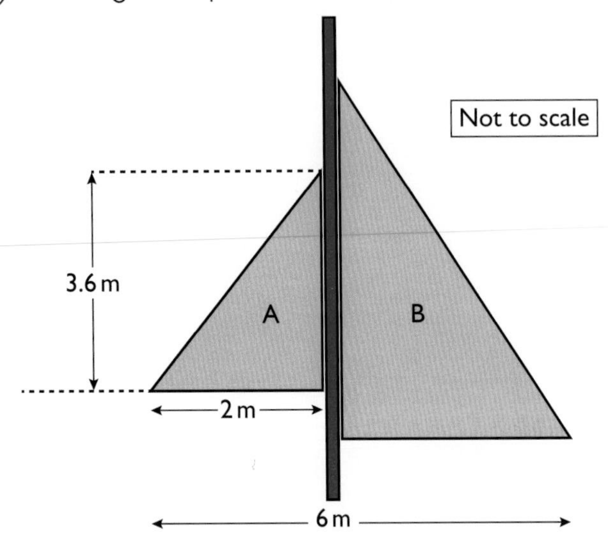

The ratio height : base is the same for both sails.
Some of the dimensions are given on the diagram.
Work out the total area of the sails.

Exam-style

2. Jim owns a café.
He wants to put this banner across a space on his door after enlarging it.
The space is 1 m wide and 36 cm high.

a What is the greatest size Jim's banner can be?

b What is the scale factor of the enlargement?

c What is the area of the banner?

Exam-style

3. A and B are two cubes. B is bigger than A.
The ratio of their sides is 2 : 5.
Block A has sides of length 40 cm.

a Work out the lengths of the sides of block B.

b Work out the surface area of each block.

c Show that the ratio of their surface areas is 4 : 25.

4 Bill makes a model garage to the scale of 1 : 40. It has a flat roof.

The length of the real garage is 8 metres.

a Work out the length of the model garage. Give your answer in cm.

The width of the garage is $\frac{3}{4}$ of the length.

The height of the garage is $\frac{2}{5}$ of the width.

b Write down the width and height of the real garage.

c Find the volume of the real garage.

d Find the width, height and volume of the model garage.

e Show that the ratio of the volumes of the model and the real garage is $1 : 40^3$.

5 Olga makes wooden blocks for children's toys.

Here are two cubes.

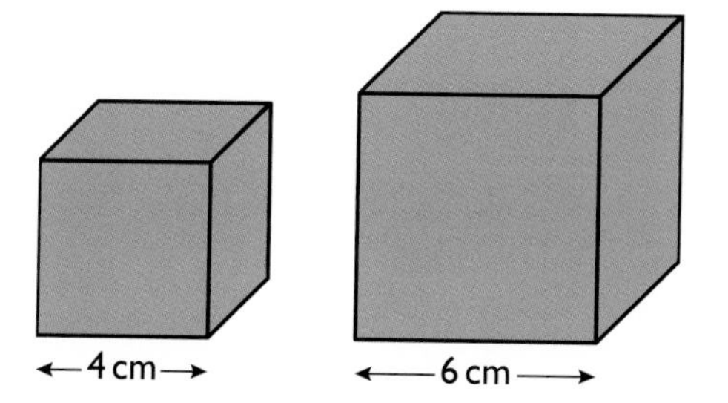

a Write down the ratio of the length of the sides.

b George thinks that the ratio of the surface areas of the two cubes is the same as the ratio of the lengths of their sides.

Show that he is wrong.

c The blocks have the same density, so their masses are proportional to their volumes.

Show the ratio of their masses is 8 : 27.

6 Here are two pictures.

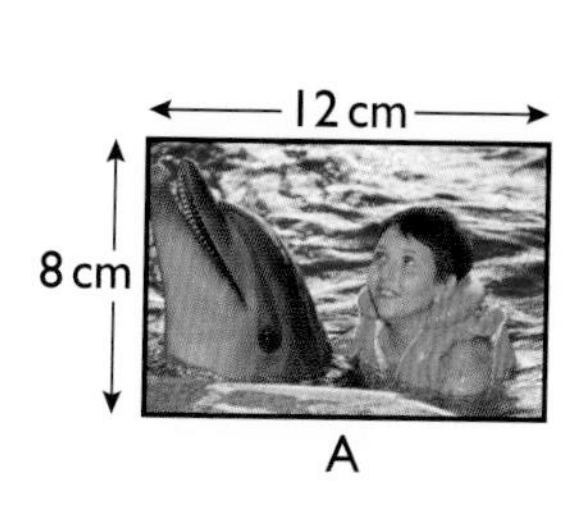

A

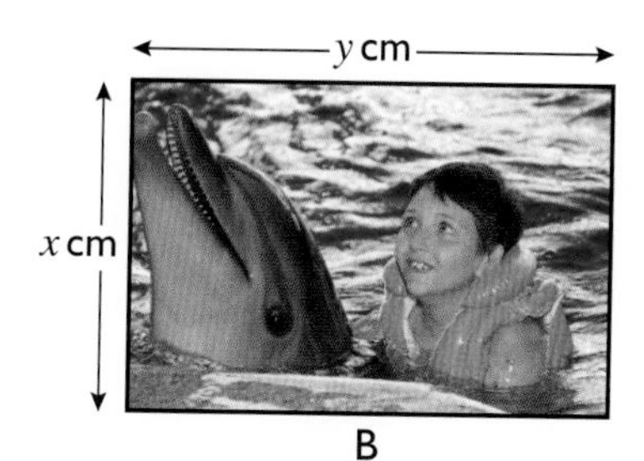

B

Picture B is a computer enlargement of picture A with scale factor 3.

a Work out the measurements of picture B.

b For the two pictures, find the ratios of

i the perimeters

ii the areas.

Gill then stretches picture B by 8 cm in width and by 8 cm in height.

c Is the new picture an enlargement of picture A?

Give a reason for your answer.

Reviewing skills

1. Look at the two cuboids.

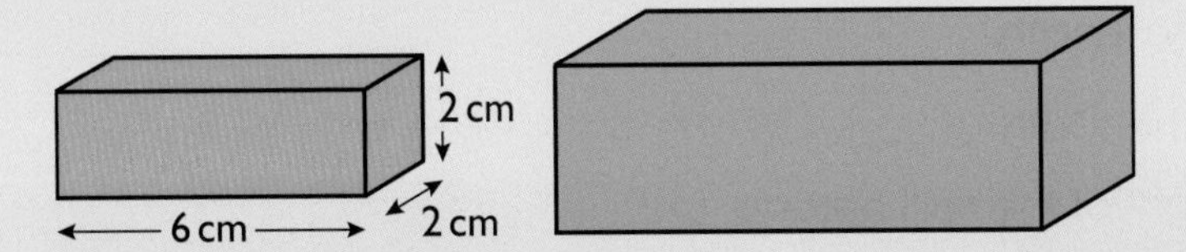

The edges of the larger one are $2\frac{1}{2}$ times the lengths of those in the smaller one.
The smaller cuboid has dimensions of 2 cm × 2 cm × 6 cm.

a Write down the dimensions of the larger cuboid.

Calculate

b the surface area of the smaller cuboid

c the surface area of the larger cuboid

d the volume of the smaller cuboid

e the volume of the larger cuboid

f the ratio of the surface areas of the two cuboids in its simplest form

g the ratio of the volumes of the two cuboids in its simplest form.

2. Jim has to paint some cubes.
He knows that 1 litre of paint is enough to paint 200 cubes of edge 3 cm.
How many cubes of edge 4 cm can he paint with 1 litre of paint?

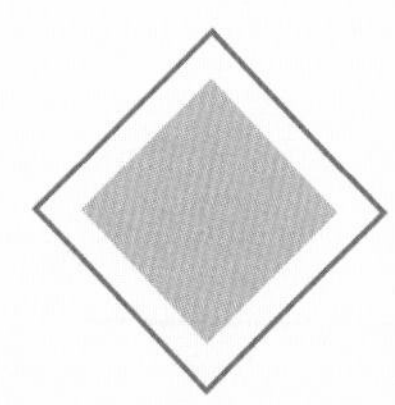

Statistics and Probability
Strand 1 Statistical measures

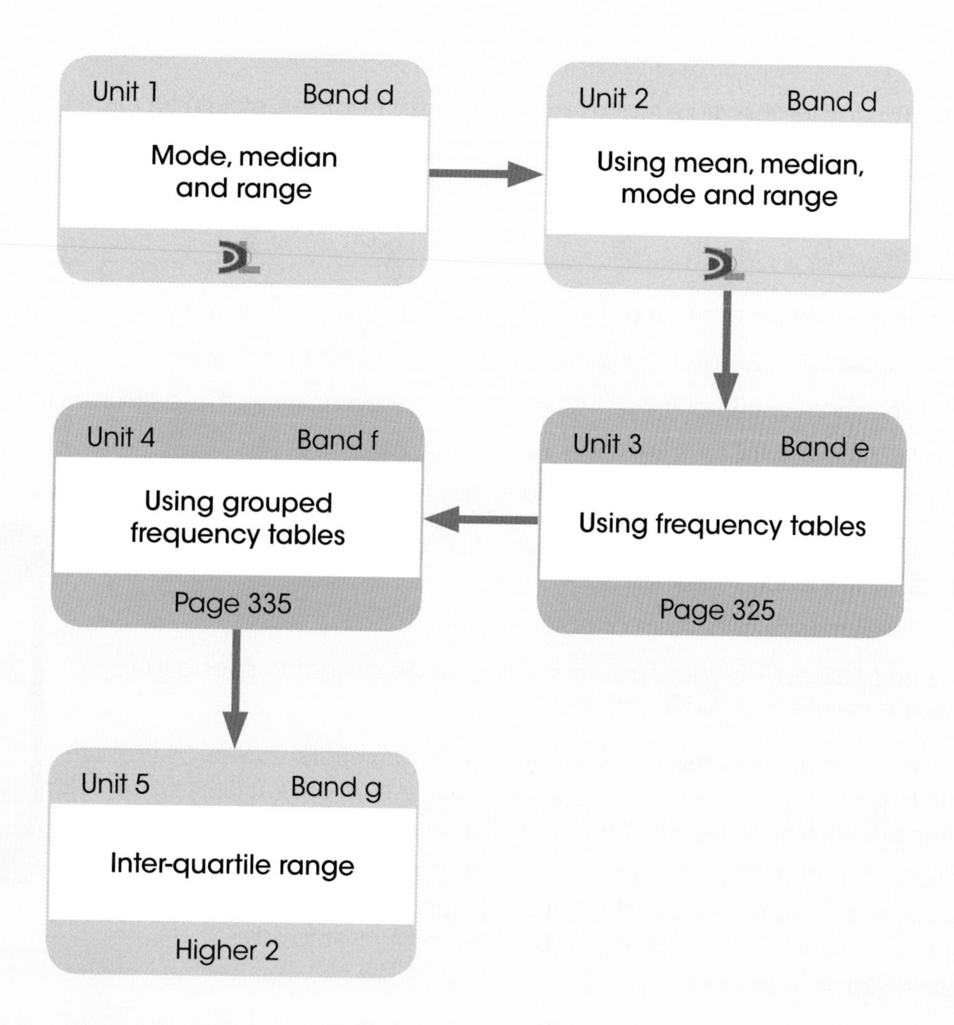

Units 1–2 are assumed knowledge for this book. They are reviewed and extended in the Moving on section on page 324.

Units 1–2 • Moving on

1. Here is a seven day temperature record (to the nearest degree Celcius) for Geneva and London.

Temperature (°C)	Mon	Tues	Wed	Thurs	Fri	Sat	Sun
Geneva	−7	−5	−3	0	1	−1	1
London	−1	−2	2	4	2	−1	−1

   a Find the mean temperature in both Geneva and London in this week.

   b Find the median temperature in both Geneva and London this week.

   c Which is the more useful of these two measures for comparing these temperatures?

   d Calculate the range of temperatures for both places.
   Explain how this helps you compare the temperatures.

2. Frank is a delivery driver. For two weeks he kept a note of how much he paid for a cup of coffee each morning.

   a Find the mean price of his cup of coffee.

   b Frank works 230 days a year. How much should he expect to spend in total on his morning coffee each year?

Coffee
£2.20
£1.95
£2.45
£1.36
£1.10
£1.95
£2.20
£2.10
£1.95
£2.20

3. Entry to a local museum is free to the public. They ask visitors for donations.
   One Saturday morning there are 7 visitors between 11 a.m. and 12 noon.
   The museum curator notes how much each person donates.
   To encourage donations, the curator wants to put up a notice showing the average donation. What would you advise the curator to say?
   Give reasons for your answer.

£2
£5
£5
£5
£5
£0
£0

4. A car service at Molly's Garage costs £200 plus parts. One day, the final bills were as shown here.

	A	B	C	D	E	F	G	H
1	£227	£265.50	£285.30	£212.70	£202.60	£212.70	£217	£215.30
2								

   a Find the mean cost of a service.

   b Find the median cost of a service.

   c Which is more useful for a customer to know, the mean or the median?
   Give a reason for your answer.

   d What other statistic about these data would it be useful for a customer to know? Explain why.

Unit 3 • Using frequency tables • Band e

Outside the Maths classroom

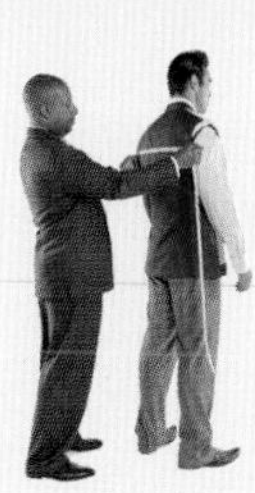

Clothing manufacture

How do clothing manufacturers know what sizes to make their clothes?

Toolbox

A **frequency table** has two rows (or columns).
The first shows all the different values that the data can take.
The second shows the frequency of each value.
It is useful to add an extra column when calculating the mean:

No. of packets	Frequency	Total no. of packets
0	9	0
1	11	11
2	9	18
3	4	12
4	11	44
5	10	50
Total	54	135

$0 \times 9 = 0$

$1 \times 11 = 11$

$2 \times 9 = 18$

$$\text{Mean} = \frac{135}{54} = 2.5$$

When deciding which average to use, remember that the mean is the only one which uses all of the data. The median shows the data split into two halves and the mode only uses the most common value.

Example – Finding the median from a table

The table shows how many goals a team scored in their 16-match season.

Number of goals	0	1	2	3	4	5
Frequency (no. of matches)	1	2	6	4	1	2

Find the median of the scores.

Solution

There are 16 scores altogether, so the median will be midway between the 8th and 9th values.
The 8th and 9th values are both 2, so the median is 2 goals.

Example – Finding the mean from a table

Max wants to be a percussionist. He needs to make both his hands work well.
He tests them by throwing a dart 20 times with each hand.
The rings on the target score 1 to 7 measuring from the centre.
His right hand scores were:

3, 2, 4, 6, 2, 5, 2, 1, 3, 3, 3, 2, 2, 3, 4, 3, 3, 3, 2, 1

a Put these data into a frequency table. Find the mean and range of the scores.

b For Max's left hand, the mean was 3 and the range was 4.
Which hand performed better?

Solution

a Mean score = 57 ÷ 20 = 2.85
Range = 6 − 1 = 5

Score	Tally	Frequency	Score × Frequency
1	II	2	1 × 2 = 2
2	~~IIII~~ I	6	2 × 6 = 12
3	~~IIII~~ III	8	3 × 8 = 24
4	II	2	4 × 2 = 8
5	I	1	5 × 1 = 5
6	I	1	6 × 1 = 6
7		0	7 × 0 = 0
Totals		20	57

b The mean for Max's left hand was higher than for his right, so his left performed better on average. The range for his left was smaller than his right so his left hand was also more consistent.

Practising skills

1 Henry asks his class how much pocket money they get each week.
Here are his results.

Amount (£)	Frequency
2.00	1
2.50	3
3.00	2
3.50	6
4.00	3
4.50	2

a Write down the mode.

b **i** Write down all the results in an ordered list.
ii Find the median.

c **i** Work out how much money they get in total.
ii Work out the mean.

(2) Priti goes ten-pin bowling. The table shows how many pins she knocks down.

Number of pins	0	1	2	3	4	5	6	7	8	9	10
Number of turns	3	0	1	4	6	3	2	1	2	1	2

a How many turns does she have?

b Work out the total number of pins she knocks down.

c Work out Priti's mean score.

d What is her median score?

(3) The table shows the number of waterproof jackets a shop sold in a week.

Mon	Tue	Wed	Thu	Fri	Sat
3	5	8	2	6	15

a How many waterproof jackets did they sell altogether?

b Work out the mean number of sales each day.

c Why do you think the sales were highest on Saturday?

(4) A driving instructor records the number of times his students take their practical driving test before they pass.

Number of attempts	1	2	3	4	5
Number of students	8	5	2	3	4

a Why are there no zero attempts?

b How many students were there?

c How many attempts were there?

d Work out the mean number of attempts per student.

e Find the median number of attempts.

(5) Daphne is doing a biology project. She asks everyone in her class to count the number of birds they see in their garden one day.
The table summarises her data.

Number of birds	0	1	2	3	4	5	30	40
Number of students	9	5	4	3	4	3	1	1

a Write down the mode. Say why it is not very representative of the data.

b Find the median number of birds in each garden.

c Work out the mean number of birds in each garden.

d Which is more representative in this case, the mean or the median?
Explain your answer.

(6) The data shows the number of goals scored by Jimmy's football team in each game.

0	5	2	3	6	1	1	1	2	0	6	3	2
1	0	3	1	0	0	2	4	6	4	0	2	3
1	0	2	3	5	0	2	6	3				

a How many games were there?

b Construct a frequency table for this data.

c Find the mode.

d Find the range.

e Work out their mean number of goals per game.

f Why is it difficult to use these results to predict what they will score in the next game?

Developing fluency

(1) Ben and Caroline go ten-pin bowling.
Here are the number of pins they knock down each turn.

Ben	4	2	0	8	10	4	3	6	3	7
Caroline	5	8	4	6	5	7	3	4	6	5

a For each person find the

i mean **ii** mode **iii** median **iv** range.

b Describe the differences between the two sets of scores.

c Who do you think is the better bowler and why?

(2) A survey records the number of people living in the households in one street.
Here are the results.

3	8	2	4	3	2	4	3	1	1
4	5	1	4	3	3	4	2	2	4

a **i** Make a frequency table for the data.

ii Draw a bar chart to show the data.

b Find the

i mean **ii** mode **iii** median **iv** range.

c What do the figures tell you about the people living in the street?

3) Alex and Eddie are comparing how long it takes them to get to work.
They write down their journey times each day for twelve days.

Alex	15	23	18	19	27	28	22	19	20	19	25	26
Eddie	24	17	19	23	28	19	70	21	24	21	26	19

a Find the outlier in the data. Give a possible explanation for it.

b What is the difference in the range between Alex and Eddie and why do you think this is so?

c For the journey times for each of Alex and Eddie work out the

i mean

ii median

iii range.

d i Who has the shorter average journey time?

ii Which average have you used and why?

4) Mrs White gives two Year 9 classes a Geography test which she marks out of 20.
Here are the results:

Class 9A

17	5	8	12	15	14	9	3	19	20
16	17	7	15	13	8	11	10	18	19
6	14	15	14	12					

Class 9B

12	11	7	9	16	17	8	12	13	14
15	14	9	11	10	10	7	12	15	11
10	6	14	10						

a Find the mean for each class.

b Find the range for each class.

c Which class do you think has done better? Explain your answer.

5) Theresa is testing a plant compost.
She records the heights of the plants, in cm, after eight weeks.

	A	B	C	D	E	F	G
1	6	22	8	33	7	9	31
2	23	31	9	21	25	36	28
3	7	27	30	27	34	24	

a Present the data as a stem and leaf diagram.

b Describe the distribution of the data.

What does it tell you about the effect of the compost?

c Find the median and mean. Which of these is the more representative figure?

d Theresa says that most of her plants have a height of 25 cm or over. Is this correct?

6 A plant food company is testing two new products, A and B.
Two batches of identical seeds are planted and each batch is treated with one of the new plant foods.
After three weeks the height of the seedlings, in centimetres, is measured to the nearest millimetre.

Batch A

1.5	2.2	2.1	1.9	2.2	1.4	1.8	2.0	1.7
1.2	1.2	1.3	2.2	1.2	1.3	1.2	1.3	1.6
2.2	1.7	1.2	1.2	1.5	2.2	2.1		

Batch B

1.9	2.4	1.9	1.8	1.4	1.4	1.6	1.4	1.4
2.3	1.4	1.5	2.4	1.6	1.5	2.2	1.4	2.4
2.4	1.4	1.5	1.6					

a Display the data for each batch in a stem and leaf diagram.

b Compare the two batches of plant food to decide which is more effective.
Give reasons for your answer.

Problem solving

Exam-style

1 The Human Resources department of a small company record the number of absences for each employee.
Here is a frequency table of their results:

Number of absences	Frequency
0	20
1	6
2	8
3	12
4	9
5	5
6	0
7	2
8	1

a How many employees are there?

b Find the mode, mean and median of the data.

c Which average do you think represents the data best?
Give a reason for your answer.

Exam-style

2 The table shows the annual salaries of employees at a small manufacturing company.
Each employee is paid according to the grade they are on.

Grade	Salary	Frequency
A	£18 000	13
B	£23 500	9
C	£30 000	10
D	£34 300	11
E	£42 800	4
F	£55 800	3
Chief Executive	£262 800	1

a How many employees are there?

b Find the salary that is the

i mode **ii** median **iii** mean.

c Which average do you think describes the data best?
Give a reason for your answer.

Exam-style

3 A riding stables records the weight of each of their horses in a stem and leaf diagram.

Stem	Leaf							
39	1	9						
40	4							
41	1	2	2					
42	2	3	6	7	7	7		
43	1	2	2	5	5	8	8	9
44	1	2	2	3				
45	7	8	9					
46	2	4						

Key: 43 | 2 = 432 kg

a How many horses does the stable have?

b The vet says that a horse is underweight if it is it more than 40 kg below the mean weight.
What percentage of the horses are underweight?

Exam-style

4 The maximum daily temperature (to the nearest degree Celsius) is recorded in Reykjavik for a month.

The frequency table shows the temperature for the first 20 days.

Maximum daily temperature (°C)	Frequency
−4	5
−3	3
−2	3
−1	1
0	3
1	2
2	1
3	2

a Here is the data for the next 10 days.

−4 −3 −2 −1 −2 0 −3 −2 0 1

Complete a frequency table to represent this information.

b What is the range of the data?

c Find the

i median temperature

ii mean temperature.

The temperature is 'freezing' when it is 0 °C.

d Use your table to work out the percentage of days when the temperature is

i above freezing

ii below freezing.

Exam-style

5 A council surveys all the households in a particular street. The number of children in each household was recorded.

Number of children	Frequency
0	18
1	7
2	13
3	5
4	3
5	3
6	0
7	1

a Draw a suitable diagram for this data set.

b The council decide that if the average number of children per household is greater than 1 then they will provide a play area.

State whether the council should provide a play area.

Explain your answer fully.

Exam-style

Extension

6 James has some tomato plants.

Some plants are treated with *Grow-well* fertiliser, other plants are not.

One day James counts the number of ripe tomatoes he picks from each plant treated with *Grow-well*.

Number of ripe tomatoes	Frequency
4	8
5	12
6	15
7	20
8	15
9	12
10	8
11	4
12	6

a Find the

i mode **ii** mean **iii** median **iv** range of the data.

James counts the number of ripe tomatoes he picks from plants not grown in fertilizer. Here are his results.

mode: 7
mean: 5.4
median: 6
range: 12

b James claims that *Grow-well* increases yield.

Does the data support the claim? Explain your answer.

7 The frequency table shows the number of pets owned by a group of students.

Number of pets	Frequency
0	9
1	
2	7
3	4
4	2
5	4

The mean number of pets is 1.8.

How many students had one pet?

Reviewing skills

1. A survey was taken of the number of goals scored in 18 matches.

5	0	2	4	3	4	1	1	0
3	1	2	0	1	3	5	2	23

a How many matches did not have any goals scored?

b Which number is an outlier?

Explain how it might have happened and give a reason why it should be removed.

c Copy and complete the frequency table.

Number of goals						
Frequency						

d **i** What is the mean number of goals?

ii What is the mode?

iii What is the median?

e What is the range?

f Work out the mean, mode, median and range with the outlier reinstated.

Which of these measures is affected the most?

Unit 4 • Using grouped frequency tables • Band f

Outside the Maths classroom

Market research

What helps market researchers make sense of their data?

Toolbox

For large amounts of data use a **grouped frequency table**.

Between 5 and 10 groups (or **classes**) is usually most suitable.

Show classes for **continuous data** using less than (<) or less than or equal to (⩽).

The **modal class** is the class with the highest frequency (if the class widths are all the same).

To **estimate the mean** from a grouped frequency table, multiply the **mid-interval value** for each group by the frequency for that group, add the results and divide by the total frequency.

You should round your answer to a suitable degree of accuracy.

Example – Finding the mean of continuous data

The maximum temperature, in °C, is recorded in Buenos Aires for each day in June one year.

16.4	12.8	17.6	19.1	16.6	15.5
11.2	18.7	19.5	16.1	15.3	14.2
15.8	15.7	14.9	14.4	13.4	12.1
13.9	11.9	13.1	12.6	10.9	13.5
14.2	15.4	16.6	15.9	15.6	14.3

June

a Present these data in a grouped frequency table.

Temperature, T °C	Tally	Frequency
$10 \leqslant T < 12$		
$12 \leqslant T < 14$		
$14 \leqslant T < 16$		
$16 \leqslant T < 18$		
$18 \leqslant T < 20$		

b Calculate an estimate for the mean temperature.

Solution

a

Temperature, T °C	Tally	Frequency
$10 \leq T < 12$	|||	3
$12 \leq T < 14$	~~||||~~ ||	7
$14 \leq T < 16$	~~||||~~ ~~||||~~ ||	12
$16 \leq T < 18$	~~||||~~	5
$18 \leq T < 20$	|||	3

b

Temperature, T °C	Midpoint, m	Frequency, f	$m \times f$
$10 \leq T < 12$	11	3	33
$12 \leq T < 14$	13	7	91
$14 \leq T < 16$	15	12	180
$16 \leq T < 18$	17	5	85
$18 \leq T < 20$	19	3	57
	Totals	30	446

estimated mean = 446 ÷ 30 = 14.866 = 14.9 °C (1d.p.)

Practising skills

1 The table shows the heights of 22 footballers about to play a match.

Height (cm), h	Frequency, f	Midpoint, m	$m \times f$
$150 \leq h < 156$	3		
$156 \leq h < 162$	6		
$162 \leq h < 168$	8		
$168 \leq h < 174$	3		
$174 \leq h < 180$	2		
Totals			

Copy and complete the table and use it to estimate the mean height of these footballers.

(2) Ruth's telephone bill shows the lengths, in minutes, of her last 20 calls.

47 min	30 min
19 min	44 min
32 min	18 min
41 min	12 min
57 min	24 min
28 min	36 min
9 min	42 min
17 min	16 min
46 min	32 min
33 min	29 min

a Copy and complete the frequency table.

Length of call (minutes), l	Frequency, f	Midpoint, m	$m \times f$
$0 \leq l < 10$			
$10 \leq l < 20$			
$20 \leq l < 30$			
$30 \leq l < 40$			
$40 \leq l < 50$			
$50 \leq l < 60$			
Totals			

b In which group does the median lie?

c Use the table to estimate the value of the mean.

(3) A group of people do a puzzle as part of an aptitude test.
Here are the times (in seconds) that it takes them to solve the puzzle.

24	83	114	84	90	103	74	176	61	40	162	49
77	92	108	124	185	89	63	79	37	91	65	19

a i Find the times of the first and the last people to solve the puzzle.

ii What is the range?

b Find the median for this data set.

c Copy and complete the frequency table.

Time (seconds), t	Frequency, f	Midpoint, m	$m \times f$
$0 \leq t < 40$			
$40 \leq t < 80$			
$80 \leq t < 120$			
$120 \leq t < 160$			
$160 \leq t < 200$			
Totals			

d In which group does the median lie?

e Use the table to estimate the value of the mean.

f The fastest 25% of the group are accepted and the slowest 25% are failed. The others have to do more tests.
What can you say about the marks of those who do more tests?

4. A speed camera recorded the speed of cars, v mph, on a road through a housing estate. Here are the results.

Speed (mph), v	Frequency, f	Midpoint, m	$m \times f$
$0 \leqslant v < 10$	1		
$10 \leqslant v < 20$	12		
$20 \leqslant v < 30$	22		
$30 \leqslant v < 40$	4		
$40 \leqslant v < 50$	1		
Totals			

a Complete a copy of the table and use it to estimate the mean speed.

b What do you think the speed limit is?

Local residents say that any speed above 25 mph is unsafe.

c Estimate the percentage of cars that are travelling at 25 mph or faster.

5. A group of students record how many hours each week they spend on homework. Here are their results.

Hours, h	$0 \leqslant h < 4$	$4 \leqslant h < 8$	$8 \leqslant h < 12$	$12 \leqslant h < 16$	$16 \leqslant h < 20$
Frequency	3	15	14	9	4

a Estimate the mean number of hours a student spends doing homework.

b Most students do homework on six days. What is the average amount of homework they do each day?

Developing fluency

1. A sports centre records how many people use its facilities each day. Here are the results.

32 71 64 28 66 128 184 94 47 104
95 73 69 80 115 162 106 82 73 118

a Find the mean and the median for this data set.

b Copy and complete the frequency table.

Number of people, n	Frequency, f	Midpoint, m	$m \times f$
$0 \leqslant n < 50$			
$50 \leqslant n < 100$			
$100 \leqslant n < 150$			
$150 \leqslant n < 200$			
Totals			

c In which group does the median lie?

d Use the table to estimate the value of the mean.

e Compare your answers for parts **c** and **d** with those for part **a**.

f In what circumstances would you use an estimate of the mean and when would you need to calculate it exactly?

2 A running club records the time members take to complete a cross-country race.

Time (minutes), t	Frequency, f	Midpoint, m	$m \times f$
$0 \leq t < 10$	0		
$10 \leq t < 20$	1		
$20 \leq t < 30$	8		
$30 \leq t < 40$	14		
$40 \leq t < 50$	7		
Totals			

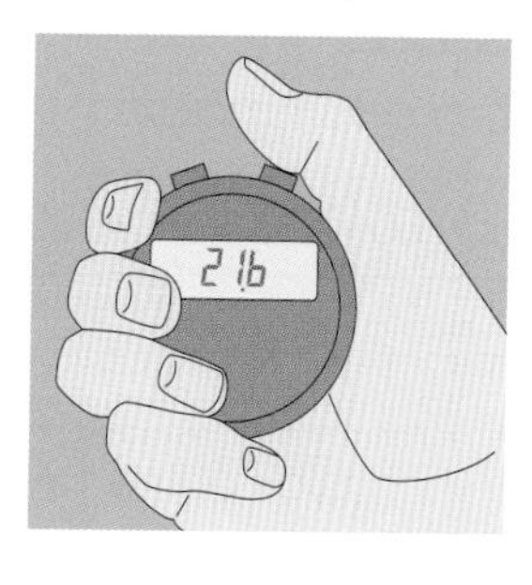

a Copy and complete the table and use it to estimate the mean time.

b What can you say about the range?

c Another club had a mean of 27.9 minutes and a range of 55 minutes.
Compare the performance of the two clubs.

3 Harry travels to work by train.

He records the time, in minutes, that he waits for the train.

Waiting time (minutes), t	Frequency, f
$0 \leq t < 5$	6
$5 \leq t < 10$	5
$10 \leq t < 15$	11
$20 \leq t < 25$	9
$25 \leq t < 30$	4
Total	

a Complete a copy of the table, with any extra columns, and use it to estimate the mean waiting time.

b How many times does Harry have to wait at least 20 minutes?

c One day the waiting time was 72 minutes.
Can you explain what might have happened? Why did Harry exclude this time from his calculation?

4 The table shows the distance jumped by long jumpers at a school sports day.

Distance (metres), d	Frequency, f	Midpoint, m	$m \times f$
$0 \leq d < 1$	1		
$1 \leq d < 2$	5		
$2 \leq d < 3$	12		
$3 \leq d < 4$	9		
$4 \leq d < 5$	8		
Total			

a Complete a copy of the table and use it to estimate the mean distance jumped.

b There were 14 jumpers and they had 3 jumps each. How many foul jumps were there?

c The qualification for the school team is a jump over 4.5 m.
What can you say about the number who qualified?

5 Alan plants two types of early potato, A and B. He weighs the potatoes, w grams, that each plant produces.

The results are summarised in this table.

	$700 \leqslant w < 750$	$750 \leqslant w < 800$	$800 \leqslant w < 850$	$850 \leqslant w < 900$	$900 \leqslant w < 950$	$950 \leqslant w < 1000$
A	5	8	4	2	5	0
B	2	4	5	7	3	3

Compare the yields of the two types of potatoes.

6 A small firm employs 25 people. The table shows their salaries.

Salary, £s	Number of people
$0 \leqslant s < 10000$	6
$10000 \leqslant s < 10000$	11
$20000 \leqslant s < 30000$	5
$30000 \leqslant s < 40000$	2
$40000 \leqslant s < 50000$	1

The managing director is considering three schemes to increase the employees' pay.

Scheme 1: Give everyone a 5% increase on their current salary.

Scheme 2: Find 5% of the mean salary and increase all salaries by this amount.

Scheme 3: Find 5% of the median salary and increase all salaries by this amount.

a Copy and complete the table below to find out how much Sandra, Shameet and Comfort would earn under each scheme.

	Current salary	Scheme 1 increase	Scheme 2 increase	Scheme 3 increase
Sandra	£8000			
Shameet	£2200			
Comfort	£3600			

b Which member of staff would benefit most from each scheme?

c Which scheme do you think the managing director should use?
Give a reason for your answer.

Problem solving

Exam-style

(1) An IT company give all applicants for a vacancy an aptitude test.
The applicants score s marks on the test.
The table shows the results.

Score, s	Frequency, f	Midpoint, m	$m \times f$
$0 \leqslant s < 10$	11		
$10 \leqslant s < 20$	6		
$20 \leqslant s < 30$	12		
$30 \leqslant s < 40$	10		
$40 \leqslant s < 50$	9		
Totals			

a Calculate an estimate of the mean mark.

b The IT company interview all applicants who achieved 5 marks more than the mean. Estimate how many applicants were interviewed.

c Explain why your answers are estimates.

Extension

Exam-style

(2) The grouped frequency table gives information about the distance that 100 commuters travel to work.

Distance travelled, d km	Frequency, f	Midpoint, m	$m \times f$
$20 < d \leqslant 30$	6	25	
$30 < d \leqslant 40$			420
$40 < d \leqslant 50$	20	45	900
$50 < d \leqslant 60$			
$60 < d \leqslant 70$	11		715
Totals	**75**		

a Copy and complete the table.

b In which group does the median lie?

c Estimate the mean distance travelled by the commuters.

d A commuter expects that their journey will take $1\frac{1}{2}$ minutes for every kilometre commuted. Estimate the mean time spent commuting.

Exam-style

3 Chloe buys an IQ testing puzzle which she tests out on her friends.
She times how long each person takes to solve the puzzle.
The frequency table shows Chloe's results.

Time, t minutes	Frequency, f
$0 < t \leqslant 5$	2
$5 < t \leqslant 10$	3
$10 < t \leqslant 15$	6
$15 < t \leqslant 20$	8
$20 < t \leqslant 25$	18
$25 < t \leqslant 30$	13

a What is the modal class?

b Estimate the range of the data.

c Estimate the mean time to solve the puzzle.

d The puzzle carries this statement:
'Are you a genius? You are if you can solve me in less than 10 minutes!'
Comment on the statement.

Exam-style

4 A doctors' surgery surveys a group of adult male patients to find out their heights.
Here is a frequency table of the results.

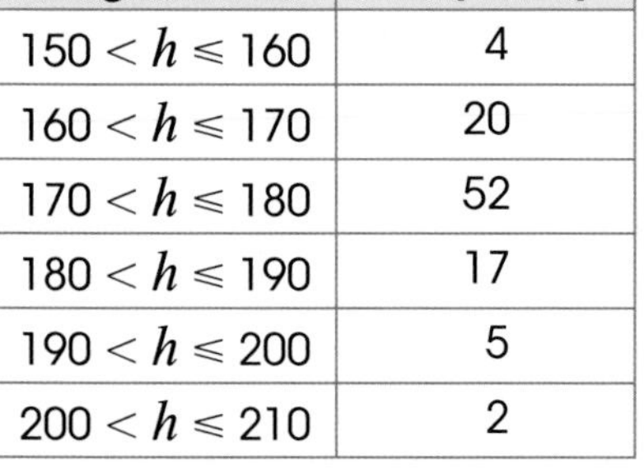

Height in cm	Frequency
$150 < h \leqslant 160$	4
$160 < h \leqslant 170$	20
$170 < h \leqslant 180$	52
$180 < h \leqslant 190$	17
$190 < h \leqslant 200$	5
$200 < h \leqslant 210$	2

a How many people took part in the survey?

b In which class does the median lie?

c Dr Smith says the range of heights is 60 cm.
Is Dr Smith correct? Give a reason for your answer.

d Estimate the mean height of the patients.

Dr Smith decides a patient is classified as tall if he is 9.5 cm taller than the mean.

e Estimate the percentage of tall patients at the surgery.

Exam-style

5 The production team for a TV quiz show run a general knowledge test to identify suitable contestants.

The table below shows the number of correct answers from a group of hopeful contestants.

Number of correct answers, c	Frequency
$30 < c \leqslant 40$	41
$40 < c \leqslant 50$	16
$50 < c \leqslant 60$	10
$60 < c \leqslant 70$	7
$70 < c \leqslant 80$	8
$80 < c \leqslant 90$	10
$90 < c \leqslant 100$	8
Total	

a What is the modal group?

b In which group does the median fall?

c Estimate the mean number of correct answers.

The test has 100 questions.

A correct answer scores 2 points, an incorrect or missing answer scores −1 point.

d Estimate the mean score.

Contestants need a score of 125 or more to appear on the television show.

e Estimate the probability that a randomly chosen person will be selected for the show.

Exam-style

Extension

6 A teacher wants to compare the performance of two different classes, X1 and Y1.

This table shows the results m% of a recent test.

Percentage	$0 < m \leqslant 20$	$20 < m \leqslant 40$	$40 < m \leqslant 60$	$60 < m \leqslant 80$	$80 < m \leqslant 100$
X1	2	5	11	9	3
Y1	6	4	3	9	8

Which class do you think is better? Explain your answer.

Reviewing skills

1 A group of people take an aptitude test for a flying school. Here are the results.

Score, s	Midpoint, m	Frequency, f	$m \times f$
$0 \leqslant s < 20$		6	
$20 \leqslant s < 40$		8	
$40 \leqslant s < 60$		15	
$60 \leqslant s < 80$		14	
$80 \leqslant s < 100$		5	
Totals			

a Complete a copy of the table and use it to estimate the mean score.

b The pass mark for the flying school is 75. Estimate how many people passed.

c The lowest $\frac{1}{3}$ of the group of people are told they cannot try again.

What can you say about the marks of this group?

Statistics and Probability Strand 2 Statistical diagrams

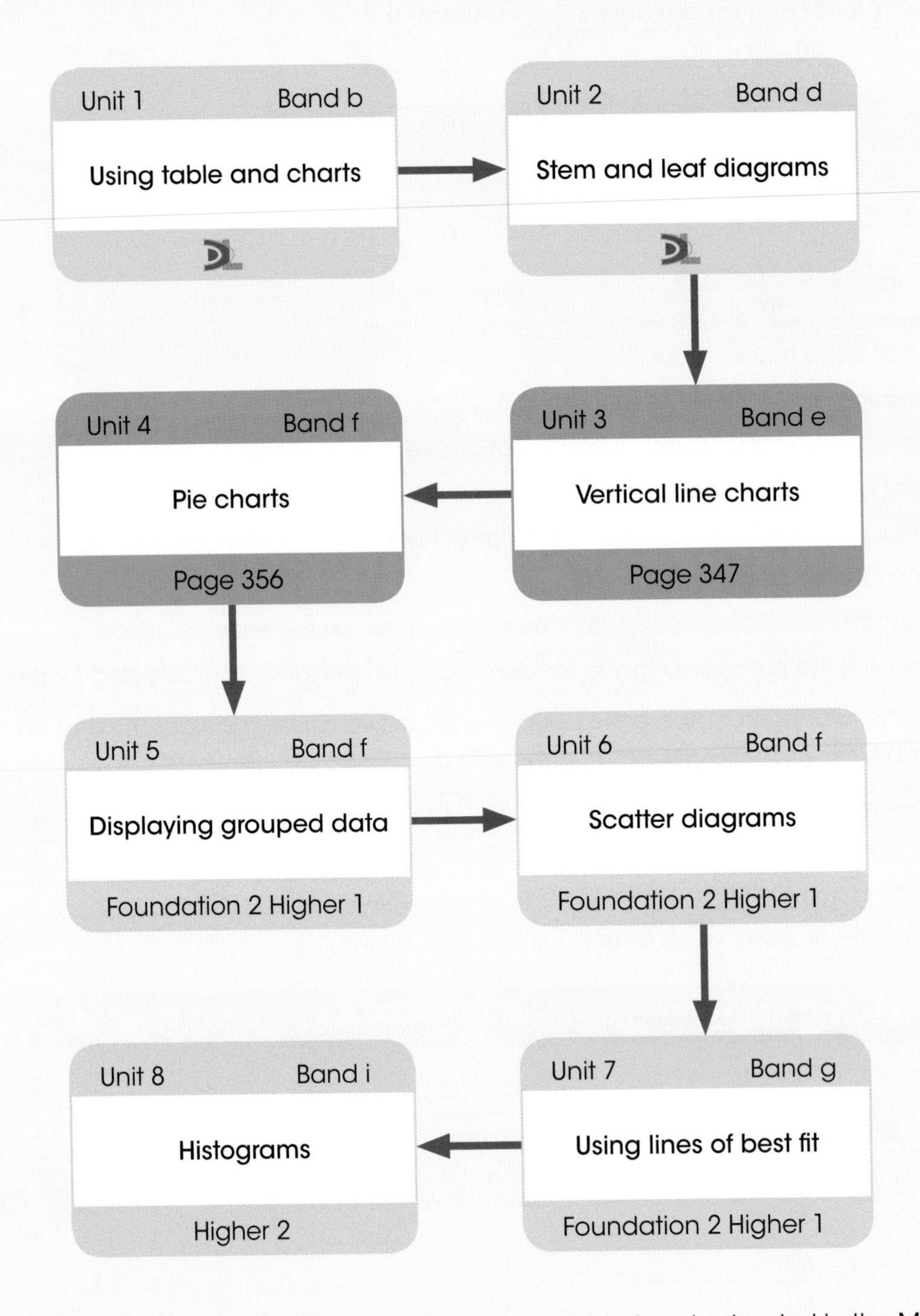

Units 1–2 are assumed knowledge for this book. They are reviewed and extended in the Moving on section on page 345.

Units 1–2 • Moving on

Exam-style

1. The manager of a restaurant collects data on the meals served for a week.
The bar chart shows her results.

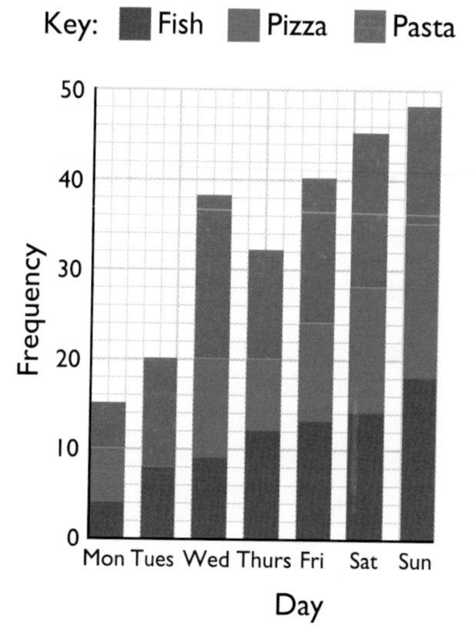

a On which day were no pizzas served?

b How many pizzas were served that week?

c What fraction of the meals served on Friday were pasta?
Give your answer in its simplest form.

d On which two days were the same number of pizzas served?

e Which dish was the least popular choice that week?

f Calculate the mean number of meals served each day.

g Why is a compound bar chart a useful way of displaying the data?

Exam-style

2. The bar chart shows the monthly mean number of daylight hours per day in London, UK, and the city of Barrow, Alaska.

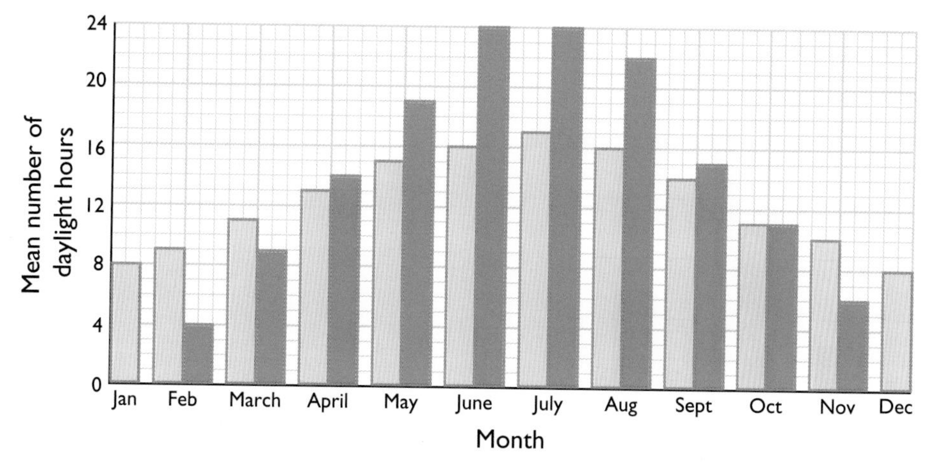

a In which month is the mean number of daylight hours the same in London and Barrow?

b Work out the range in the mean number of daylight hours in London.

c How long is the average night in

i July in Barrow

ii December in Barrow?

d Work out the total number of daylight hours in London in June.

e Why has a multiple bar chart been used to display the data?

3 Peter records the number of texts he sends and receives each day.

The stem and leaf diagram shows his results.

0	4	7	8	9					
1	1	2	3	3	3	4	5	8	9
2	0	0	2	7	9				
3	1								

Key: 2 | 0 = 20 texts

a How many days does Peter collect data for?

b Find the modal number of texts.

c Find the median.

d Find the range.

e Peter and Isobel both record the number of minutes they use their mobile phones for each day for 2 weeks.

Peter	Isobel
110, 102, 113, 127, 132, 108, 114	113, 122, 132, 128, 127, 131, 135
100, 109, 117, 128, 129, 131, 104	119, 109, 112, 122, 126, 138, 129

i Display the data in an ordered back-to-back stem and leaf diagram.

ii Who spends more time on their phone? Explain how you can tell from your stem and leaf diagram.

4 A doctors' surgery collects data to find out why patients have not attended an appointment. The bar chart shows the results.

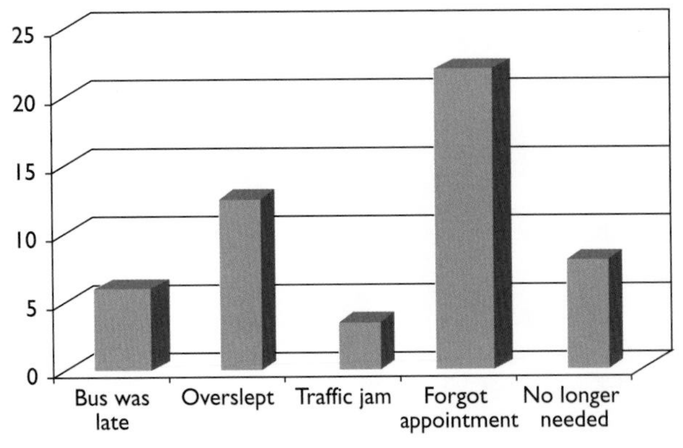

Write down three criticisms of this bar chart.

Unit 3 • Vertical line charts • Band e

Outside the Maths classroom

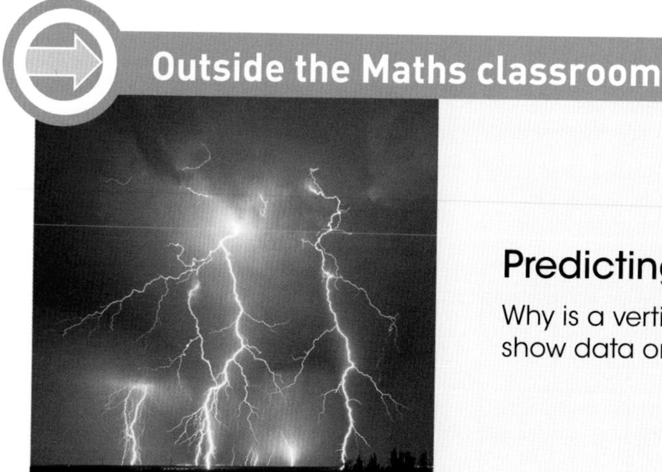

Predicting weather patterns

Why is a vertical line chart a useful way to show data on average monthly rainfall?

Toolbox

A vertical line chart is used to show data that are obtained at intervals of time or place.

The scales on the axes do not have to be the same as each other. They will usually represent quite different things.

The scale must be the same all along each axis so the numbers must be evenly spaced.

If time is involved it goes along the horizontal axis.

A company declares its profits on 1st June each year, profits for 2010 to 2014 are shown here.

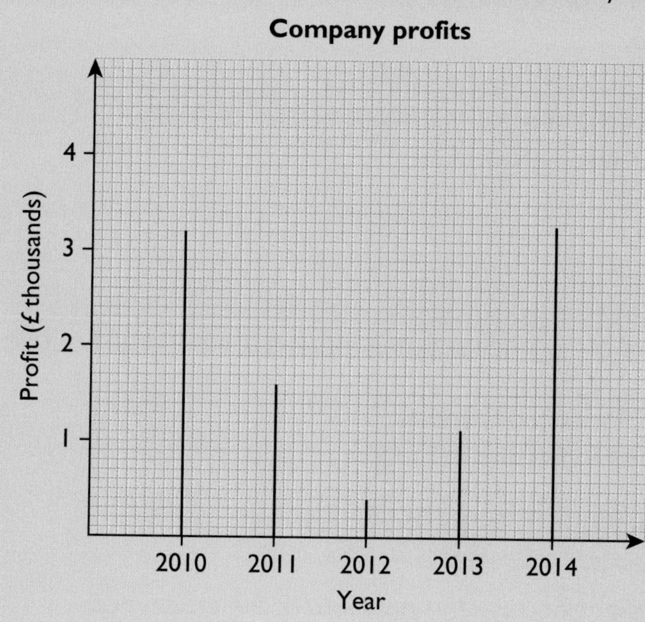

Example – Interpreting a vertical line chart

Jamila plants a shrub.

She measures its height on 1st January each year. She draws this vertical line chart.

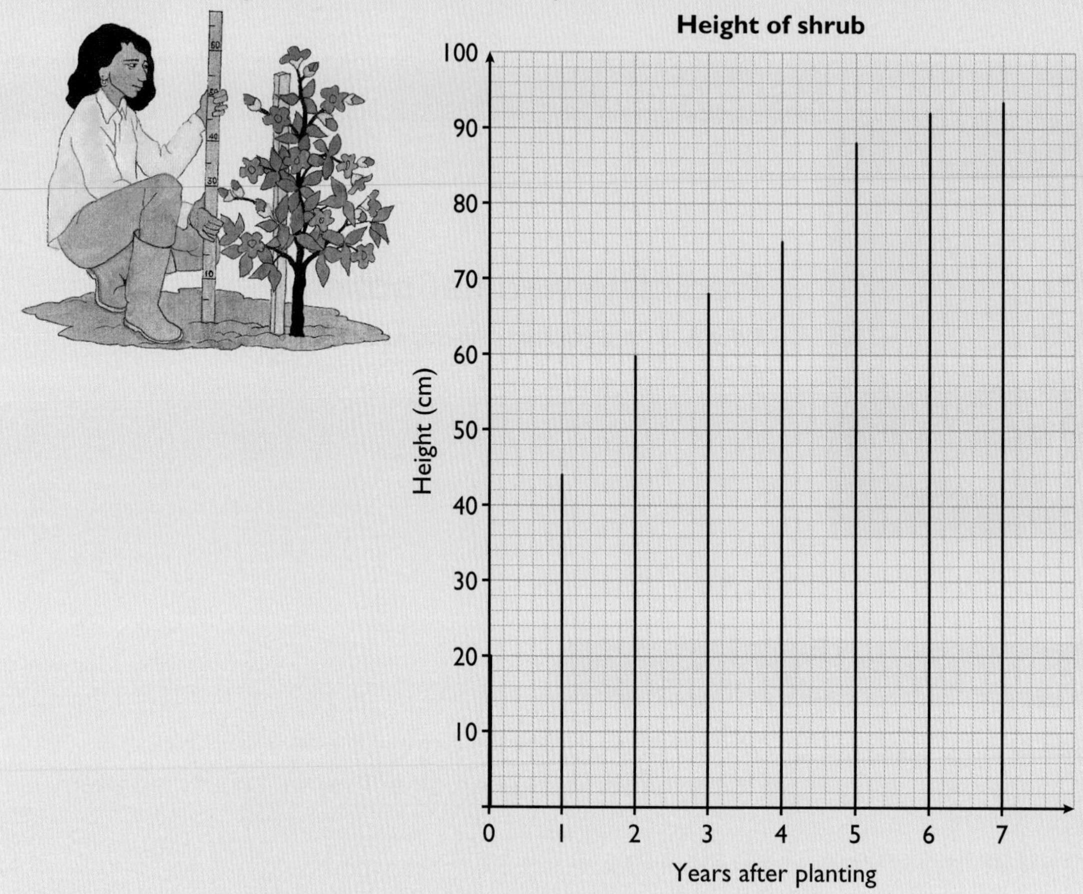

a What does each small square represent on the height scale?

b In which year does the plant grow the most? How do you know?

c In which year does the plant grow the least?

d Do you think the plant will grow much more? Explain your reasoning.

Jamila's mother says, 'You should join up the tops of the vertical lines to make a graph'.

Jamila replies, 'No, most of the year the shrub doesn't grow at all'.

e Comment on this conversation.

Solution

a Each small square represents 2 cm.

The large square represents 10 cm and there are 5 small squares making up the large square: $10 \div 5 = 2$ cm.

b The plant grows most during its first year of planting, 0–1 years.

This can be seen by the change in height from 20 cm to 44 cm, an increase of 22 cm (more than in any other year).

c The plant grows least in year 7, 6–7 on the graph.

This shows a change in height from 92 to 94 cm.

d It is unlikely that the plant will grow much taller, because the **gradient** of the graph has become flatter.

e Jamila is right. The shrub grows mostly in the summer. It hardly grows at all in winter months. So she should not join the points with a line because it is not representative of the data.

Practising skills

1) A guesthouse keeps a record of the number of guests staying there each month.

Month	Jan	Feb	Mar	Apr	May	Jun	Jul	Aug	Sept	Oct	Nov	Dec
Number of guests	24	15	28	18	36	42	52	54	37	19	23	48

a Show the data in a vertical line chart.

b In which month was there the fewest guests?

c In which month was there the most guests?

d Explain the shape of the graph.

2) Ravindra records the midday temperature each day for one week.

Day	Mon	Tue	Wed	Thu	Fri	Sat	Sun
Midday temp. (°C)	10	10	9	3	4	8	10

a Draw a vertical line chart to illustrate these figures.

b Join the top of these lines with a dotted line. What does this show?

c Between which two days was there the greatest fall in midday temperatures?

d Describe the tempreature over the week.

3) This table gives the dates when the world population reached certain milestones.

Billions of people	1	2	3	4	5	6	7
Date	1804	1927	1959	1974	1987	1999	2012

a Draw a vertical line chart to illustrate this data. Join the tops of these lines with a smooth curve.

b Describe the trend.

c Estimate the world population in 1950.

d Estimate the year when the world population was 4.5 billion.

4) Here is a misleading vertical line chart.

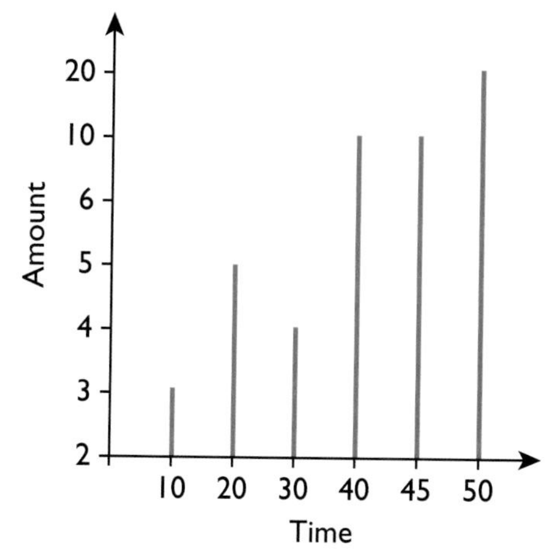

Give reasons why the graph is misleading.

Developing fluency

1) A small shop opens at 9 a.m. and closes at 5 p.m. One day the owner counts the number of people in the shop once an hour, at 10 a.m., 11 a.m. and so on until 5 p.m.

The vertical line chart shows these data.

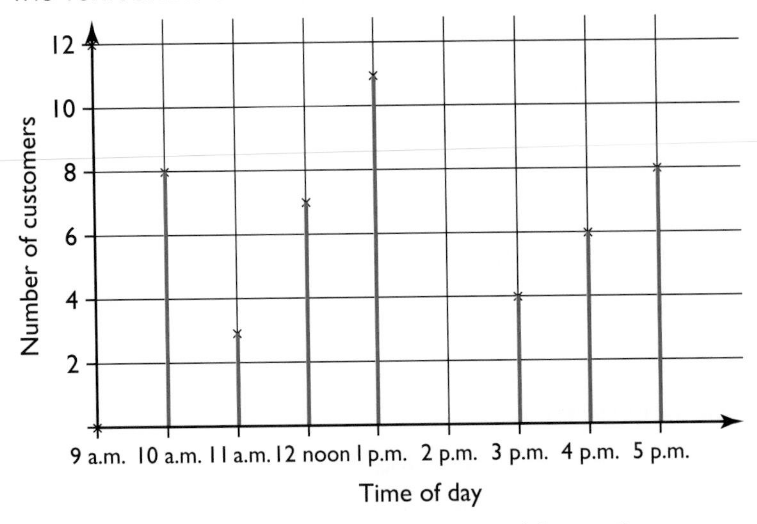

a How many people were in the shop at 3 p.m.?

b Give two possible explanations of what happened at 2 p.m.

c Which of the following statements is true for the number of people in the shop at 11.30 a.m.?

i There were 5 people.

ii There were between 3 and 7 people.

iii You have no information so can't say.

d Explain why it is wrong to join the tops of the vertical lines.

2) A car dealer records the number of new cars he sells each month.

Month	Jan	Feb	Mar	Apr	May	Jun	Jul	Aug	Sept	Oct	Nov	Dec
Number of cars	3	4	15	6	2	3	7	5	16	8	2	7

a In which months does he sell the most cars? Can you explain why this is so?

b In which months did he sell the fewest cars?

c Draw a vertical line chart to show the number of cars sold each month.

d The table and chart show the same information. Which do you think shows it more clearly?

3) A school keeps a record of the attendance of its students for two weeks.
The table shows the results for two weeks.

Day	Mon1	Tue1	Wed1	Thu1	Fri1	Mon2	Tue2	Wed2	Thu2	Fri2
% present	87	91	95	97	89	92	94	97	98	93

a Illustrate the data in a vertical line graph. Join the tops of the vertical lines with a dotted line.

b Describe the pattern of the graph and give a possible explanation.

Exam-style

4 A highway maintenance company keeps a record of the number of days that it does work on a particular stretch of motorway each month for 2 years.

Month	Jan	Feb	Mar	Apr	May	Jun	Jul	Aug	Sept	Oct	Nov	Dec
Number of days Year 1	6	4	3	0	0	5	0	2	5	3	0	6
Number of days Year 2	7	5	2	1	0	0	1	1	4	3	3	5

a Show the data for the two years in a vertical line chart.

b During which months were there no repairs done?

c Which months had the most days spent doing repairs?

d Join the top of the vertical lines with a dotted line.

e Describe

i any pattern in the data

ii any trend in the data.

5 Halley is a keen cricketer.
This vertical line graph shows the number of runs he scored when he was different ages.
(His birthday is in the winter.)

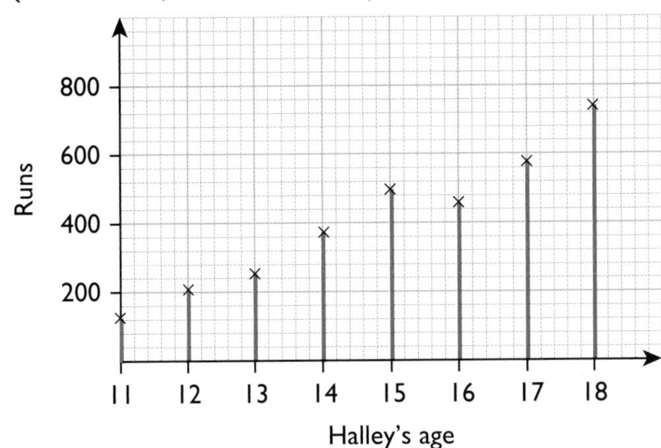

Say whether the following statements are true or false. Give your reasons.

a There is an upward trend, overall, in Halley's run scoring.

b Halley scored 400 runs when he was $14\frac{1}{2}$ years old.

c Halley had an exceptionally good season when he was 15.

d He will definitely score more than 720 runs when he is 19.

Exam-style

6 This vertical line chart shows Francesca's temperature at certain times.
She is ill in hospital.

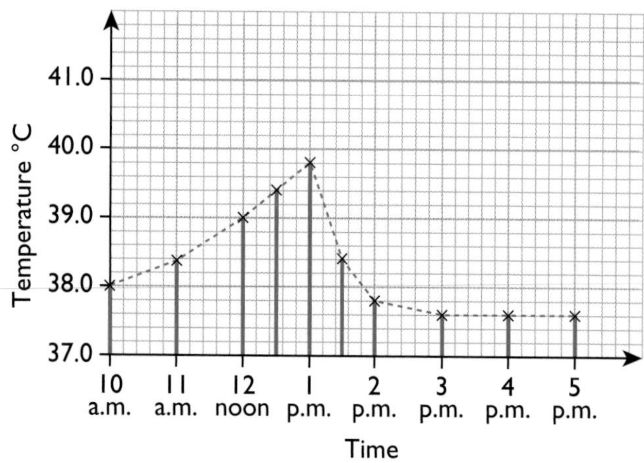

a What is Francesca's temperature at
 i 11 a.m.
 ii 3.30 p.m.?

b When was her temperature highest? What was it at its highest?

c How often was Francesca's temperature taken?

d Normal body temperature is 37.0 °C. What do the lengths of the vertical lines represent?

e Write a short paragraph describing Francesca's temperature during her illness.

7 Here is a graph showing the temperature at 3 p.m., in degrees Celsius, for ten days in August in Sudbury.

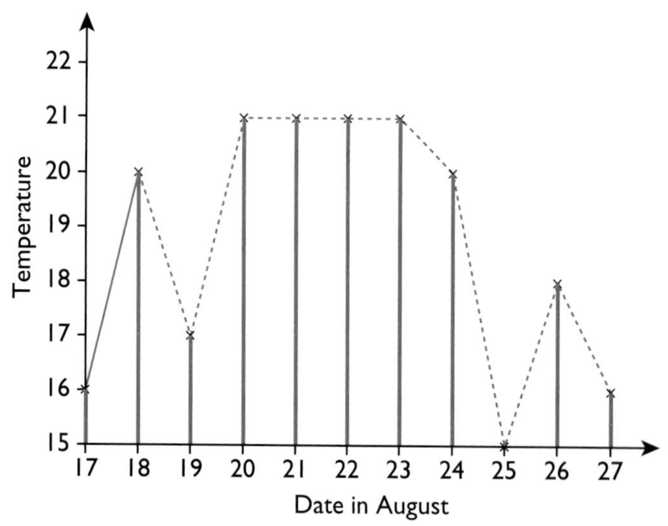

Describe what is right and what is wrong with this graph.

Problem solving

Exam-style

1. Toby records the temperature inside and outside his house every 2 or 4 hours over a 24-hour period.

   The graph shows his results.

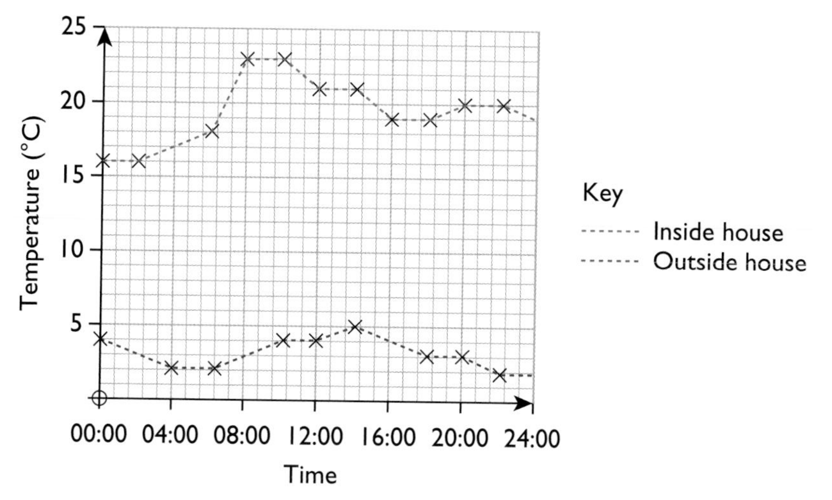

   a What was the temperature inside Toby's house at 4 p.m.?

   b How much warmer was it inside Toby's house at 00:00 than outside?

   c Estimate the range in temperature outside Toby's house that day.

   d Estimate the time that Toby's heating turns on in the morning.

   e At what time was the difference between the inside and outside temperatures the greatest? What is the temperature difference at this time?

Exam-style

2. An estate agency records their total house sales for the last twelve months.

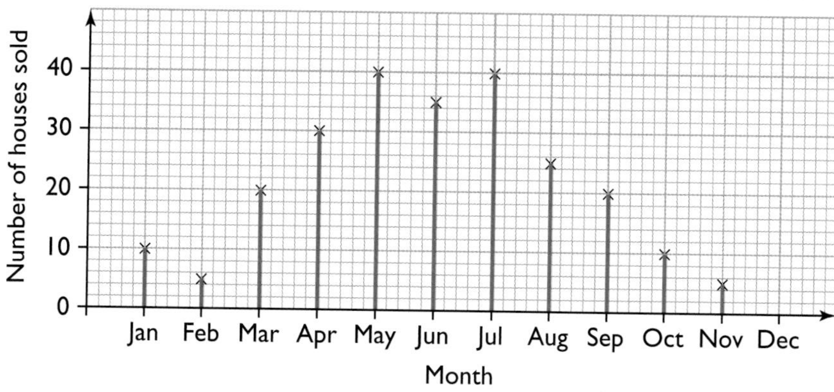

   a How many houses were sold in

   i June **ii** December?

   b What is the average number of houses sold in a month?

   c In which months did the estate agents sell

   i more than the average

   ii fewer than the average

   iii the average number of houses?

   d At what time of year do the estate agents sell the least number of houses?

Exam-style

3 Alfie has an internet bookshop.
He records the number of books he has in stock each month.
The graph shows the amount of stock Alfie has on the first of each month.

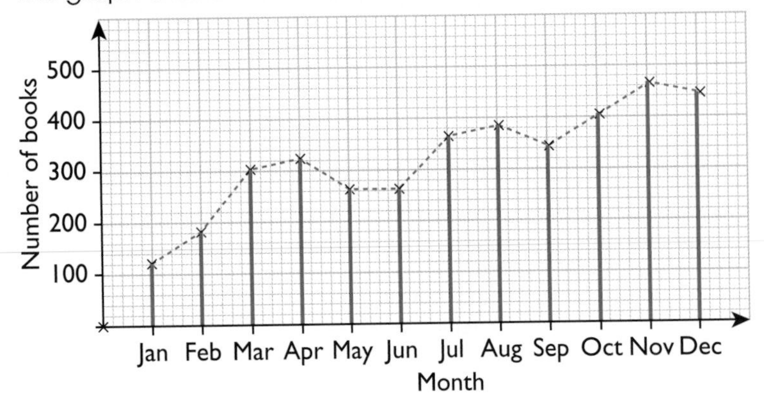

a How many books does Alfie have on 1 September?

b When does Alfie have
  i the most
  ii the fewest books in stock?

c On the 1st of which month does he have 320 books in stock?

d In which month is the greatest change in the number of books in stock?

e Between which months is there no change?
Does this mean Alfie sells no books that month? Explain your answer.

Exam-style

Extension

4 The midday temperature in Brighton was recorded each day for a week.

Day	Mon	Tue	Wed	Thur	Fri	Sat	Sun
Temperature (°C)	21	23	25	21	19	18	22

a Draw a suitable diagram to show this information.

b Sam says, 'The hottest day was Wednesday.'
Present an argument to show this may not be true.

c What can you say about the temperature on Thursday?

Exam-style

5 Neil records his best time to run 100 metres over eight weeks.
The table shows this information.

Week	1	2	3	4	5	6	7	8
Time (s)	17.4	17.1	16.9	16.8	19.4	16.7	16.4	16.5

a Plot a vertical line graph for this information.

b Which week did Neil run the fastest?

c Give a reason why Neil's time in week 5 might be higher than any other week.

d Can you draw a trend line from week 1 to week 8 to predict future times?
Explain your answer.

Exam-style

(6) Katy plants two dwarf fruit trees. She records the heights of each tree each year.

Year	1	2	3	4	5	6	7	8
Tree A (cm)	20	63	125	158	201	219	223	224
Tree B (cm)	20	36	58	73	83	90	97	101

a Draw a suitable chart to show this information.

b During which years did the trees grow the most?

c Estimate the maximum height of the trees.

d Do you think the trees are the same species?
Explain your answer.

Reviewing skills

(1) Jenny is in hospital and her body temperature is monitored every hour.

Time	8 a.m.	9 a.m.	10 a.m.	11 a.m.	12 noon	1 p.m.	2 p.m.	3 p.m.	4 p.m.
Temp. (°C)	38.0	38.2	38.5	39.0	38.7	38.4	38.1	37.9	37.7

a Show the data in a vertical line chart. Join up the tops of the vertical lines.

b During which time period did her temperature rise?

c When do you think she was given some medicine to bring her temperature down?

d When is Jenny's greatest temperature change?

e Can you say that at 8.30 a.m. her temperature was definitely 38.1 °C?

Unit 4 • Pie charts • Band f

Outside the Maths classroom

Budgeting

Why is a pie chart a good way to illustrate how you spend your money?

Toolbox

A pie chart shows the parts of a whole.

This pie chart shows that $\frac{1}{2}$ of the crisps sold were cheese and onion and $\frac{1}{4}$ were ready salted. The other two flavours are both the same size and take up $\frac{1}{4}$ of the pie chart in total. This means that they are each half of $\frac{1}{4}$, that is $\frac{1}{8}$.

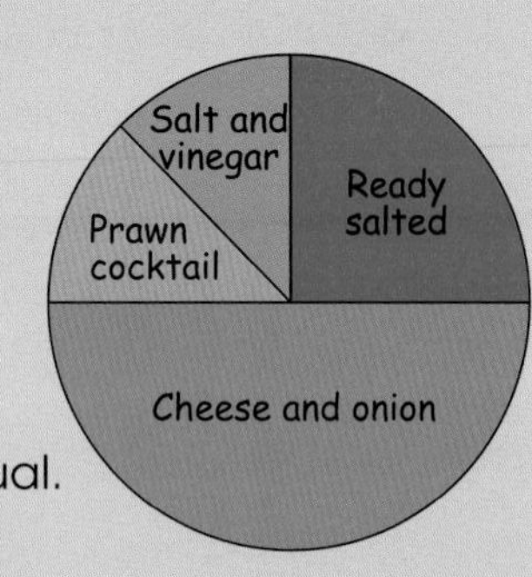

To find the size of each sector, find the angle that represents one individual. That is 360° divided by the total. Then multiply by the frequency for each category.

The data show which of four television channels thirty people were watching on Monday evening.

	BBC1	BBC2	ITV1	ITV3	Totals
Frequency	11	6	5	8	**30**
Pie chart angle	132°	72°	60°	96°	**360°**

$360 \div 30 = 12°$ for one person, $12 \times 11 = 132°$

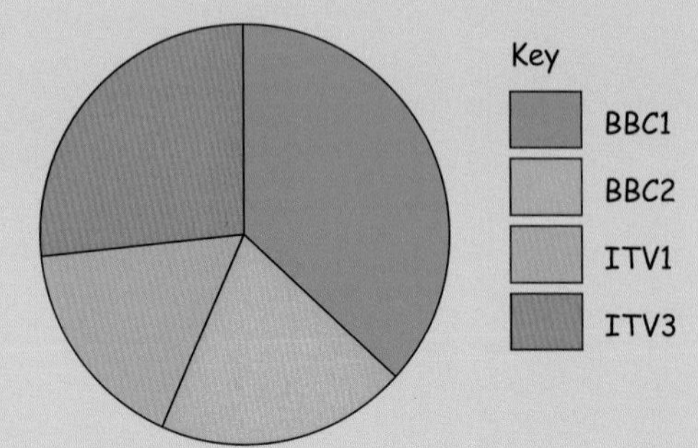

Example – Reading a pie chart

These pie charts show information about two football teams in the same season.

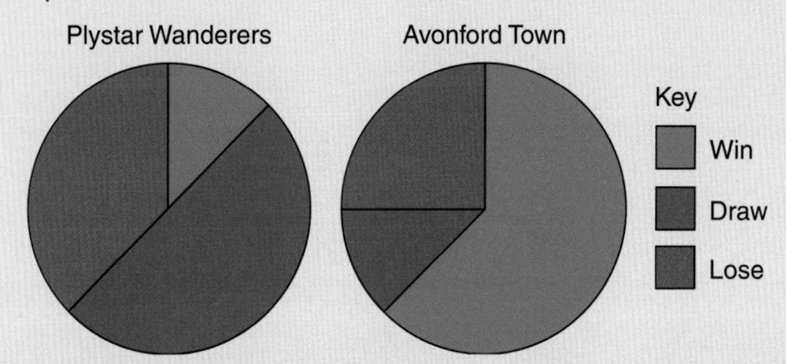

a Which team played better?

b What fraction of their matches did Plystar Wanderers draw?

c **i** What fraction of their matches did Avonford Town lose?

ii Which angle is used to show this?

d Each team has played 32 matches. How many matches did Avonford Town lose?

Solution

a Avonford Town, as their pie chart shows a greater proportion of wins.

b $\frac{1}{2}$

c **i** $\frac{1}{4}$ **ii** $\frac{1}{4}$ of $360° = 90°$

d $\frac{1}{4}$ of $32 = 8$ matches

Example – Drawing a pie chart

Mark has made a list of the favourite type of music of his friends.

Dance	Pop	RnB	Dance	RnB	RnB	Dance	RnB
RnB	RnB	RnB	Garage	RnB	Pop	Dance	
Garage	RnB	Garage	Dance	RnB	Dance	RnB	
RnB	RnB	RnB	Garage	Pop	RnB	Pop	
RnB	Dance	RnB	Dance	Pop	RnB	Dance	

a Make a tally chart for Mark's data.

Mark draws this pie chart to show the data.

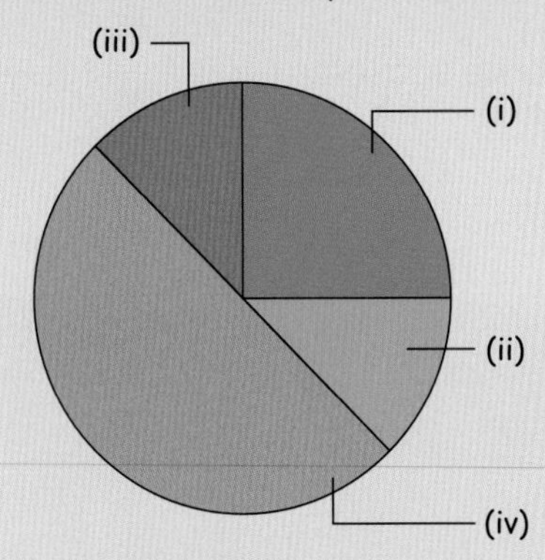

b What angle represents one person?

c Copy the pie chart and label each section with the correct type of music.

d What fraction chose Dance?

e What is the angle for Pop?

f What is the angle for RnB?

Solution

a

Favourite music	Tally	Frequency
Dance	~~IIII~~ IIII	9
Pop	~~IIII~~	5
RnB	~~IIII~~ ~~IIII~~ ~~IIII~~ III	18
Garage	IIII	4
Total		**36**

b There are 36 people. The angle for each person is $360 \div 36 = 10$

c **i** Green - Dance

ii Orange - Pop

iii Red - Garage

iv Blue - RnB

d $\frac{9}{36} = \frac{1}{4}$

e $10° \times 5 = 50°$

f $10° \times 18 = 180°$

Practising skills

1. Students in Avonford School Year 13 were asked about their plans for next year.

   The pie chart shows their replies.

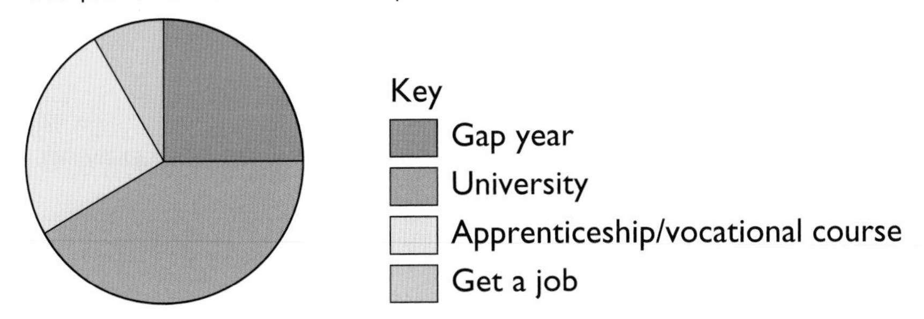

   a Which option is the most popular choice?

   b What is the angle for gap year?

   c What fraction of the whole year choose gap year?

   d There are 168 students in Year 13.
   How many choose each of the four options?

2. A sample of people were asked which of the following best described their diet.

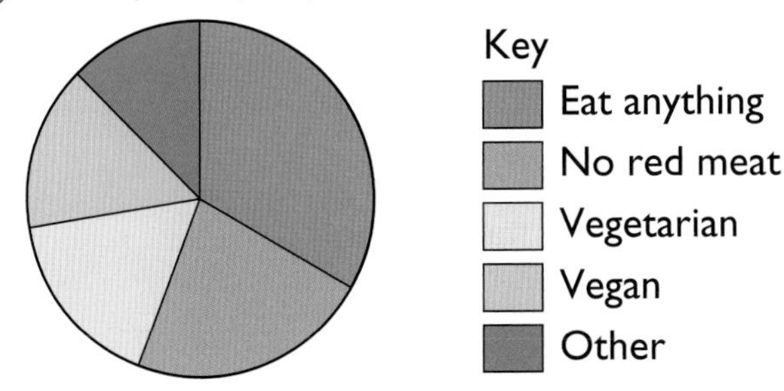

   a Which is the most popular option?

   b What is the angle for vegetarian?

   c What fraction chose the vegetarian option?

   d There were 120 people in the sample. How many chose each of the options?

3. A rail user group do a small survey of how people get to their local station.
   Here are their results.

Method	Walk	Bus	Taxi	Own car	Lift
Frequency	10	5	6	14	5

   They want to show the results on a pie chart.

   a How many degrees should they use for one person?

   b Draw the pie chart.

   c The rail user group are campaigning for more car parking space.
   1200 people use the station each day. How many of them do you expect to need a parking space?

4) Jamini is a cat. Her owners watch how she spends her day.

Activity	Sleep	Prowling	Eating	Grooming
Time (hours)	16	5	1	2

They show this on a pie chart.

a How many degrees do they use for one hour?

b Draw the pie chart to show how Jamini spends her day.

c Jamini's owners watch her some more. They decide that she spends $\frac{1}{2}$ an hour playing, $6\frac{1}{2}$ hours prowling and less time sleeping.

Describe the changes they must make to the pie chart.

5) The bar chart shows the types of drink that Sally sold in her café one day.

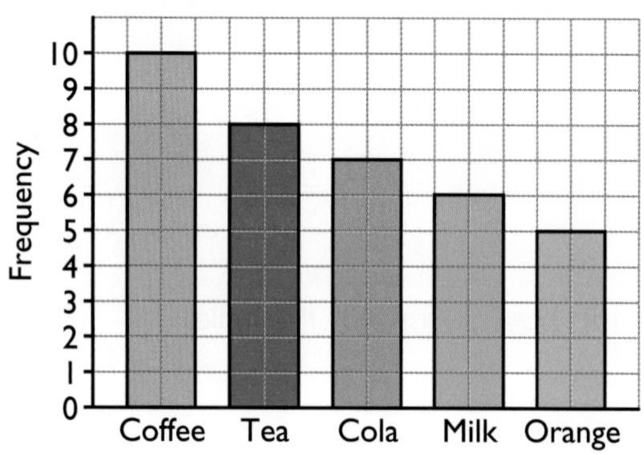

a Draw a pie chart to show this information.

b Which diagram do you find more helpful, the bar chart or the pie chart?

Developing fluency

1) This pie chart shows the favourite sports of 30 students.

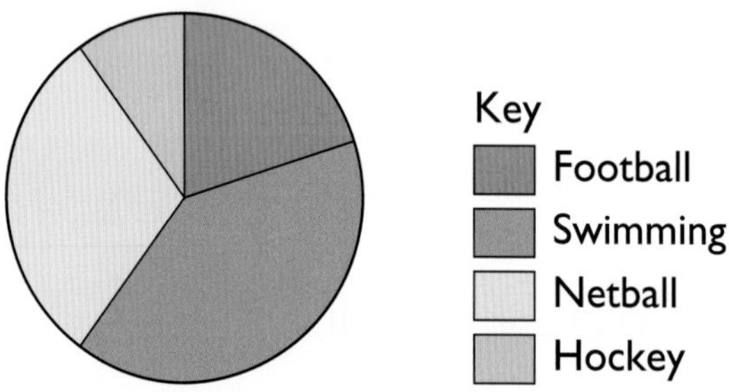

a Which sport is the most popular choice?

b What is the angle for the choice of football?

c What fraction of the 30 students chose football?

What is this as a percentage?

There were actually 40 students in the group. Three said they didn't like any sports and the rest gave other sports (such as rugby).

d What are the angles for a pie chart that shows the choices of all the students in the group?

2 A travel operator carried out a survey of British people in a bar in a Spanish holiday resort. She asked them how they had travelled there.

Here are her results.

Method	Air	Drive	Coach	Live here	Other
Frequency	32	28	20	9	1

a How many people were in the survey?

b The travel operator draws a pie chart to show the information. How many degrees does she use for one person?

c Draw the pie chart.

d Draw a bar chart showing the same information.

e Which do you personally find most helpful, the table, the pie chart or the bar chart? Explain why.

Exam-style

3 A fruit seller carries out a survey on people's favourite fruit, with these results.

Fruit	Apple	Orange	Peach	Grapefruit	Pineapple	Banana	Other
Frequency	11	8	9	6	4	12	40

a How many people were in the survey?

b The information is to be shown as a pie chart. How many degrees represent one person?

c Draw the pie chart.

d Comment critically on the data collection and the display of the results as a pie chart.

4 Emma works for an animal charity. It re-homes pets.

She keeps a record of the types of animals they re-home.

$\frac{3}{8}$ are dogs, $\frac{2}{9}$ are cats, $\frac{1}{4}$ are rabbits.

The rest (for example, snakes) come in the category of 'others'.

Emma's records covered 72 animals.

a Work out the angles for a pie chart to show this information.

b Draw the pie chart.

The next week she re-homes 12 cats, 4 rabbits and 2 snakes, which she adds to her record.

c What should the angles in the pie chart be now?

Exam-style

Extension

5 Sam draws this diagram to show the different types of vehicles in a car park at 12 noon one day. He says it is a pie chart.

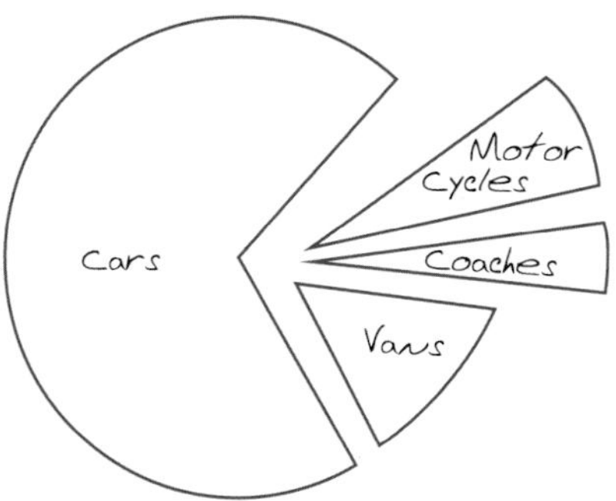

What things can you find that are wrong with it?

Problem solving

Exam-style

1 A cinema manager records the number of different types of tickets sold on Monday.

Tickets	Senior	Adult	Student	Child
Number sold	42	10	6	14

The manager draws a pie chart to represent these figures but it is wrong.
What mistakes has the manager made?

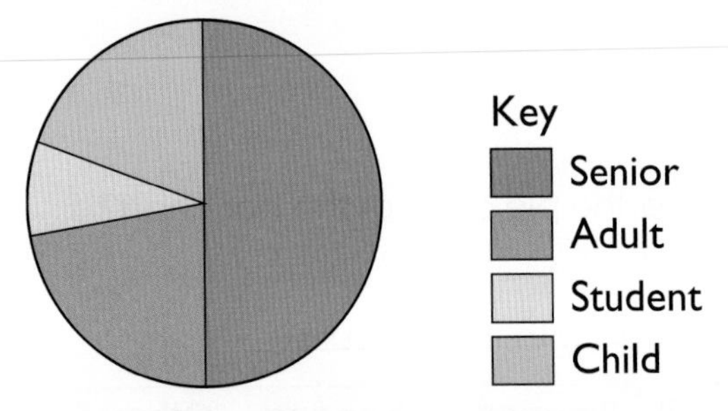

Exam-style

Extension

2 In an election there were five candidates: Mrs Taylor, Mr Hussain, Ms White, Mr Clift and Ms Treble.
The table shows the number of votes each candidate received.

Candidate	Taylor	Hussain	White	Clift	Treble
Number of votes	2567	1850	980	6751	2252

a Draw a pie chart to illustrate the data.

b Work out the percentage vote each candidate received. Give your answers to the nearest 1%.

c Mr Clift claims he got a majority of the vote.
How can you see that this is false from
i the pie chart
ii the percentages?

Exam-style

3 A careers adviser surveys a group of students to find out their career aims.

$\frac{7}{12}$ of the students plan to go to university.

$\frac{1}{4}$ of the students plan to start an apprenticeship.

The remainder plan to take a gap year.

a What fraction of students plan to take a gap year?

b Draw a pie chart to illustrate this data.

c 140 students plan to go to university.
How many students were surveyed altogether?

Exam-style

4 A clothes company collected data to find out why customers returned products to one of their stores last month. These are the results.

Reason	Frequency
Wrong size	27
Faulty	4
Wrong colour	16
Unwanted gift	5
Changed mind	8

a Draw and label a pie chart to represent this data.

b The clothes company has 15 000 returns annually.
Estimate how many items are returned because they are faulty.

Exam-style

5 The pie chart shows how Mel spent her allowance last month.

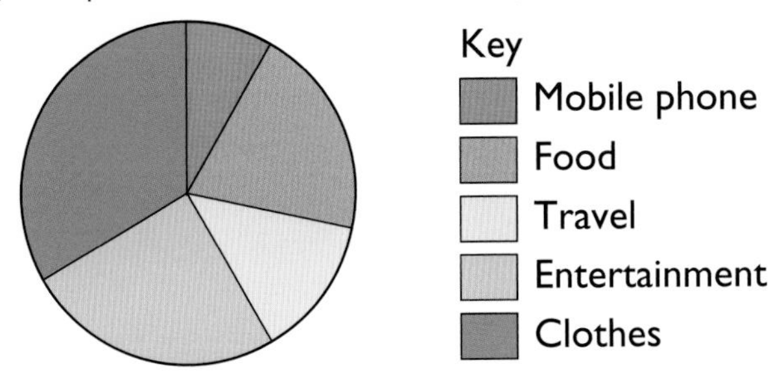

Mel spent £7.50 on her mobile phone.

Work out how much Mel spent

a on entertainment

b altogether

c on travel.

Exam-style

6 Ted records the annual cost of running his car in a pie chart.

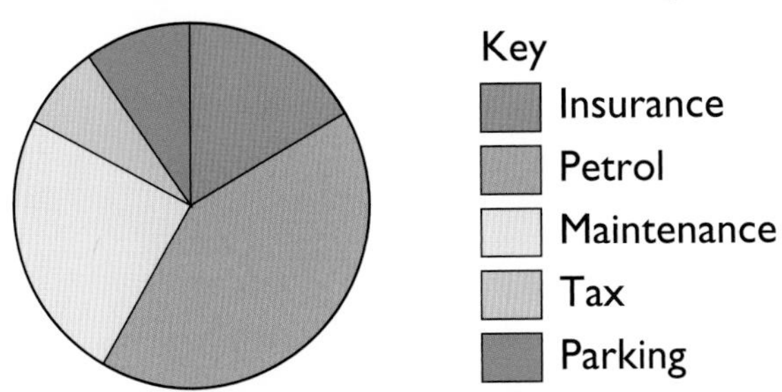

a What fraction of Ted's running costs are spent on petrol? Give your answer in its simplest form.

b Ted spent £160 on parking in this year.
Calculate the total annual running cost of the car.

Reviewing skills

1. There are 224 children at a primary school.

   $\frac{1}{8}$ of the children go to art club.

   $\frac{1}{4}$ of the children go to football club.

   a What fraction of the children go to neither art club nor football club?

   b What angle would this be represented by on a pie chart?

   c Draw the pie chart.

   d How many children do not go to either art club or football club?

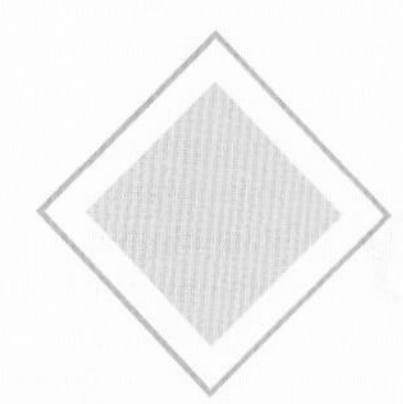

Statistics and Probability
Strand 4 Probability

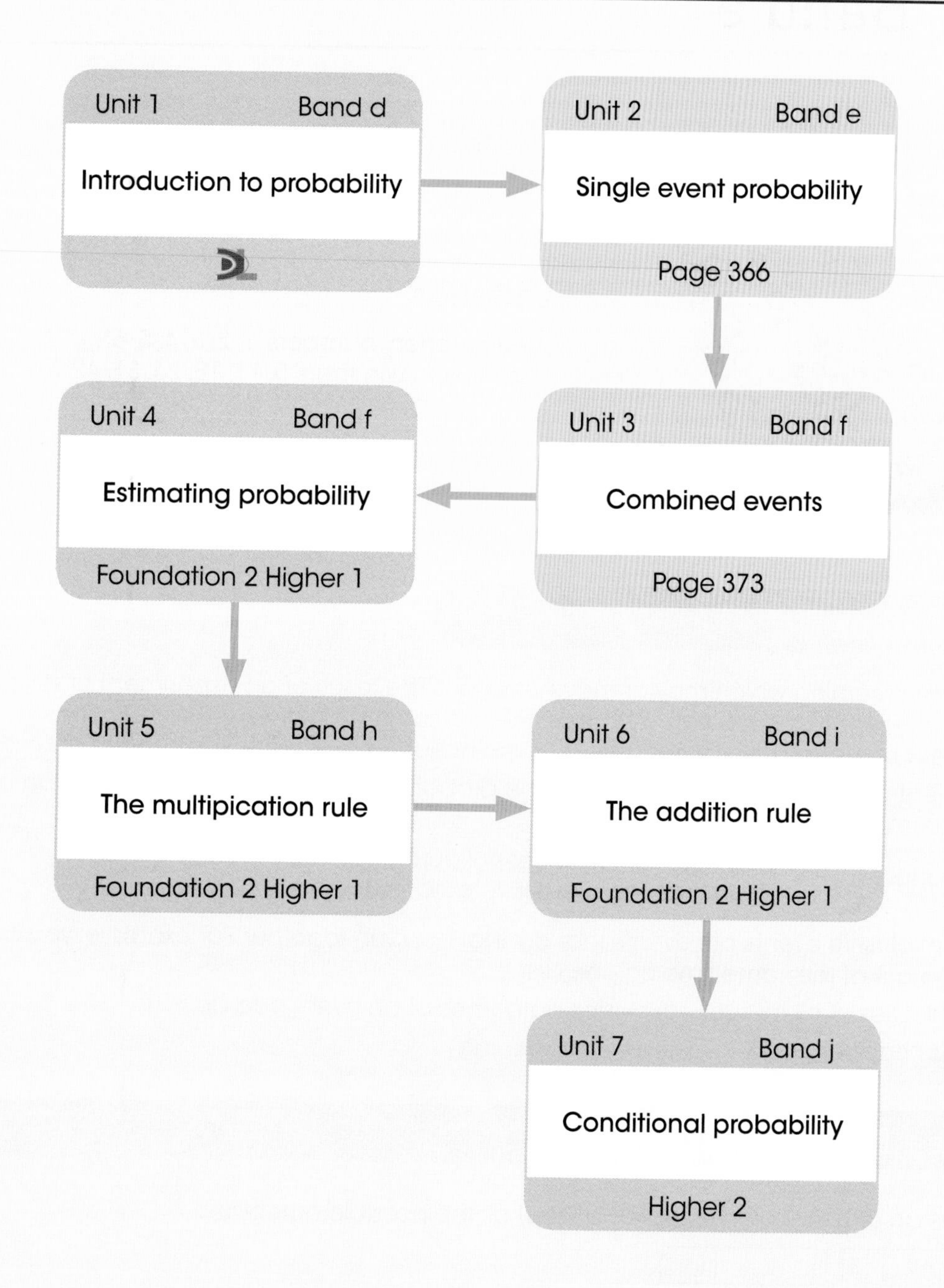

Unit 1 is assumed knowledge for this book. It is reviewed and extended as part of Unit 2.

Unit 2 • Single event probability • Band e

Outside the Maths classroom

Lotteries

Are the lottery numbers '1, 2, 3, 4, 5, 6' less likely to win than '5, 11, 18, 24, 31, 42'?

Toolbox

An **event** is something which may or may not occur. The result of an experiment or a situation involving uncertainty is called an **outcome**, like the score on a die.
The word **event** is also used to describe a combination of outcomes, like scores 5 or 6 on a die.

For any event with equally-likely outcomes, the probability of an event happening can be found using the formula:

$$\text{P(event happening)} = \frac{\text{total number or successful outcomes}}{\text{total number of possible outcomes}}$$

Mutually exclusive events are events that cannot happen together. For example, you cannot roll a 2 and a 5 at the same time on one die!

The probabilities of all mutually-exclusive outcomes of an event add up to 1.

P(event not happening) = 1 – P(event happening)

Example – Equally likely outcomes

Kyle throws an ordinary die. He makes a list of all the possible outcomes.

1	2	3			

a Complete Kyle's list.

b Find the probability that Kyle gets

i 6

ii not a 6

iii an even number

iv 5 or more

v less than 4

vi a prime number.

Solution

a All possible outcomes: 1 2 3 4 5 6

b **i** P(6) $= \frac{1}{6}$ ← 1 possible throw out of 6 equally likely possibilities

ii P(not a 6) $= 1 - P(6) = 1 - \frac{1}{6} = \frac{5}{6}$

iii P(even) $= \frac{3}{6} = \frac{1}{2}$ ← 3 possible throws: 2, 4 and 6

iv P(5 or more) $= \frac{2}{6} = \frac{1}{3}$ ← 2 possible throws: 5 and 6

v P(less than 4) $= \frac{3}{6} = \frac{1}{2}$ ← 3 possible throws: 1, 2 and 3

vi P(prime) $= \frac{3}{6} = \frac{1}{2}$ ← 3 possible throws: 2, 3 and 5

Example – Explaining probabilities

A letter is picked at random, from the word PROBABILITY.

Find:

a P(letter B is chosen)

b P(a vowel is not chosen)

c P(letter C is chosen)

Solution

The letter is chosen at random. So all letters have equal chance of being chosen.

a There are 11 letters in PROBABILITY, 2 of which are B.

P(letter B is chosen) $= \frac{\text{total number of letter Bs}}{\text{total number of letters}} = \frac{2}{11}$

b There are 4 vowels out of the 11 letters in the word PROBABILITY, and 7 letters that are not vowels.

If the letter is chosen at random, all letters have equal chance of being chosen.

P(vowel is chosen) $= \frac{\text{total number of vowels}}{\text{total number of letters}} = \frac{4}{11}$

P(a vowel is not chosen) $= 1 - \text{P(vowel is chosen)}$

$= 1 - \frac{4}{11} = \frac{7}{11}$

This is the same as $\frac{\text{total number of non-vowels}}{\text{total number of letters}}$.

c The letter C does not occur in PROBABILITY so it is impossible to choose it.

The probability of picking a C is 0.

Practising skills

1 Here are 10 shapes.

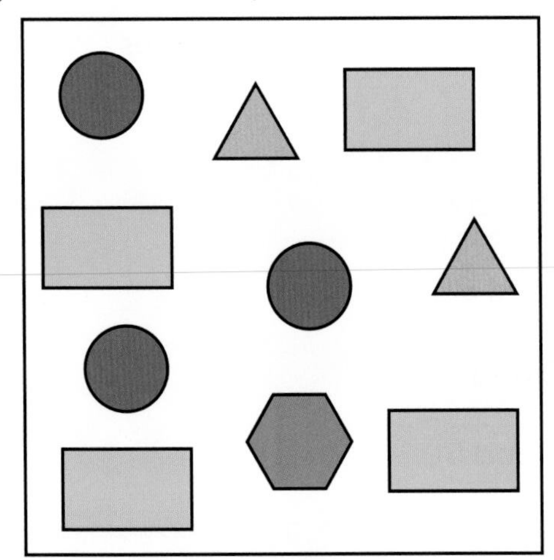

One is selected at random.
What is the probability that the shape selected is:

a a rectangle

b a circle

c a hexagon

d a shape with straight sides

e a square?

2 Howard rolls a normal six-sided die.
What is the probability that he rolls:

a a six

b a three

c an even number

d a prime number

e not a six?

3 Gemma has a pack of 52 playing cards.

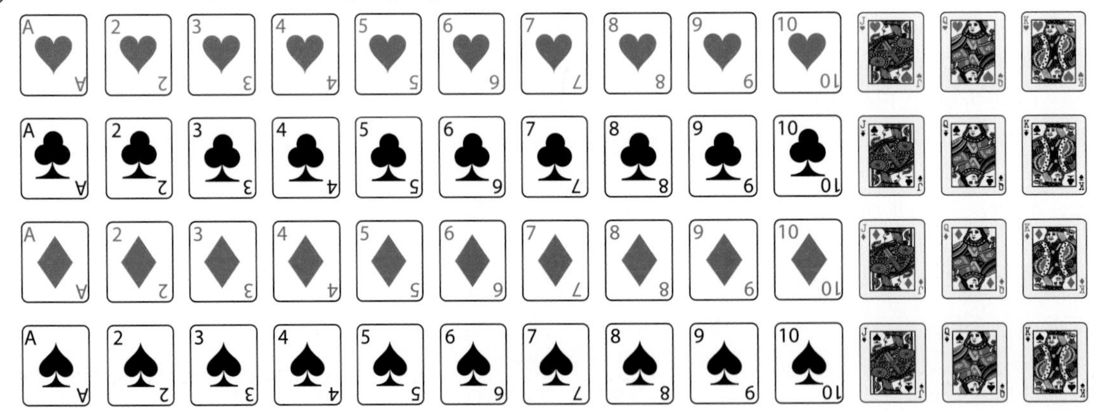

She picks one card at random.
What is the probability that the card she picks is:

a **i** a Queen

ii a red Queen

iii the Queen of Hearts

iv not the Queen of Hearts?

b **i** a five

ii a black five

iii the five of Clubs

iv not a five?

4. Here are some letter tiles.
Richard picks a tile at random.
What is the probability that he picks:

a the letter P
b the letter O
c the letter B
d the letter I
e the letter Z
f a blank tile?

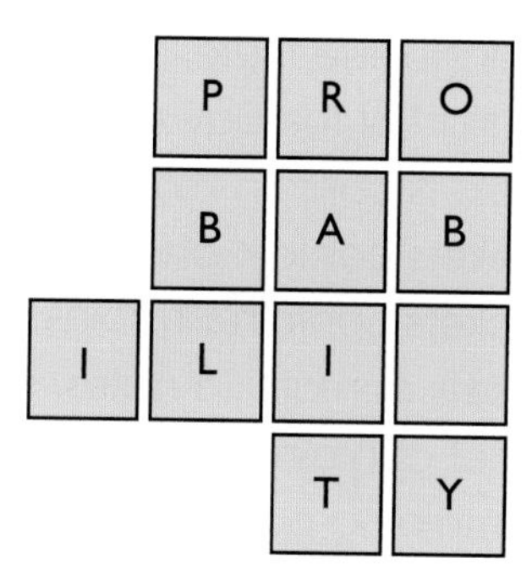

5. Bree has a spinner with 9 equal sides numbered 1 to 9.
What is the probability that:

a on the first spin the number is 1?
b on the second spin the number is 2?
c on the ninth spin the number is 9?
d on the tenth spin the number is 10?

Developing fluency

1. Here is Ahmed's password to his savings account: BADBOY6527.
Ahmed enters the first nine characters correctly but he has forgotten the last one.
He enters a character at random.
What is the probability that he is right if:

a he chooses from all the letters and numbers
b he remembers it is a number and chooses one of them
c he thinks it is a letter and chooses one of them?

Exam-style

2. A lottery has 100 balls numbered 1 to 100.
To enter the lottery you 'buy' certain balls at £1 each.
One ball is then selected at random. The person who has this number wins £50.
Jasmine buys all the square numbers and Mark buys all the prime numbers.

a What is the probability that:
  i Jasmine will win
  ii Mark will win?

b Jasmine suggests that they form a syndicate. She says, 'If one of us wins, we will share the £50 equally, £25 each'.
Is this a fair arrangement?
Should Mark accept her offer?
Explain your answer.

3 At a village fete there is a 'Lucky Dip' stall which has a barrel with packets in it.
At the start of the day there are 400 packets.
60 packets contain key rings.
20 packets contain chocolate coins.
16 packets contain £1 coins.
3 packets contain £5 notes.
1 packet contains a £20 note.
The remaining packets are empty.

a How many empty packets are there?

b On the first draw what is the probability that the packet contains:
i nothing **ii** a key ring **iii** a chocolate coin
iv a £5 note **v** a £20 note?

c Each draw costs £1.
Jules says she is going to keep on buying tickets until she gets the £20 note.
Do you think this is a good strategy?

4 A die with 6 faces is believed to be biased. It is rolled many times and the results are as follows.

Number	1	2	3	4	5	6
Frequency	70	38	45	30	52	15

a How many times was the die thrown?

b Estimate the probability of each of the outcomes 1, 2, 3, 4, 5 and 6.
Show your answers in a table.

c How can you check the figures in your table?

d The die is rolled again.
What is the probability that the number rolled is:
i a 6
ii a number less than 6
iii a number greater than 6?

Exam-style Extension

5 Martina has a bag of counters which are either red, green or blue.
She chooses one at random.

a The probability that Martina gets a red counter is $\frac{5}{12}$ and for a green it is $\frac{1}{3}$.
Find the probability that she gets a blue.

b Martina says, 'I think there are 18 counters in the bag.'
Explain why she must be wrong.

c What is the fewest possible number of counters in the bag?

Exam-style Extension

6 Here is a spinner.
Hamid says, 'If I spin the spinner four times I will get a 1, a 2, a 3 and a 4'.
Is Hamid right?
Explain your answer.

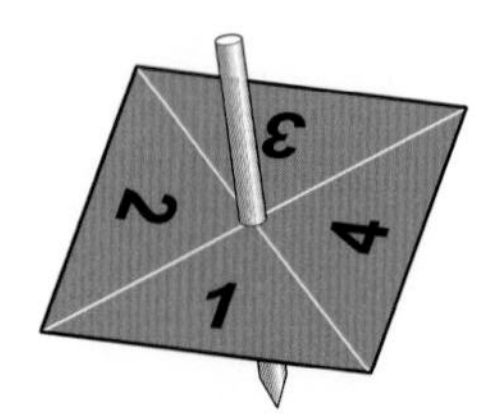

Problem solving

Exam-style

1. A bag contains red balls and blue balls.

   The probability of taking a red ball from the bag at random is $\frac{2}{3}$.
   There are 12 red balls in the bag.
   How many blue balls are in the bag?

Extension

Exam-style

2. The pie chart shows the sport last played by each member of a health club.
   The manager of the club displays this information using a pie chart.

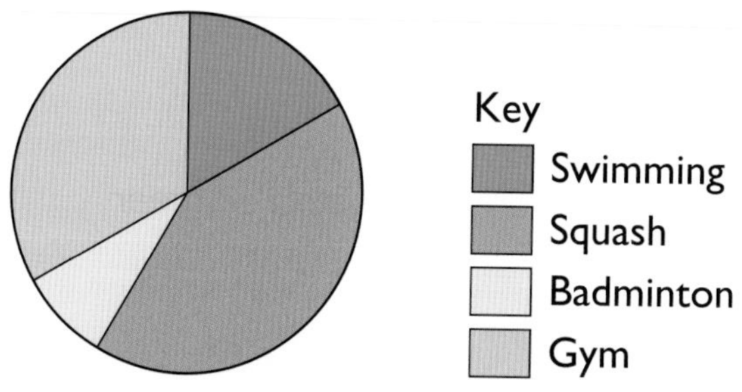

   Work out the probability that a randomly selected club member:

   a played badminton

   b went to the gym or played squash

   c went swimming

   d didn't go swimming.

Exam-style

3. Kyle and Isabelle are playing a game with a set of 21 cards, numbered 1 to 21.
   Kyle selects a card a random, it is the card with number 15.
   Isabelle shuffles the rest of the cards and picks one card at random.

   a What is the probability that the number on Isabelle's card is higher than 15?

   They put their cards back and play again.
   Kyle takes out a card at random.
   Isabelle shuffles the rest of the cards and picks one card at random.
   The probability that the number on this card is higher than the number on Kyle's card is 0.6.

   b What is the number on Kyle's card?

   c What is the probability that the number on Isabelle's card is lower than 15?

Extension

Exam-style

4. Ellie throws a biased die.
   The die is equally likely to score a 2 or 4.
   It is twice as likely to score a 1 as a 2.
   It is twice as likely to score a 6 as a 1.
   Find the probability that Ellie throws:

   a a 3

   b a 6

   c an even number.

Extension

Exam-style

Extension

5 James has a bag of toffees and pieces of fudge.
He takes a sweet at random from the bag.
The probability that the sweet is a toffee is $\frac{7}{12}$.
James eats four of the sweets and then chooses a sweet at random.
The next sweet is equally likely to be toffee or fudge.

a How many sweets were in the bag to start with?

b How many toffees were there?

c How many sweets of each type did James eat?
Find both possible answers.

Exam-style

Extension

6 A spinner has:

- eight equal sections and four different colours
- the probability of landing on red is greater than the probability of landing on green
- landing on yellow or blue has equal probability.

Draw the two possible designs.

Reviewing skills

1 Oliver throws a die with 8 faces, numbered 1 to 8.
What is the probability that Oliver rolls:

a **i** an 8

ii a number between 1 and 8

iii a number between 0 and 8

iv zero?

b The die is described as 'fair'. What does this mean?

2 A cafe for long distance lorry drivers offers these breakfast dishes.

Egg, sausage, beans and chips
Egg, bacon, sausage, tomato and toast
Bacon buttie
Scrambled eggs on toast
Sausage roll

a If a driver chooses his breakfast at random, what is the probability that his breakfast contains

i egg

ii sausage

iii tomato?

b If his breakfast contains egg, what is the probability that it also contains bacon?

Unit 3 • Combined events • Band f

Outside the Maths classroom

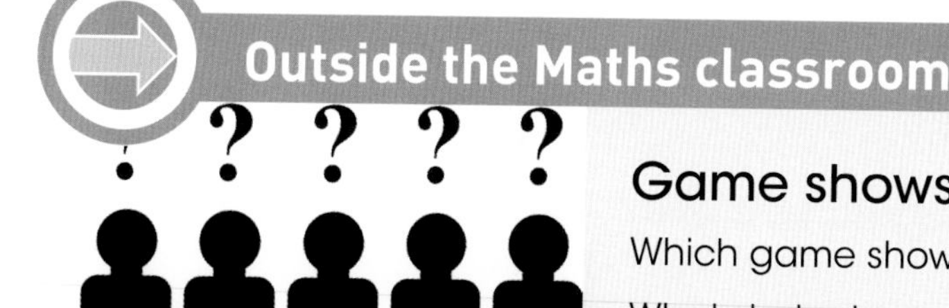

Game shows

Which game shows involve probability?

What strategies can help you win?

Toolbox

You can use the formula that for any equally-likely event

$$P(\text{event}) = \frac{\text{total number or successful outcomes}}{\text{total number of possible outcomes}}$$

to find how likely a combination of events is.

To find the total number of combinations, list all of the outcomes if this is possible.

Be systematic and change one item at a time.

If it is not possible to list all the outcomes, you can use a **possibility space** diagram or a **Venn diagram**.

Example – Listing all possible outcomes

Sam is choosing her breakfast. She can choose one cereal and one drink.

Cereals: Wheatamix, Cornflakes or Sugarloops

Drinks: tea or coffee

a Draw a diagram to show all the possibilities for Sam's breakfast.

b Sam selects her cereal and drink at random. What is the probability that she has Sugarloops and coffee?

c How would the list change if Sam had three options for drinks, for example tea, coffee and orange juice ?

Solution

a

Cereal / Drink	Wheatamix	Cornflakes	Sugarloops
Tea	T&W	T&C	T&S
Coffee	C&W	C&C	C&S

b There are 6 possible outcomes and 1 successful outcome.

Probability = $\frac{1}{6}$

c There will be an extra row in the table in answer **a**.

Example – Probability space diagrams

Hannie throws two dice, one red and one green.

a Draw and complete a table to show all the possibilities for their total scores.

b What is the probability the total score is

  i exactly 3

  ii 3 or less

  iii greater than 12

  iv a prime number?

c Would the answers be the same with two red dice?

Solution

a

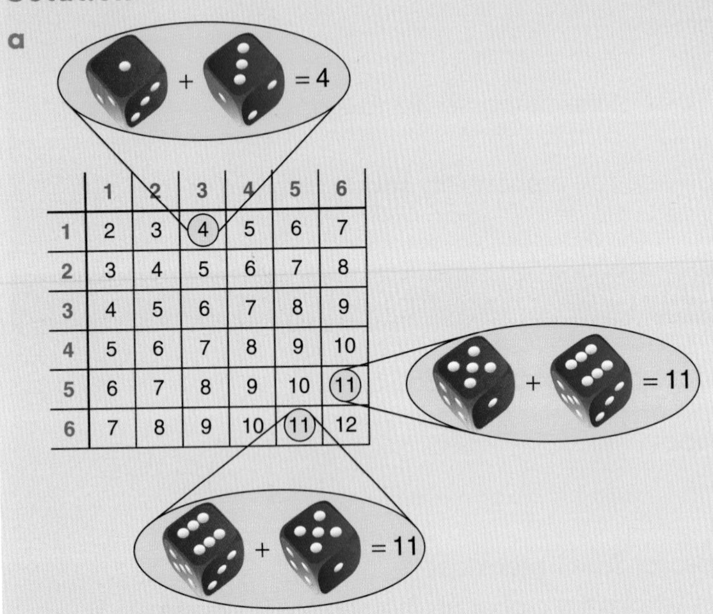

	1	2	3	4	5	6
1	2	3	4	5	6	7
2	3	4	5	6	7	8
3	4	5	6	7	8	9
4	5	6	7	8	9	10
5	6	7	8	9	10	11
6	7	8	9	10	11	12

b There are 36 possible outcomes.

i $\frac{2}{36} = \frac{1}{18}$ — There are 2 ways of getting 3, 1 + 2 and 2 + 1, so 2 favourable outcomes.

ii $\frac{3}{36} = \frac{1}{12}$ — There are 3 ways of getting 3 or less.

iii 0 — It is impossible to get more than 12.

iv $\frac{15}{36}$ — Prime numbers 2(1way), 3(2 ways), 5(4 ways), 7(6 ways), 11(2 ways)

c The answers would be the same.
The colour of the dice makes no difference. — Sometimes it is easier to see what is going on with different coloured dice.

Example – Venn diagrams

Nick has some cards with different shapes printed on them.
This Venn diagram describes the shapes printed on the cards.

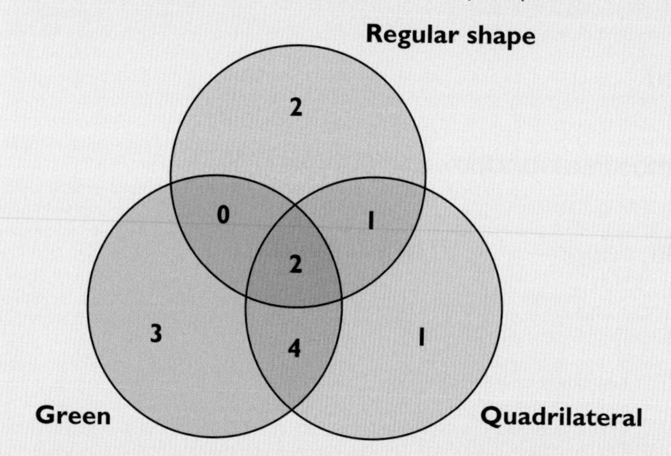

a How many cards are there in Nick's pack?

b Nick picks a card at random. What's the probability that his card is

 i a quadrilateral

 ii a green quadrilateral

 iii not green?

c Draw the shape in the central intersection.
 Write a sentence about the probability of picking it.

Solution

a There are 13 cards. ← Add the number of cards in all the different regions.

b i P(quadrilateral) = $\frac{8}{13}$ ← $4 + 2 + 1 + 1$

 ii P(green quadrilateral) = $\frac{6}{13}$ ← $4 + 2$

 iii P(not green) = $\frac{4}{13}$ ← $2 + 1 + 1$

c

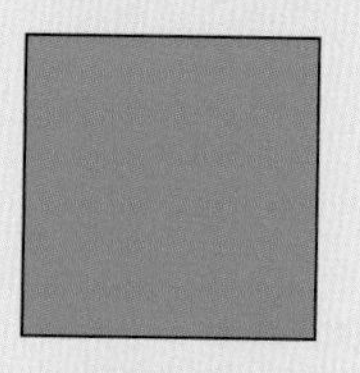

← A regular quadrilateral is a square.
The intersection requires a green one.

The probability of picking a card with a green square at random is $\frac{2}{13}$.

Practising skills

1 Anwar is a keen cricketer. He is a bowler.
So far this season he has got 30 batsman from other teams out.
The table shows how they were out.

Bowled	Caught	LBW	Stumped
9	12	3	6

a One of the batsman Anwar got out is chosen at random.
What is the probability that he was:
i bowled **ii** LBW **iii** caught **iv** stumped?

b Add your answers to part **b** together.
What does your answer tell you?

c So far Anwar has played 10 matches.
He has 20 matches more to play.
How many more batsman can he expect to get out by being bowled?

2 A restaurant has 4 starters and 3 main course meals as shown on the menu card.

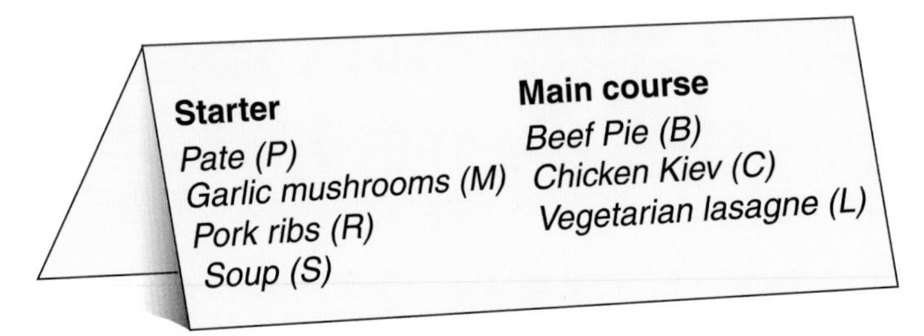

a Make a list of the possible combinations of starter and main course.

b How many meals are there in your list?
How could you work it out quickly without listing them all?

c Jamie says, 'Get me anything'.
If the starter and main course are selected at random, what is the probability he gets soup and beef pie?

3 Two ordinary fair dice are rolled.
One die is red and the other is blue.
The numbers that come up are multiplied together.

a Copy and complete this table to show all the possible outcomes.

		Red die					
	×	1	2	3	4	5	6
Blue die	1						
	2						
	3						
	4						
	5						
	6						

b What is the probability that the product is:
i 8
ii 25
iii 7
iv a number below 37?

c Do the colours of the dice make any difference?

d What types of numbers between 1 and 36 have a probability of $\frac{1}{36}$?

4 Gloria has two spinners: the blue one has four equal edges numbered 1 to 4 and the green one has 6 equal edges numbered 1 to 6.
Gloria spins them and adds the two numbers together.

a Make a table to show all the possible outcomes.

b What is the probability that Gloria gets a score of:

i 5 **ii** 12 **iii** 7?

c Which other numbers between 1 and 20 have the same probability as:

i 5 **ii** 12 **iii** 7?

5 Members of a sports club may play golf (G) or bowls (B).
Some play both and some play neither.
The numbers are shown on the Venn diagram.

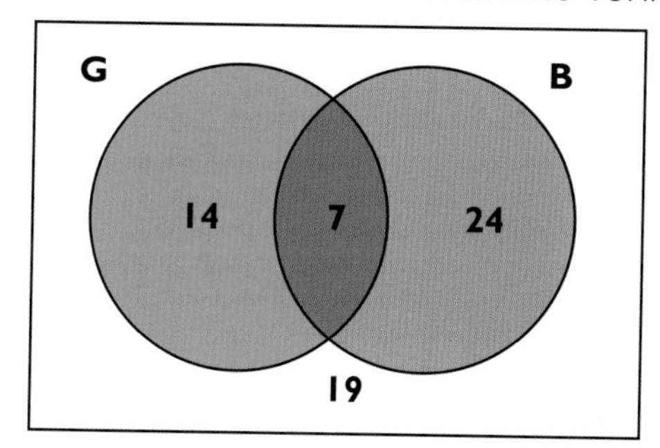

The club holds a raffle. Every member has a ticket and the winner is chosen at random.
Find the probability that the winner

a plays golf

b plays bowls

c plays both golf and bowls

d does not play golf

e does not play bowls

f does not play either bowls or golf?

Developing fluency

1 A mother cat has had a lot of kittens.
Her owner, Veronica, classifies them as follows.

	Tabby	Ginger	Black & White	Tortoiseshell
Female	3	0	4	3
Male	3	4	3	0

a Veronica has a photo of each of the kittens.
She selects a photo at random. What is the probability that the photo is of:

i a female kitten **ii** a tabby kitten **iii** a female tabby?

b Veronica says, 'The kittens prove conclusively that all ginger cats are male and all tortoiseshell cats are female.'
Comment on this statement.

2. Geoff is on holiday.

   In his wardrobe he has four shirts and three pairs of shorts.

Shirts	Shorts
Green	Black
Red	Grey
Black	Cream
Cream	

   a Make a list of all the possible combinations to wear.

   b Geoff grabs one shirt and one pair of shorts in the darkness.

   What is the probability that he has:

   i a red shirt and grey shorts

   ii a shirt and shorts of the same colour?

3. Amanda has two bags with coloured balls in each.

   Bag 1 has 2 red and 3 blue balls.

   Bag 2 has 4 red and 2 blue balls.

   a Copy and complete this table to show all possible outcomes when Amanda selects a ball at random from each bag.

		Bag 1				
		R	**R**	**B**	**B**	**B**
Bag 2	**R**					
	R					
	R					
	R					
	B		RB			
	B					

   b She selects a ball from each bag at random.

   What is the probability that they are:

   i one red and one blue

   ii two red

   iii two blue?

4. The Venn diagram shows the membership of a rugby club.

   There are 30 members. Some of them are backs, others are forwards and there are some non-playing members.

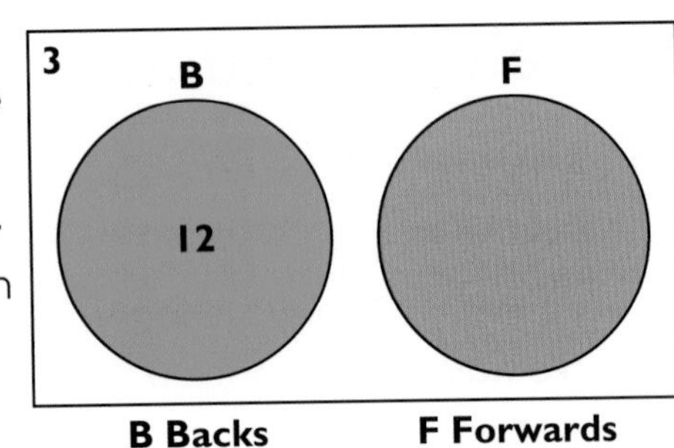

B Backs **F** Forwards

   a Fill in the missing number on a copy of the Venn diagram.

   b A team of 15 players (8 forwards and 7 backs) is chosen for an away match.

   Another member of the club is chosen at random from the remaining players to be the travelling substitute.

   What is the probability that the substitute is

   i a forward

   ii a back?

   c A new player joins the club. He can play as a forward or as a back.

   Draw a new Venn diagram to illustrate the changed membership of the club.

5 Tiffany and Neil are playing a game with a red spinner and a yellow spinner.
The red spinner is numbered 1 0 1 0 1.
The yellow spinner is numbered 0 1 0 1 0.
They are spun together.

a Make a table showing the total of the numbers on the two spinners.

b What is the probability that the total is:
i 0 **ii** 1 **iii** 2?

c Now make a table for the product of the numbers on the two spinners (when they are multiplied together).

d What is the probability that the product is:
i 0 **ii** 1 **iii** 2?

Problem solving

Exam-style

1 Sam and Chloe are playing a game.
They each throw a die and the scores are multiplied together.

a Complete the possibility space diagram.

	1	2	3	4	5	6
1						
2						
3				12		
4						24
5		10				
6						

b Sam wins if the product is odd. Is the game fair? Explain your answer fully.

c Find the probability of getting a product of
i 6
ii a multiple of 6
iii a factor of 60.

Extension

2 A factory manager records whether employees arrived late or on time one Friday.
She also records how employees got to work that day.
The table shows the results.

	Walk	Bus	Car	Cycle
Late	5	28	6	7
On time	95	85	142	32

a An employee is selected at random. Find the probability that they
i were on time
ii walked and were late
iii were late and came by bus.

b The manager says that the results show that employees are not making enough effort to arrive at work on time. Comment on the manager's statement.

3 Baby Ben has lots of different bricks.

	Cube	Cuboid	Cylinder	Total
Red	14	40		71
Green	21		12	60
Blue		33		69
Total		100	42	

a Copy and complete the table.

b Find the probability that a randomly chosen brick is

i a cuboid

ii red

iii a blue cube.

c Find the probability that a randomly chosen

i cuboid is green

ii green brick is a cuboid.

Exam-style

4 Isobel and Peter are playing a game.
They each spin a spinner with sides numbered 1, 2 and 3.

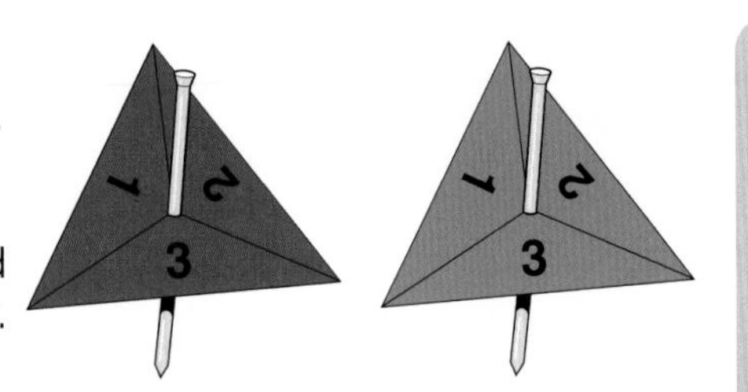

a List all the possible outcomes of the two spinners.

b Isobel wins the game when one spinner shows an odd number and the other spinner shows an even number.
Work out the probability that Isobel wins the game.

c How can they make the game fair?

Extension

5) A group of scientists find a remote population in the Amazon rainforest.
They estimate there are 1500 adults living in 30 villages.
The scientists spend time in one village and notice that many of the people are left-handed.
They record these data for the adults in that village.

	Women	Men
Right-handed	16	14
Left-handed	11	9

a An adult is chosen at random from the village. Find the probability that the person is

i a woman

ii left-handed

iii a left-handed woman.

b One of the scientists says that the women in the village are more likely to be left-handed than the men.

Comment on this statement.

c The data from the village is used to estimate the number of left-handed adults in the complete remote population.

i Calculate this estimate.

ii Give a reason why this estimate may be very inaccurate.

6) A school has 200 Year 11 students. Of these, 136 study French, 96 study German and 8 students study neither French nor German.

a Copy and complete the Venn diagram.

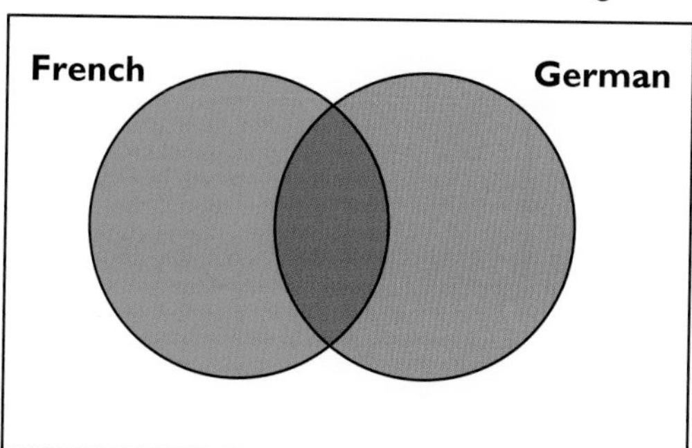

b Find the probability that a student selected at random studies

i French

ii French and German

iii German but not French.

7) A café records what their customers order one morning.

The table shows the results.

	Coffee	Not coffee
Cake	82	35
No cake	90	43

a A customer is chosen at random.

Find the probability that they order:

i coffee and a cake

ii neither coffee nor a cake

iii coffee.

b The café is open 6 days a week and expects to have 400 customers a day.

i Estimate how many customers have a coffee in a week.

200 g of coffee beans make 24 cups of coffee.

The café owner wants to order enough coffee to last for the next four weeks.

ii How many kilograms of coffee should she order? Give your answer correct to the nearest kilogram.

Reviewing skills

1) Steve and Bella are playing a game with a green die and a blue die.

The green die is numbered 1 3 3 4 6 6.

The blue die is numbered 2 3 4 4 5 5.

The dice are rolled and the scores that come up are added together.

a Make a table showing the possible outcomes.

b Find the probability of getting

i a 3 on both dice

ii a total of 7

iii a double

iv a total of at least 6

v a number on the blue die that is 1 more than that on the green die.

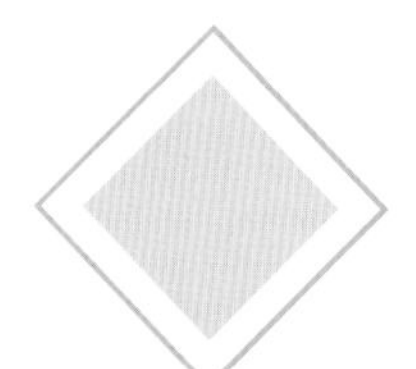

Index